High Temperature Electronics

ELECTRONIC MATERIALS SERIES

Electronic Materials Series:
This series, devoted to electronic materials subjects of active research interest, provides coverage of basic scientific concepts as well as relating the subjects to the electronic applications and providing details of the electronic systems, circuits or devices in which the materials are used. The series will be a useful reference source at senior undergraduate and graduate level as well as for research workers in industrial laboratories who wish to broaden their knowledge into a new field.

Series editors:

Professor A.F.W. Willoughby,
Department of Engineering Materials,
University of Southampton,
UK.

Professor R. Hull,
Department of Material Science,
and Engineering,
University of Virginia,
USA.

High Temperature Electronics

Edited by

M. Willander

Department of Physics, Chalmers University of Technology
and Göteborg University, Sweden

and

H. L. Hartnagel

Technische Hochschule Darmstadt, Germany

CHAPMAN & HALL

London · Weinheim · New York · Tokyo · Melbourne · Madras

Published by Chapman & Hall, 2–6 Boundary Row, London SE1 8HN, UK

Chapman & Hall, 2–6 Boundary Row, London SE1 8HN, UK

Chapman & Hall GmbH, Pappelallee 3, 69469 Weinheim, Germany

Chapman & Hall USA, 115 Fifth Avenue, New York, NY 10003, USA

Chapman & Hall Japan, ITP-Japan, Kyowa Building, 3F,
2-2-1 Hirakawacho, Chiyoda-ku, Tokyo 102, Japan

Chapman & Hall Australia, 102 Dodds Street, South Melbourne,
Victoria 3205, Australia

Chapman & Hall India, R. Seshadri, 32 Second Main Road, CIT East,
Madras 600 035, India

First edition 1997

© 1997 Chapman & Hall

Typeset in Ireland by Doyle Graphics
Printed in Great Britain T. J. Press (Padstow) Ltd., Padstow, Cornwall

ISBN 0 412 62510 5

A catalogue record for this book is available from the British Library

Library of Congress Catalog Card Number: 96-86291

∞ Printed on permanent acid-free text paper, manufactured in accordance
with ANSI/NISO Z39.48-1992 and ANSI/NISO Z39.48-1984 (Permanence
of Paper).

Contents

Contributors

R. B. Brown
University of Michigan College of Engineering

G. Burbach
Fraunhofer Institute of Microelectronic Circuits and Systems

R. G. Carter
Delco Electronics Corporation

D. W. Christenson
United States Air Force, Ogden Air Logistics Center,
   Technical and Industrial Support Directorate

J. C. Erskine
General Motors Research and Development Center

T. Fallet
SINTEF Instrumentation

H. L. Fields
Delco Electronics Corporation

G. Forre
SINTEF Instrumentation

K. Fricke
Fachhochschule Fulda

J. Gakkestad
SINTEF Instrumentation

H. L. Hartnagel
Technische Hochschule Darmstadt

J. A. Hearn
Delco Electronics Corporation

J. M. Himelick
Delco Electronics Corporation

M. A. Khan
APA Optics

V. Krozer
Technische Universität Chemnitz

W.-Y. Lee
Technische Hochschule Darmstadt

M. Schüßler
Technische Hochschule Darmstadt

M. Shur
University of Virginia

Y. M. Tairov
St Petersburg State Electrical Engineering University

R. Werner
Fraunhofer Institute of Microelectronic Circuits and Systems
(Now: National Semiconductor)

M. Willander
Chalmers University of Technology and Göteborg University

K. Wu
University of Michigan College of Engineering

J. A. Yurtin
Delphi Packard Electric Systems

Preface

There is a growing demand for electronic signal processing at elevated temperatures. A number of approaches have been used to develop this capability. Silicon circuits could be developed and fabricated with an appropriate technology to cover increased temperature ranges. In a search for semiconductors with a wider energy gap to avoid leakage currents at high operating temperatures, one developed compound semiconductors such as GaAlAs on GaAs substrates. Efforts to use GaN are also useful, although difficult due to the lack of a suitable substrate material for lattice-matched epitaxial growth. Other work concerns electronic component and circuit developments with SiC. Preliminary results have proved interesting.

This book attempts to present the possibilities of such circuitry. Some of the solutions obtained so far are directly usable for the many applications where high environmental temperatures exist. Other concepts, particularly the more demanding ones, such as operation above 500 °C, still need much more researching. This also concerns estimates of device lifetimes for continuous high temperature operation.

This book may help the potential user of such circuitry to find a suitable solution. It should also stimulate more research groups to enter this demanding effort. And finally, it should stimulate a broad awareness of the need and the solutions for this type of electronics. That is why Part One is devoted to high temperature applications.

M. Willander and H. L. Hartnagel
Göteborg and Darmstadt
September 1996

PART ONE

Users

High temperature electronics in aircraft and space systems

D. W. Christenson

1.1 INTRODUCTION

Since men and women began creating products, they have faced, accepted and made design compromises associated with materials limitations. An engineer's primary responsibility is to make the 'best' compromises and create a product that can reliably operate within the criteria for which the product is designed. It was by balancing materials and technology limitations that the Wright brothers were able to create and fly their first aircraft. From their first efforts, designers have found materials limitations unique to flight.

1.2 AIRCRAFT ENVIRONMENT

The focus of this text is the examination of component characteristics that will support extending the operable temperature range of electronic components. An examination of the aircraft environment will provide understanding that will further define the need for extension of the temperature range of electronic products.

Several characteristics of aircraft and space systems make some of their problems unique from land vehicles:

1. The cost of maintaining a kilogram of anything in the air for the life of a commercial aircraft is measured in significant portions of a typical

High Temperature Electronics. Edited by M. Willander and H.L. Hartnagel. Published in 1997 by Chapman & Hall, London. ISBN 0 412 62510 5.

engineer's yearly salary. For high performance aircraft, weight is a major design criterion and is measured in grams. On a typical aircraft program, engineers will be offered as much as one hour of salary per gram weight loss for suggestions that results in a design change that lowers aircraft weight. This concern is unique from land vehicles because it costs so much to create the power to lift the weight.

2. Current aircraft experience leading-edge temperatures in excess of 130 °C for short periods during operations. Future aircraft will lengthen the exposure, and perhaps raise the temperatures significantly.

3. In general, the causes of heating in aircraft also produce vibration, a major contributor to component failures. The primary source of vibration is the physical act of pushing the aircraft through the atmosphere. Additional vibration sources in an aircraft include acceleration-induced vibration, engine vibration/noise, actuators and generators.

The current and future aircraft environment is one of higher temperatures, longer exposures and vibration.

1.3 CUSTOMER NEEDS

The current customer demand is for higher performance avionics that operates without degradation in the aircraft environment with increased functionality and reliability.

1.4 CURRENT AVIONICS CAPABILITIES

Aircraft utilize a wide range of electronics. Most of the equipment is microprocessor based, using processors capable of several millions of instructions per second. Bus bit sizes range over 8, 16 and 32 bit microprocessors with 8 and 16 being the most prevalent today. Bus sizes of 64 and 128 bits and greater are being investigated for support of ever increasing needs to process data for future aircraft applications. Memory sizes range from 32 K bytes and upwards with 128 K bytes utilized to create an average system. Sensors of many types (radar, infrared, radio frequency, etc.) are required in an avionics suite. To aid with sensor processing, digital signal processors (DSPs) are the norm. These components create systems which accomplish the following functions as well as others: flight controls, weapons control, sensors, navigation, electronic countermeasures, threat warning, attack radar, etc.

1.5 THE LEADING EDGE

Pursuit of the leading edge of technology has been and will continue to be a road filled with disappointment and partial success. A major cause for disappointment in this pursuit is the lack of knowledge. The idea that the leading edge of technology is a multifaceted concept appears to be generally accepted; but approaching the leading edge in any one area may require a loss of capability in another.

The following leading-edge areas are known to exist: (1) high temperature performance, (2) reliability, (3) testability, (4) support ability, (5) additional and/or unique functionality, (6) processing speed, (7) cost, (8) power consumption, (9) radiation hardness, (10) electromagnetic interference (EMI) and related areas, e.g. ESD, EMP susceptibility, (11) EMI generation and (12) weight.

Potential performance with respect to each area is constantly being improved. For example, industry can put more functionality into a component today than ever before. Industry knows how to create devices which are capable to reliable operation at higher temperatures. The factors that control each of these aspects of performance are similar, but inversely related. The space that is created by using leading-edge technology can be used to provide additional functionality or additional temperature capability; however, both capabilities cannot be maximized simultaneously.

This technological balancing act is complicated by politic decisions which may impede performance improvements in a given area. Today the performance characteristics of electronics are stalled at 125 °C due to an artificial barrier placed upon 'standard' components. The 125 °C barrier, although political, is a stable operational point around which significant improvements in functionality have been made. Significant efforts will have to be made throughout industry if this barrier is to be broken. The barrier has been in place for over 10 years. During that time the real capability to operate outside (or even within) the 125 °C parameter has actually decreased. In those same 10 years, military requirements have continually derated performance within this specification. Initially 125 °C case temperature operation was allowed. Later 110 °C junction maximum was mandated. Today many aircraft requirements mandate 60 °C junction temperature or less.

Maintaining the required temperatures has become a major burden on the airplane. Supplying cooling to the wing areas is very difficult. It may require as much as 10% of the aircraft total weight and 50 kW to power it. Continuing in this manner, the pilot will find that the weight and power requirements of the electronics significantly impact the flight characteristics of the airplane. Thus, the airplane will be electronically more functional but the flight performance will be degraded.

If the engineer were allowed the freedom to develop products that operate at 150 °C case temperature, much of the environmental control system (ECS) would no longer be required. Several component characteristics known to be impacted are size, speed, cost and power consumption. In turn, a real improvement in temperature and/or reliability performance can be realized.

An interesting side note is that all new technologies will suffer initially from a projected poor reliability until performance is proved, and a level of technological maturity has emerged. This is true even of components whose design was driven from the need to improve reliability.

1.6 MIL STD COMPONENTS

For the past 30 or more years, the Military Standard for electronics has been 125 °C. This temperature has become the standard for component development. It has also gained the reputation as the temperature at which electronic components become 'unreliable'.

There are at least three major characteristics associated with reliability that must be examined. Each of these characteristics can be considered 'failure' characteristics:

1. The following examples are characteristic descriptions of the environment for which a component no longer operates properly but will recover when returned to its operational environment:
   (a) raising or lowering the temperature of a component to the point where it no longer operates, then returning it to an ambient that is within the operational performance range of the device so it resumes proper operation;
   (b) exposure of electronics to a nuclear blast such that EMP or radiation causes a temporary upset but following the exposure the electronics resume operation without degradation.
2. There are characteristics of the environment that cause a component to fail permanently. An example might be heating a component to the point that a package failure occurs.
3. The environment can cause a component to wear out. An example might be operation of a component at an elevated temperature for a sufficient period of time so that metalization electromigration occurs. At the point that the component fails from electromigration, one may say that the environmental exposure 'wore out' the device.

In general, each of these failure characteristics can be modeled with sufficient accuracy to allow design parameters to be derived. Providing these models and making them part of the engineer's design tool makes it possible for the engineer to develop products utilizing these characteristics as design variables instead of constants.

1.7 PRACTICAL DESIGN OF AIRCRAFT AND SPACECRAFT

There are three significant reasons for the high temperature exposure in aerospace applications:

The speed with which the vehicle pushes through the atmosphere results in frictional heating. Supersonic atmosphere flight has long been recognized as a cause of high skin temperatures for aircraft and spacecraft. One of the highest temperatures experienced in spaceflight occurs during atmosphere braking during reentry. Frictional and compressive heating effects from passing through the atmosphere create a heat source localized near the leading edges which are 'cutting through' the atmosphere. The friction-induced heating effects associated with motion through the atmosphere are widely known and well characterized in aerospace engineering texts. As aircraft and spacecraft continue to increase supersonic performance, the temperatures and length of exposure will continue to increase.

There are several methods of containing the temperatures associated with flight. The ablative and ceramic heat shields on space systems have the primary responsibility to protect the rest of the spacecraft from exposure to the heat generated during atmospheric reentry and braking. Many techniques are being evaluated to alleviate aircraft heating and vibration effects, including making the aircraft skin smoother, making the aircraft more aerodynamic and using 'lubricants' on the aircraft skin during flight (such as pumping air out of the leading edge).

Operation in a stagnant environment results in little or no ability to remove heat by conduction or convection. There is no more stagnant environment than a vacuum. If a small power source is operated in such an environment with no thought of how to eliminate heat, the terminal temperatures of the devices may cause electronics to cease operation. Operation in space or rarefied atmospheres must address methods to remove the power utilized by electronic components.

To understand how important this is, consider a tungsten filament encased in a vacuum chamber – an incandescent light. A relatively small amount of power is able to cause the tungsten to achieve a temperature at which the lightbulb creates a bright light. Tungsten becomes incandescent at temperatures near 1000 °C. Similar effects will be observed in electronic systems if a thermal path to some lower temperature point is not provided. This makes space operations problematic. It is unlikely that operation in space without cooling is possible. A thermal path must therefore be provided and radiators operated such that the heat is dissipated.

There are three ways that heat can be transferred from one place to another: conduction, convection and radiation. The lack of density of the atmosphere at high altitudes and space means that the cold plate has little or no ability to conduct or utilize convection to remove heat from the vehicle. This makes the craft rely on radiation to support heat loss in place of conduction and convection.

The vehicle contains multiple thermal sources. A good example is the engines. The reason for wanting to place electronics on an engine is to make engine controls and maintenance simpler and more efficient. First, if all the engine controls are on the engine and only fuel lines and electrical connections need be made, the maintenance and manufacture of the vehicle are much simpler. Second, most engines have combustion chambers which operate at several thousand degrees Celsius. These chambers require monitoring to operate efficiently and to allow proper control of the combustion process. Current technology accomplishes this monitoring by indirect means. If sensors capable of operating in the environment provided by the engine were available, the control systems of the vehicle could be made more effective.

1.8   ENVIRONMENTAL CONTROL

Higher temperature electronics lowers the work required by the environmental control system (ECS). This is due to the ECS being required to create a temperature environment which is much closer to ambient. A relatively small improvement will often remove heat load from the ECS entirely because the component is capable of operating in its ambient conditions.

Some current aircraft designs transfer the heat generated by the electronics into the fuel, conduct the heat through the body of the aircraft or exhaust it with the fuel through the engine, and eventually dissipate the heat into the environment around the aircraft. This method has several drawbacks. If the heat is transferred to the fuel in the fuel tanks, there is a finite amount of fuel that must be maintained in the aircraft to support the ECS. This fuel reserve has an impact on aircraft range.

The amount of heat generated is also a concern. The electronics power load is continually increasing on current aircraft. Typical new fighter aircraft utilize approximately 50 kW of power. Almost all of this power ends up as heat. This heat must be dissipated by the ECS. In comparison, the pilot, for whom the ECS was created, generates only a few hundred watts. Nearly all of the ECS cooling capability is utilized by the electronics. The power generation system which creates the 50 kW of power needed by the avionics suite is approximately 50% efficient, requiring about 100 kW to create 50 kW. Dissipating the 100 kW of power translates into 100 kW that the ECS must create (assuming 100% efficiency in the cooling system). This results in approximately 200 kW of power taken from the engine.

The cooling system is a major component in aircraft and spacecraft, sometimes weighing as much as 5–10% of the total aircraft weight. The ECS cannot be made redundant without massive weight gains. The ECS is a mechanical system; therefore, is has reliability and mechanical wear-out problems. The ideal solution is for the aircraft electronics to stand on their own and not require cooling; this would remove the major burden on the ECS and would consequently provide options for redundancy and/or weight improvements.

In general, using ECS to cool the electronics has always been an expensive way to accomplish the required function.

1.9 RELIABILITY DIRECTIVES

Rome Air Development Center (RADC) has reduced field-failure data into a design guideline known as MIL-HDBK-217. One of the findings is that the reliability of components is heavily driven by the junction temperature at which the components are operated. This relationship has been translated into Navy, DOD (US Department of Defense) and program directives. It has been stated in several directives from the DOD that junction temperatures of less than 110 °C must be maintained. These directives include the Navy's Willoughby Templates and the DOD's tri-service memorandum. Each of these documents requires junction temperatures to be maintained at less than 110 °C. Again, these directives are based on implications from the RADC data – that junction temperature is directly related to reliability performance on the aircraft.

The true interest of the United States Air Force (USAF), and other military and space services, is in making the equipment they operate more reliable. This has created pressure on developers to produce more reliable systems. A method which has been considered is the use of more rugged components which can be operated in higher temperatures, and to utilize them in cooled compartments to provide even greater reliability.

The directive to build an aircraft with the reliability assessments which today's components achieve when operated at 60 °C has become the baseline for many designs. However, it has generated many negative implications on aircraft performance. The first, best answer is to gain an understanding of the reliability considerations which would allow equipment to operate in the required environment with the required reliability assessment.

Although the information that was collected by RADC is unquestioned, the implications associated with the data are seriously questioned. After much study of the methodology utilized to gather the data, it has become apparent to the community at large that the temperature relationship might not be the only contributor to the reliability problem. In general, as noted elsewhere in this text, the location that is hottest on an aircraft also experiences other energies which contribute to hardware failures. Major known contributors to wear-out are vibration and rapid, wide temperature swings. Generally, the hotter locations on an aircraft also have more vibration and wider, more rapid temperature swings. The temperature swings impact the standard packaging systems by putting stress on the printed circuit boards (PCBs) and the solders they use to manufacture the systems. Temperature induced stresses can break transformer cores and destroy electronic packages. The implication from this data means that, even though higher temperature electronics may be provided, if the packaging is not made more resistant to vibration and thermal shock, the reliability problem will be unsolved.

1.10 CUSTOMER CONCERNS

Several published reports indicate that reasonable increases in temperature, in some cases as little as 25 °C over a military temperature range, would result in components that could be used in the aircraft ambient environment without requiring cooling. These reports have suggested a temperature range of −55 °C to 150 °C ambient for aircraft.

Space applications would benefit from this increase in temperature performance; however, most space applications would prefer a wider range in temperatures with increases in peak ambient operating temperature being most beneficial. Controlling peak ambient operating temperature in space requires some form of cooling system. The greater the operating temperature of a component, the smaller the temperature difference between it and the radiator system that removes heat from the spacecraft.

It appears that little work has been done in the past and is currently being accomplished to create standard components which could be used to increase the temperature performance of the aircraft, increase the reliability of the aircraft and/or lower the weight of the aircraft. The explanation seems to be that it has not been identified as either feasible or desirable.

There are at least two major aircraft subsystems that could benefit from development of high temperature electronics; they are identified in the following paragraphs.

1.11 WEAPONS CONTROLS

In a military aircraft the stores management system (SMS) electronically controls all weapons and other 'stores' (e.g. fuel tanks) attached to the wings or fuselage of the aircraft.

The portion of the SMS which directly interfaces with the store is ideally located at the store interconnect. This allows a modification of the electronics to add additional stores to the weapon suite with minimal impact upon the aircraft. This design technique also lowers the wire mass and bulk of the aircraft. The result is sophisticated, mission-critical electronics with a premium on weapons safety – electronics located on or in the wings or the centerline, etc., with little availability for cooling.

There are at least two reasons why cooling the components in aircraft pylons is very difficult, and perhaps impossible: inability to insure safe sealing of the environmental control system in emergency conditions and a small wing-root with little room for plumbing and wiring.

Stores have been added to the wing pylons which house the electronics and provide mechanical support for the store, but this disturbs the ideal flight characteristics of the aircraft. And these stores tend to be highly volatile weapons or fuel. It is therefore desirable that the stores and pylons are jettisoned in emergency conditions. This 'cleans' the wing, allows the aircraft to be unburdened and makes it safe from the carriage of these external elements, elements which add additional weight, disturb the lifting characteristics, raise the aircraft's drag coefficient and provide potential danger to pilot and crew.

Were jettisonable pylons to have cooling, a faulty jettison could cause the rest of the cooling system to be opened – a highly undesirable failure that might cut off cooling to critical aircraft systems, including the pilot. The electronics associated with jettisonable pylons cannot therefore be cooled.

1.12 SMART SKINS

It is desirable to put sensors in the skin of the aircraft, specifically within the leading edge of the aircraft. The reason for wanting to place sensors or weapons in these areas is also obvious, it is the same reason the pilot faces forward. Most interesting events are occurring in the direction in which the aircraft is pointed. The addition of these sensors imbedded in the skin has resulted in the term *smart skins*. A major electronic element of such a system would be some form of data collection and data reduction computer to compress the data which must travel to the fuselage. This arrangement again creates a need for high temperature electronics without available cooling.

1.13 TECHNICAL CONCERNS

There appears to be no technical reason why electronics cannot be made to operate reliably at temperatures greater than would support uncooled aircraft use. In fact, commercially available industrial applications are currently using components rated to operate at temperatures of 200 °C. Several of the companies have published test results on components which were operational at temperatures above 300 °C using silicon-based devices. Some companies indicate reliable operation at over 500 °C using devices based on silicon carbide (SiC). Experts agree that, at high enough temperatures, reliability problems plague the metalization. As the temperatures increase, problems occur with the packages. Finally, the semiconductor no longer operates. These effects can be modeled and used as design parameters to provide reliable products for the aircraft environment.

1.14 SUMMARY

The impact of high temperature electronics is dramatic. It holds the promise of improving aircraft performance in many ways: increasing the reliability, lowering the flyaway cost, lowering the life-cycle cost, lowering the weight and increasing the range. In the final analysis, the goal of the high temperature electronics expert is to provide the design engineer with the opportunity to add one additional variable to the design process – the temperature range. The engineer can then choose the temperature range to suit a particular application, which may help to optimize the design. The views presented are those of the author and do not represent the views of the DoD or its components.

High temperature automotive electronics

2

J. C. Erskine, R. G. Carter, J. A. Hearn, H. L. Fields,
J. M. Himelick and J. A. Yurtin

2.1 INTRODUCTION

Until the end of the 1950s the only electronic component in the automobile was the radio. Everything else was mechanical, electromechanical or electrical. At the end of the 1950s, the first major solid-state electronic component, the alternator, was introduced. The first units were for trucks and buses and used selenium rectifiers, but silicon diodes were used in car applications. The voltage regulator was transistorized in the early 1960s.

With the success of the alternator, suppliers began to offer other automotive electronics. The 1960s saw the first solid-state ignition systems, tachometers, automatic speed controls, automatic headlight controls, windshield wiper controls, turn signal and emergency flashers, all-solid-state radios and the first electronic engine control system. The new solid-state components were mostly used in noncritical applications, so their failure would not result in loss of drivability, they were generally more expensive, and thought to be not as reliable as electromechanical systems.

Growth in the use of solid-state automotive electronics slowed in the 1970s as other issues (federal emissions, fuel economy and vehicle integrity requirements, plus an oil crisis) commanded the industry's attention. Next came the introduction of the catalytic converter, solid-state ignition systems, massive downsizing of the fleet and rapid growth in fuel efficiency. Electronic displays appeared briefly and automatic heating, ventilating and air conditioning (HVAC) systems were introduced. Electronic engine control and fuel injection were used on some European cars. The 1970s, however, were the staging point for a host of new electronic components that would burst on the scene in the 1980s.

High Temperature Electronics. Edited by M. Willander and H.L. Hartnagel. Published in 1997 by Chapman & Hall, London. ISBN 0 412 62510 5.

In 1979, 1980 and 1981 the microprocessor-based engine control module (ECM) was phased in to control the operation of the engine to meet emissions requirements with acceptable drivability. With the ECM came the solid-state manifold pressure sensor and exhaust oxygen sensor. This was a major undertaking for the industry and its success gave confidence that solid-state electronics had matured to the point where it could be used in primary vehicle functions. Later in the 1980s, there appeared electronic applications such as antilock braking systems, air bags, electronic displays, all-electronic instrument panels, head-up displays, wide use of fuel injection, suspension control, electronically controlled transmissions, traction control, multiple microprocessors, data links between microcontrollers and limited multiplexed signal and power distribution. The value of the electronic components in the average North American automobile has corresponding-ly increased from about $20 in 1955 to $782 in 1993 and to a projected $900 in 1995 [1].

The next decade will most likely see the introduction of automotive electronic systems, subsystems and components at least at the same pace as in the previous decade. Automotive trade magazines offer lengthy lists of possible new and/or improved automotive electronic systems such as electric power steering, electrically multiplexed power distribution, optical multi-plexing, door air bags, active suspension control, traction control, collision avoidance systems, vision enhancement systems, route guidance and navi-gation systems, high intensity discharge headlamps, electrically darkenable windows for solar heat load control, active noise cancellation, interactive voice recognition systems, adjustable vehicle surfaces for aerodynamic drag reduction, higher power/higher voltage power generation systems and the electric vehicle. Customer acceptance, cost and world economic climate will determine which are successful.

The importance and value of the electrical and electronics (E/E) content of the typical North American automobile has thus increased dramatically over the past 25 years. Electronic controls are now indispensable in meeting legislated emissions and fuel economy requirements as well as providing drivability, safety and comfort. Today the value of the E/E system in a high-end vehicle is comparable with that of the powertrain or the body-in-white, and the function and value represented by the E/E system will continue to increase for the next decade at least before saturating.

In our discussion of high temperature automotive electronics, we focus on ceramic substrate hybrid electronics and the cables and connectors which distribute the signals and power in the E/E system. Ceramic substrate hybrid electronics is currently extensively used in automotive applications and it is our belief that, of the proposed approaches to higher operating temperature electronics [2], this well-developed, cost-effective circuit tech-nology offers the best opportunity for improvement to meet the more stringent specifications predicted for the future. We do not discuss actuators because, with the exception of the alternator, they are primarily electromagnetic devices with no internal electronic content.

This chapter first describes the automotive environment and outlines the various electrical and physical specifications that automotive grade E/E components must meet. After reviewing present and projected future specifications, we move on to discuss the reliability issues associated with increasing the maximum operating temperature of the major components of the E/E system: (1) semiconductor devices and integrated circuits, (2) hybrid circuit technologies, (3) hybrid circuit packaging technology and (4) cables and connectors. These components constitute the E/E system, and each part of the system must meet all reliability specifications in order for the system to meet specifications. The time horizon for these projections is 10 years into the future. This work expands on a shorter and less complete discussion of high temperature automotive electronics which we published previously [3].

2.2 THE AUTOMOTIVE ENVIRONMENT

Automotive components must operate under very difficult conditions with outstanding reliability. Our vehicles must operate satisfactorily in any environment they may encounter anywhere in the world. The operating life of the automobile is about 8000 hours and the nonoperating or shelf life is 8–10 years or more.

2.2.1 Nonthermal environment

The nonthermal (mechanical, chemical and electrical) environment for automotive E/E components is quite rigorous. Automotive electronics parts must sustain substantial shock and vibration, exposure to a wide variety of chemical agents and a relatively hostile electrical environment over their operating life. Very high quality parts are required to survive this severe environment. Even though automotive E/E specifications are akin to those required for military and aerospace application, they must be met with parts that have prices more typical of consumer electronic parts. SAE standards J1211 [4] and J1455 [5] provide a detailed description of the environmental conditions under which vehicular components must operate.

2.2.2 Thermal environment

Table 2.1 summarizes the currently required baseplate operating temperature range for specific component locations and the maximum baseplate

Table 2.1 Existing and projected operating temperature for principal vehicle locations

Component location	Operating temperature range (baseplate)	Projected future maximum operating temperature (baseplate)
Passenger compartment	−40 °C to 85 °C	Unchanged
Engine compartment	−40 °C to 125 °C	160–165 °C
On-engine and on-transmission	−40 °C to 140 °C	160–165 °C
Wheel-mounted components	−40 °C to 250 °C	Unchanged

temperature projected for future applications. The baseplate temperature is the temperature of the heat transfer surface of the module. Internal temperatures will be higher due to power dissipation. Typical junction temperatures for ICs are 10–15 °C and for power devices 25 °C higher than the baseplate temperature. The temperature specification for nonoperating conditions is −40 °C to 85 °C. At present, automotive electronic modules are designed for a maximum operating temperature of 125 °C. Thus, there are few electronic components mounted on the engine or transmission.

2.2.3 Future trends

We expect that some thermal specifications will become more demanding in the future, as indicated in Table 2.1. Cars are lasting longer and lifetime range is increasing, which may require operating and shelf-life specifications to lengthen. Underhood temperatures may increase above today's levels due to factors such as the reduction in the size and lower radiator opening position, adoption of 'cab forward' styling and greater engine compartment 'fill factor', all of which can reduce the amount of cooling air moving over the engine. Electronic components may be mounted directly on the engine and transmission and within the actuators to reduce costs and simplify assembly [2]. The mechanical, chemical and electrical specifications are likely to remain much the same, while the maximum operating temperature specification is increased. The increasing automobile electronic content and the lengthening of the new vehicle warranty period requires that the already admirably low defect rate and high reliability of automotive electronic components be significantly improved while the operating specifications become more stringent.

2.3 SEMICONDUCTOR DEVICES AND INTEGRATED CIRCUITS

It is well known that silicon devices have the potential to operate at temperatures higher than their current maximum temperature ratings. The band gap of silicon is large enough that, with the suitable choice of doping concentrations, bulk CMOS ICs can operate up to the vicinity of 250 °C, albeit with greater leakage current [6, 7]. In recent years considerable research on raising the operating temperature of silicon ICs has been published. It now appears that silicon-on-insulator ICs, with their dramatically reduced p–n junction cross-sectional area, can be successfully operated at over 300 °C [8].

Higher operating temperature silicon power transistors, however, have not received the same research investment as ICs. Thus they lag behind ICs in demonstrated potential for increased maximum operating temperature, although it is likely that improvements can also be made to increase their operating temperature. Figure 2.1 shows the reliable operating temperature domain of the silicon semiconductor devices and device materials discussed. The reliable operating temperature domain is a graphical approach to illustrating the material or structural changes that can be used to increase the reliable operating temperature of automotive components. We use it for each of the four categories discussed to provide the reader with easy insight

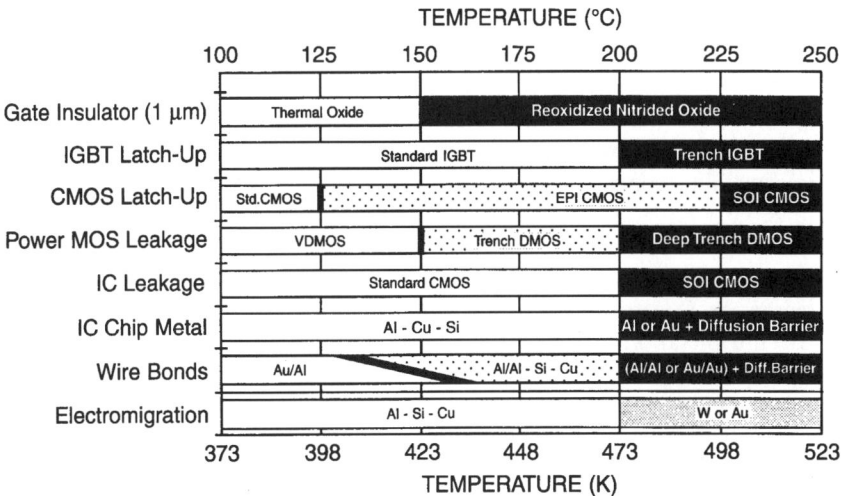

TEMPERATURE (°C)

	100	125	150	175	200	225	250
Gate Insulator (1 µm)	Thermal Oxide		Reoxidized Nitrided Oxide				
IGBT Latch-Up		Standard IGBT			Trench IGBT		
CMOS Latch-Up	Std.CMOS	EPI CMOS				SOI CMOS	
Power MOS Leakage	VDMOS		Trench DMOS		Deep Trench DMOS		
IC Leakage	Standard CMOS				SOI CMOS		
IC Chip Metal	Al - Cu - Si				Al or Au + Diffusion Barrier		
Wire Bonds	Au/Al		Al/Al - Si - Cu		(Al/Al or Au/Au) + Diff.Barrier		
Electromigration	Al - Si - Cu				W or Au		
	373	398	423	448	473	498	523

TEMPERATURE (K)

Figure 2.1 Reliable operating temperature domain for silicon semiconductor devices and materials.

into the aspects which limit the increase in reliable operating temperature of automotive E/E systems and the options available to increase the maximum operating temperature.

2.3.1 Integrated circuits

Various integrated circuit technologies are utilized in automotive electronic designs, including CMOS, bipolar, BiCMOS and smartpower. Devices in each of these technologies have a temperature sensitivity that must be understood with respect to the application. In addition to the thermal properties of base silicon, the other materials making up the device (gate insulator, interconnect metal, contacts, isolation layers, external contacts, etc.) must have a correspondingly high maximum operating temperature. The research literature indicates that there are improvements in the materials and processes that will permit higher maximum operating temperatures.

SiO_2 gate insulators experience problems with charge trapping and breakdown above about 250 °C (200 °C for the thinner insulators) [9, 10]. Reoxidized nitrided oxides [11] are reported to be more reliable than conventional gate oxides at 250 °C.

Above about 200 °C, the aluminum or Al/Si/Cu chip metalization alloys with the silicon at the contacts and the consequent spiking through and shorting of p–n junctions becomes a problem. Inserting a diffusion barrier, such as niobium, between the aluminum metal and the silicon can suppress alloying and give contacts which are stable to 300 °C or higher [12]. Other chip metals, such as Ti/Pt/Au, have also shown good reliability at high temperatures [13].

Aluminum wire bonds to Al chip metal have been reported to be reliable to 300 °C [14]. The commonly used gold thermocompression wire bond to Al chip metal has well-known problems with the formation of brittle Au–Al intermetallic compounds. Reliable operation to the 130–160 °C range is possible with good control over material purity, contamination and the bonding process. Above this temperature range, it is appropriate to change to a monometal bond system, such as Al to Al chip metal with a diffusion barrier or Au to Au chip metal with a diffusion barrier. Al bond wires and Al chip metal with a diffusion barrier probably represent the most cost-effective choice for the bond metal system of high temperature automotive silicon devices.

Electromigration, which produces open circuits in chip-level interconnects, is accelerated by temperature and current density [15]. For temperatures above about 200 °C, aluminum metalization can fail due to electromigration in a few thousand hours. Electromigration can be controlled by reducing the maximum allowed current density or using a more electromigration-resistant interconnect metalization such as W or Au.

2.3.2 Power transistors

Increased operating temperature power transistors have received relatively little attention from the research community and yet the output function they supply is essential in many electronic systems. Silicon power transistors have large p–n junction areas and thus have substantially larger leakage currents than ICs. Planar vertical DMOS power transistors are widely used in automotive applications because of the ease of interfacing them with microcontrollers, their low on-resistance and their relatively low cost. Since the electron mobility has a negative temperature coefficient, achieving low on-resistance at elevated temperatures will require larger area dies than used currently but with the penalty of increased leakage current.

Within the last two years, a new power MOSFET structure has been introduced into the marketplace. It uses a trench structure where the gate electrode is placed in trenches etched in the surface of the wafer [16]. Not only does the trench DMOS (TDMOS) power MOSFET offer a much larger channel width-to-length ratio for lower specific on-resistance, it reduces the area of the p–n junctions inherent in the vertical DMOS structure and thus offers the potential for lower leakage current at high temperature.

2.3.3 Sensors

Sensors are the most recent addition to the automotive E/E system catalog of parts. Because of their newness, they tend to be expensive and offer the greatest uncertainty about their potential for higher temperature operation. The trend is toward microsensors due to the promise of low cost inherent in microelectronic batch fabrication, their small size makes location in a crowded engine compartment easier, microsensors can use some of the packaging technology that has already been developed for other microelectronic devices, and the possibility of integrating the interface electronics on the sensor chip can reduce component count and cost and increase reliability.

Microsensors use some materials which are not traditionally found in semiconductor devices and some microsensors have micromachined features such as diaphragms and cantilevered elements. Capacitance and piezoresistance – mechanisms commonly employed to sense physical parameters such as acceleration, pressure [17] and yaw rate [18] in microsensors – make these devices sensitive to interference from differences in the thermal expansion of the sensor materials and from the difference in thermal expansion of the sensor chip and the packaging.

Some devices, such as absolute pressure manifold vacuum sensors and yaw rate sensors, require vacuum packaging which must retain its integrity over the life of the vehicle. Many of the temperature sensitivity issues discussed in section 2.3.1 also apply to microsensors. Microsensor technology may require the most development, in the basic device and device materials technology as well as in the packaging, to attain the higher projected maximum operating temperatures listed in Table 2.1.

2.3.4   Other semiconductor materials

Semiconductor devices based on semiconductors other than silicon, such as GaAs, SiC, GaN or diamond, have been proposed and some research reported [19]. The most mature of these alternate semiconductor device materials is GaAs which is used extensively in high frequency applications. GaAs devices and circuits are more expensive than equivalent silicon devices and circuits. Because GaAs does not have an adherent native oxide like silicon, the circuit technology is built around depletion mode MESFET structures. Generally, automotive electrical engineers do not favor depletion mode devices and circuits because off or leakage currents can increase the parasitic current load on the battery in applications where the power bus is not switched off when the vehicle is not in operation. Published studies of high temperature GaAs devices report degradation of the gate contacts and ohmic source–drain contacts and high leakage currents [19]. Additional development is usually required to improve the operating life at 250 °C and above. It is not likely that devices and circuits based on III–V compounds will see more than niche application in future vehicles. Light-emitting diodes (LEDs) made from III–V semiconductors will, however, continue to be widely used as indicators.

SiC has a much larger band gap than GaAs and has the potential for much higher operating temperature. SiC's large band gap and high breakdown field offer exciting potential to replace Si, especially for power transistors [20]. There has been considerable progress in developing silicon carbide (SiC) devices over the past few years. Now that device quality SiC wafers with lower defect densities than in the past are becoming available, device and fabrication process research can move forward. However, much work remains to be done before SiC devices can satisfy more than niche markets: wafer size must be increased considerably beyond the present 2 in. diameter, defect densities must be very substantially reduced, and shallower activation energy n- and p-dopants must be developed that do not begin to freeze out at the lower end of the automotive operating temperature range. Because of the severe cost constraints of automotive application, we are skeptical that SiC active devices will see significant automotive use in the next 10 years.

Diamond and some nitrides are semiconductors with wide band gap and electrical properties attractive for high temperature application [19]; however, electronic technologies for these materials are in the embryonic stage and it seems unlikely they will see automotive application in the near future.

2.4   HYBRID CIRCUIT TECHNOLOGIES

The typical hybrid circuit today consists of a ceramic (alumina) substrate containing a circuit formed by thick film printing with passive and active devices attached by a soldering process [21]. Resistors, capacitors, thermistors and dielectric are also printed on the substrate. Table 2.2 presents an abbreviated list of materials, components and attachment systems used in automotive hybrid circuits. Figure 2.2 is a photograph of a hybrid electronic intake manifold pressure sensor [17] which has been extensively used in

Table 2.2 Materials, components and attachment systems used in automotive hybrid circuits

Thick film inks	Surface mount components	Attachment systems
Conductor	Capacitors: ceramic, electrolytic	Solder
Dielectric	Resistors	Conductive adhesive
Resistor	Varistors	Wire bonding
Thermistor	Resonators and crystals	Flip chip
Capacitor	ICs and power transistors	Tab tape

Figure 2.2 Hybrid intake manifold pressure sensor. Note the use of heavy wires to interconnect the circuit and package terminals and the flip-chip integrated circuit. The cavity is filled with a silicone gel prior to attaching the cover. This unit, first introduced in 1979, has been supplanted by smaller and more advanced hybrid pressure sensors.

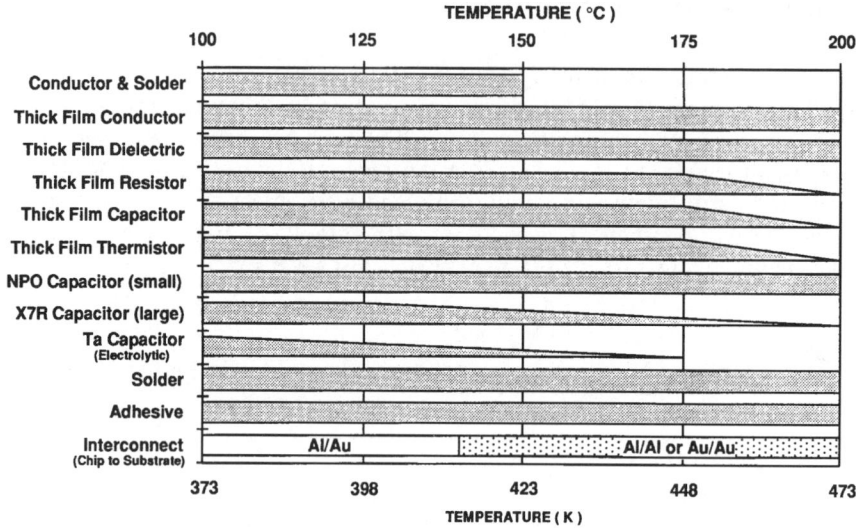

Figure 2.3 Reliable operating temperature domain for hybrid circuit technologies.

£TTT∠𝔼#£ General Motors automobiles. It illustrates the basic features of the automotive hybrid electronic module.

The present underhood automotive hybrid module is designed for a 125 °C maximum baseplate temperature. As Table 2.1 indicates, this is not as high as some applications may require in the future. In examining an increase in the operating temperature, there are the components and two major systems to consider: the thick film inks and the attachment system. Both systems have a temperature sensitivity related to solid-state diffusion of their elements. Raising the operating temperature will increase the amount of diffusion over the life of the circuit, which may change the material structure and physical properties, creating new failure modes. The goal is to increase the operating temperature while increasing the useful life expectancy to 10 years or more, which will require selecting combinations of materials with lower diffusion constants. Figure 2.3 shows the reliable operating temperature domain for the hybrid circuit technologies considered.

2.4.1 Ceramic substrates

Traditionally, ceramic substrate electronics have been used for underhood automotive applications and principally on the engine. The early applications were small circuits with one or two ICs on a substrate of less than 1 in.^{2} (6.5 cm^2) in area. Today the size has grown to as large as 12 in.^{2}

($77.4\,\text{cm}^2$) with many ICs, each containing very complex circuits. Alumina ceramic was chosen as the substrate material for several reasons. For use on the engine, stability at high temperature and resistance to vibration is mandatory. Alumina provides both of these with its very high melting temperature and large modulus. The high melting temperature of alumina allowed the development of thick film inks that are fired at high temperatures around $800\,^\circ\text{C}$. These high melting temperature inks provide reliable, very stable circuits at the underhood operating temperatures.

Other major advantages of using alumina compared to organic printed circuit boards are its low thermal coefficient of expansion and high thermal conductivity. The low thermal coefficient of expansion provides a mounting surface for silicon ICs and chip components; this produces low thermally induced mechanical stresses in the part and the interconnect material, improving reliability and allowing a greater operating temperature range for the circuit. The high thermal conductivity of alumina allows the substrate to be used as a heat sink for moderately high thermal flux densities. This simplifies the thermal structure for ICs that dissipate heat, allowing smaller packaging and simpler, lower cost manufacturing.

Other high temperature substrate materials are available today with properties as good or better than alumina [22]. BeO, AlN and SiC all have higher thermal conductivity than alumina and as low or lower thermal coefficient of expansion. Some of these materials are used today in very small substrate sizes. For example, BeO is commonly used as a high thermal conductance dielectric isolation for packaging high power ICs. These materials have the same problem – higher cost than alumina. Also, there is very little infrastructure for manufacturing large-area substrates and associated materials in high volume. The prospect of alumina being replaced by these materials is very small at this time. However, there will be special applications where the properties of these materials make them worth the additional cost.

2.4.2 Thick film inks

There are five types of inks, and for each type there are many available inks from which to choose. In order not to limit the electrical design, there must be a complete system of compatible inks with a wide range of parametric values. The selection and/or development of a **set of inks** is a difficult task because the diffusion constants are not well known and thus most development must be accompanied by extensive testing. The required parametric values are solderability, bondability, low sheet resistance (conductor inks), wide resistance range (resistor inks), no migration, excellent adhesion, fine line printable and no parameter change with aging. Given the complexity of the system, finding the optimum set of inks is a difficult task. Most ink systems are stable to $200\,^\circ\text{C}$ with some small parameter variations. The compatibility of any system should be tested at the higher temperature. However, the compatibility of the commonly used solders and conductor inks will have problems above 150 °C. Compatibility of inks could become an issue at higher temperatures.

2.4.3 Attachment systems

Solder and conductive adhesives are the attachment systems in use today. The reliable operating temperature domain of the adhesives is shown in Fig. 2.3. The adhesive systems are useful at 150–200 °C depending on the base resin. They are being used on ceramic, rigid and flexible circuit substrates. To use these materials at high temperatures, it is necessary to choose a resin that has a glass transition temperature above the maximum operating temperature. The required elasticity will be a function of the substrate and the components being attached. The adhesion properties must give high strength over the required lifetime.

The solder attachment system has a solid-state diffusion problem which is system (conductor, solder and component metalization) oriented. Since the solders currently used are low temperature melting materials (e.g. 63–37 lead–tin eutectic, high lead content solders, indium–lead solders), the solidus point must be maintained well above the operating temperature. There are many solders available with a wide range of melting points. However, solder selection is a trade-off of strength, ductility and compatibility with the attached components. Also, the conductor, the solder and the metalization must have a desirable mutual solubility for good surface wetting in order to give high reliability and good yields. Finding the most desirable system will require extensive testing to assure long-term reliability.

There may be little flexibility in the choice of metalization of the components due to the relatively low automotive volumes for surface mount components compared to the total market. The automotive market is not a strong driver for the improvement of these components. Developing a proprietary metalization for automotive application could generate a significant increase in component cost.

2.4.4 Surface mount components

The major electronic components are listed in Table 2.2. Of the many available components, very few are designed to operate above 150 °C and some may have parameters derated to zero at 125 °C. NPO ceramic capacitors have a very low temperature coefficient of capacitance and function at 200 °C without difficulty. X7R capacitors are advertised to function at 200 °C but at this temperature they are derated to 35% of room temperature capacitance. Another example concerns very high capacitance parts where electrolytics are normally used. The voltage rating at 175 °C is

25% of the room temperature rating and the reverse voltage is derated to zero. These parts may work for some circuits but not where the capacitance or voltage rating is critical.

Adding to the parametric problems is the predictable loss of reliability with the increase in operation temperature. With increasing temperature, the reliability of any component or material system will decrease unless the component or material system is modified to offset this reduction. Some components can be designed to operate at higher temperatures, but others have parametric problems which are a function of the material properties so improvements must come from changing the materials. Some users desire parts with higher temperature ratings than can presently be supplied by the manufacturers.

2.5 HYBRID PACKAGING TECHNOLOGY

The typical technologies used in packaging the hybrid circuits consist of a housing, heat sink, terminals, insulators, adhesives, passivants and interconnections. The present packaging systems function quite adequately in a 125 °C environment. There is a very large database of field experience to support applications in this temperature range. Very few if any temperature-related failure modes exist with the packaging. This does not mean the temperature may be increased without developing failure modes. It implies that moderate increases could be accommodated with minimum development.

Table 2.3 lists the materials used in hybrid packages. As the temperature is increased, materials and technologies may become unsatisfactory and require improvement or changing. The present technology is adequate up to 150 °C. Above 150 °C the first items to require development are plastic housings, plastic insulators and copper alloy terminal materials. Figure 2.4 shows the reliable operating temperature domain of the package components and materials considered.

Table 2.3 List of package components and their constituent materials

Package components	Materials
Housings	Plastic, aluminum
Heat sink	Aluminum, copper
Contacts and terminals	Copper alloy, steel
Insulators	Plastic
Passivants	Silicones
Adhesives	Silicones, epoxies, acrylics
Interconnects	Ultrasonic bonding, soldering, welding

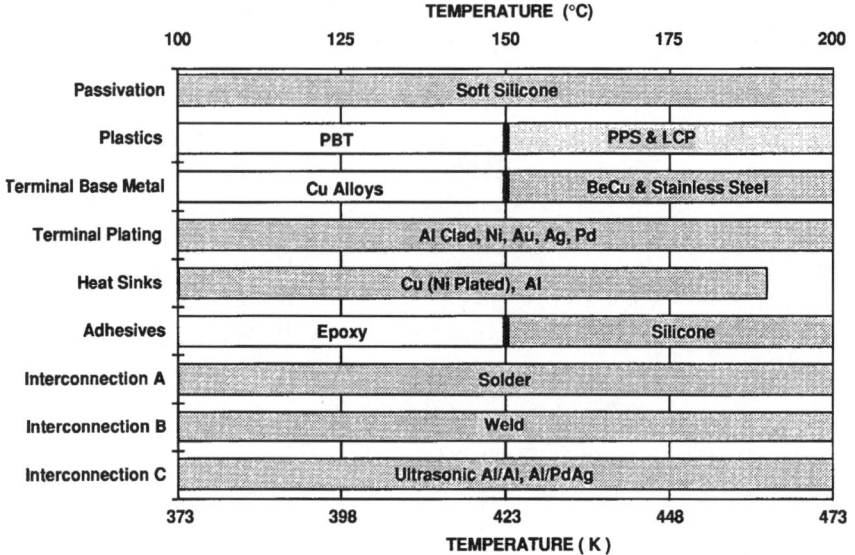

Figure 2.4 Reliable operating temperature domain for hybrid packaging technologies.

2.5.1 Housings and insulators

Plastic housings and insulators become suspect at temperatures above 150 °C. These are typically molded from polybutylene teraphthalate (PBT) today and can be molded from a different material with higher temperature capability, provided it meets all other requirements. Some of these requirements are moldability, strength, resistance to fluids and environmental contamination, bond strength of adhesives, ionic contamination, compressive and tensile creep, and cost. As noted in Fig. 2.4, polyphenylene sulfide (PPS) is a material that meets the temperature requirements above 150 °C. It also meets many of the other requirements. The penalty for using this or other options is cost, not a major factor with PPS, but would be more of a problem with materials such as liquid crystal polymers (LCPs). Aluminum die castings are widely used as housings for hybrid circuits. Aluminum housings are acceptable in the ranges being considered.

2.5.2 Heat sinks

Heat sinks can be made of copper or aluminum – die cast, extruded or wrought. There is no temperature problem with aluminum or copper but the copper surface must be protected from oxidation by plating or other means, and in some designs anneal-resistant alloys may be desirable.

2.5.3 Connectors and contacts

Contacts in many shapes and sizes made from various materials are used. The contact material must be compatible with the internal interconnect system as well as the mating connector. For this reason, the contact end is normally plated. The internal interconnect end may be plated, clad with metal or solder coated with the proper compatible metals. Base materials are typically brass, bronze or copper. With proper barrier materials, these materials will function in 175 °C environments. The mating connector to the contact requires a spring action to generate a normal force on the contact and this force must remain stable for the life of the product. For temperatures above 150 °C, a replacement base material, such as beryllium–copper or stainless steel alloys, will be necessary.

2.5.4 Passivations

A soft silicone passivation material is almost universally used in automotive modules today. It minimizes the mechanical stress on the components and wires in the package and also wets the surfaces of the materials in the package to protect the circuit from condensed water and prevent the interfacial migration of moisture. As shown in Fig. 2.4, the temperature capabilities of these materials exceed the requirements of underhood applications if properly selected.

2.5.5 Adhesives

Adhesives are used to assemble the package: attaching covers to housings, housings to heat sinks, connectors to housings and hybrid circuits to housings or heat sinks. Silicone adhesives are widely used for many of these applications. Epoxy formulations are used in limited applications because they do not have high elasticity, even though they are typically stronger than silicones.

2.5.6 Interconnections: circuit to housing

There is more variety in this area than most other parts of packaging. Various welding, soldering and ultrasonic bonding systems are being employed. Ultrasonic wire bonding is usually done with an aluminum wire bonded to a housing terminal and to a conductor on the substrate. The concerns here are the compatibility of the interface materials and temperature effects of the interface. The typical terminal would be aluminum clad or nickel plated. Neither Al–Al nor Al–Ni interfaces should be a problem at 175 °C. The substrate connection is an aluminum bond pad or a PdAg thick film conductor. The Al–Al bond would not be a problem, as discussed in section 2.3.1, but there would be a limit to the operating temperature of the Al–thick film interface.

Welding is a high temperature bond between two materials. Assuming the materials are metallurgically stable in the temperature range of interest, there should be no temperature-related problems. Interfaces of Ni/stainless steel (SS) or Ni/bronzes should be stable in the temperature range of interest. There are other interface combinations that would be satisfactory. Soldering is not the most widely used interconnection process, partly due to the low strength of solders and the requirement for a very low mechanical stress design on the interconnect conductor.

If the soldering process is such that the substrate is heated when forming the interconnections, the component attachment solder must have a higher

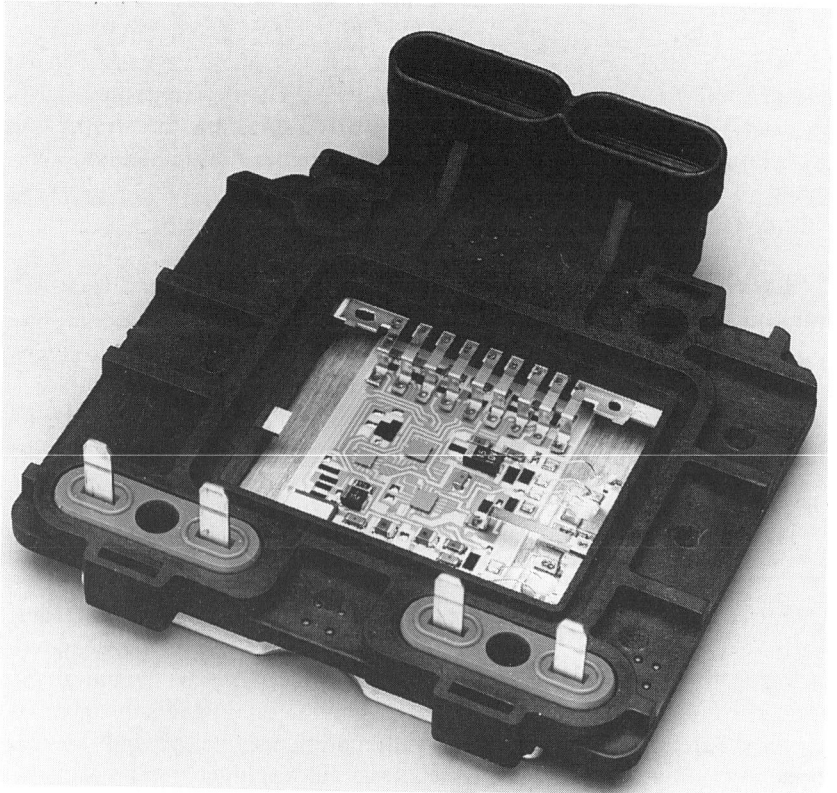

Figure 2.5 Hybrid direct ignition system module. The complete unit has one ignition coil for every two cylinders and eliminates the need for the mechanical distributor. This unit is mounted on the engine.

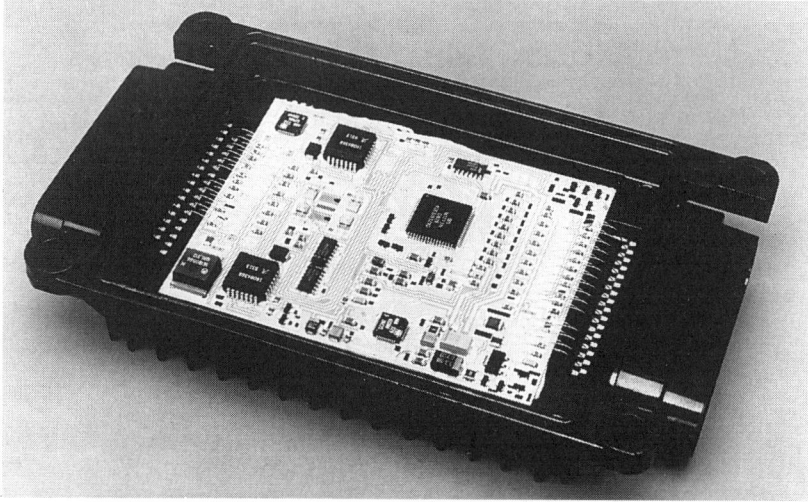

Figure 2.6 Hybrid engine control module developed for marine use. Packaged integrated circuits are used in this product because of the relatively low production volumes.

melting temperature than the interconnect solder to avoid reflow. This may reduce the choice of available solders. These solders have the same strength, ductility and diffusion problems as component attach solders. Depending on the application, some presently used solders may work at around 150 °C if the conductor-induced mechanical stress in the solder is kept low. At higher temperatures there are many choices, all of which will require extensive development to determine the compatibility of materials.

Figures 2.5 and 2.6 are photographs of the direct ignition system module, used extensively on General Motors six-cylinder engines, and a hybrid engine control module for marine use. They clearly illustrate the rugged construction and the features of automotive hybrid electronic modules as discussed in sections 2.4 and 2.5.

2.6 CABLES AND CONNECTORS

Cables and connectors provide the vital interconnects between the various modules that make up the power and signal distribution system. Figure 2.7 shows a typical engine wiring assembly. The issues associated with increasing the maximum continuous operating temperature are especially important as downsizing of interconnects follows downsizing of electronic packages.

Many of today's auto makers require specific underhood components to be validated to continuous operating (air) temperatures of greater than

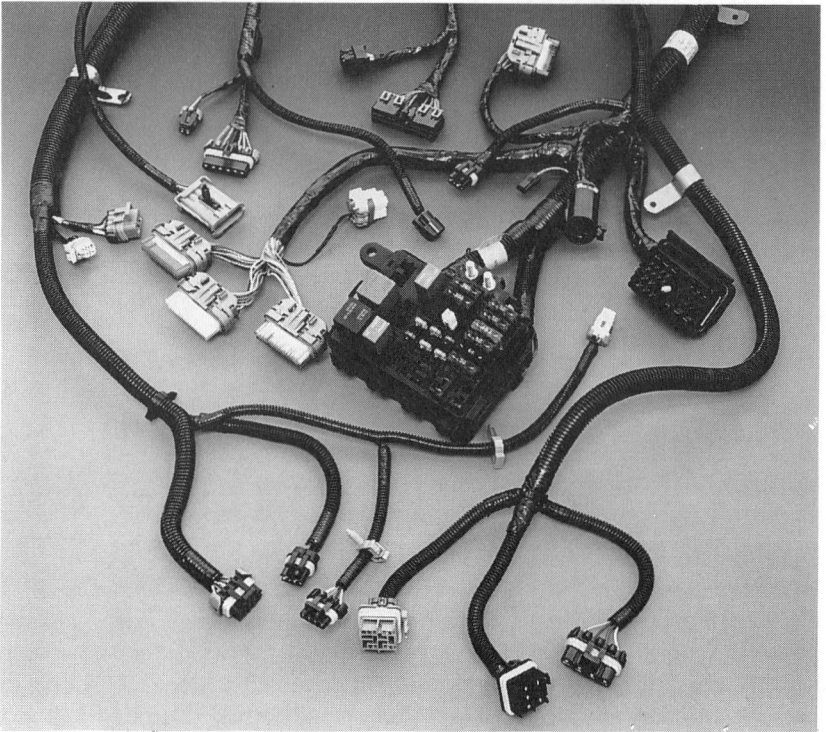

Figure 2.7 Typical vehicle engine wiring assembly with electronic content.

150 °C. If the old rule of thumb that peak engine block temperatures are 25 °C higher than peak air temperature applies, this could imply 175 °C engine block temperatures. The following discussion is divided into three sections that make up the connection system: cable (core and insulation), terminals (base metal, plating and crimps) and housings. Figure 2.8 shows the reliable operating temperature domain for cables and connectors.

2.6.1　Cables

Unplated copper conductors (core) are usable in continuous operating temperatures of up to about 150 °C, beyond which hydrogen embrittlement proceeds rapidly, reducing its useful flex life. Plating the core with silver can extend service to 200 °C and, with nickel plating, to 260 °C. Thicker platings designated by ASTM specifications can increase temperature resistance, e.g. Class 2 (2% by weight) can reach 260 °C and Class 27 (27% by weight) can reach 450 °C. Obviously, the trade-off is cost and electrical performance. If

AMBIENT AIR TEMPERATURE (°C)

	100	125	150	175	200	225	250	275	300
Cable Core	copper		Ag plated copper		Ni plated copper		stainless steel		
Cable Insulation	cross-linked polyethylene		PETFE		PTFE				
Terminal Base Metal	copper alloys		BeCu, BeNi, stainless steel, spinoidal Alloys						
Terminal Platings	tin	Ag,Au,PdNi	Au, Ag, Pd						
Terminal Crimps	crimp only	soldered or welded			welded				
Housings	PA,PBT	PPS,PEI,DAP		LCP					

373	398	423	448	473	498	523	548	573

AMBIENT AIR TEMPERATURE (K)

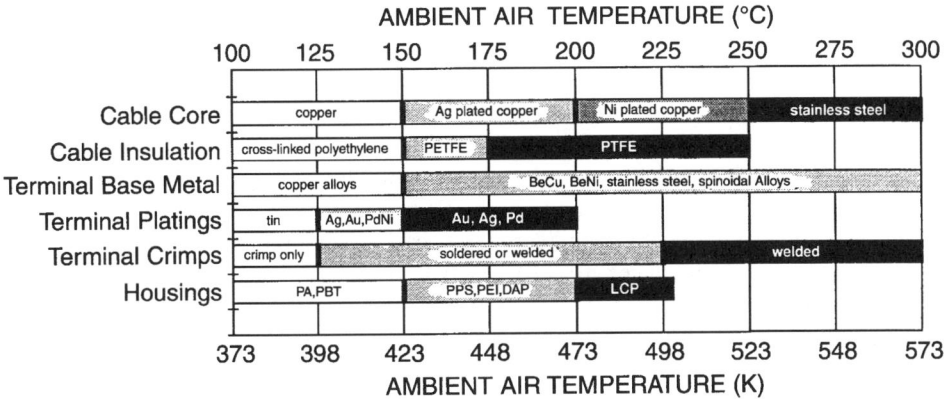

Figure 2.8 Reliable operating temperature domain for cables and connectors.

it ever becomes desirable to exceed a 260 °C ambient, a stainless steel core must be used. Since stainless steel has a higher resistivity than copper, a larger gauge size must be used to attain the same current ratings, and this may have an adverse affect on connector size due to the centerline spacing required.

The cable insulation can be conventional cross-linked polyethylene up to 135 °C and, with some proprietary formulations, up to 150 °C continuous exposure. Above 135 °C most cross-linked polyethylene becomes brittle. Tefzel (PETFE) insulation can be used to about 175 °C and Teflon (PTFE) insulation can be used to about 260 °C. Use of either insulator material entails a significant cost penalty. Further investigation is required for approved insulating materials for use above 260 °C.

2.6.2 Terminals

Most electronic terminations consist of a base metal, which provides mechanical as well as electrical integrity, and a plating to provide electrical interface integrity as well as corrosion protection.

Conventional copper alloys can be used to 150 °C, but above 150 °C high copper content materials rapidly lose their strength. Conventionally, beryllium coppers and beryllium nickels can be used but with a significant cost increase. Spinoidal alloys which are phase dispersed have advantages at higher temperatures and have emerged as an alternative terminal material. Clad metal technology opens up the possibility to use copper-clad stainless steels to allow the best mechanical and electrical properties in a single material. The other option is two-piece terminals using copper base alloy to maximize the electrical properties and stainless steel to maximize the mechanical properties. These two-piece terminals are significantly more expensive to produce.

Tin platings may be used reliably up to 125 °C, and tin-plated terminals are performing in many higher temperature applications in today's automobiles with apparent success. From 125 °C to 200 °C platings of precious metal alloys of Pd, Au, Ag and Pt can be used with a significant cost increase. Platinum or palladium alloys can be used above 200 °C, possibly in combination. A significant part of the requirement for precious metal platings is the need to provide low resistance contact with a minimum normal contact force and an extremely small terminal. Above 200 °C, however, there is little information related to connectors. Significant research and development is needed to develop reliable terminations for ambient temperatures above 200 °C. And, above 150 °C, lubricants must be solid, e.g. gold. Liquid lubricants are not recommended.

Conventional 'crimp only' technology is reliable up to 125 °C (some select circuits may require soldering). New crimp technologies are being developed

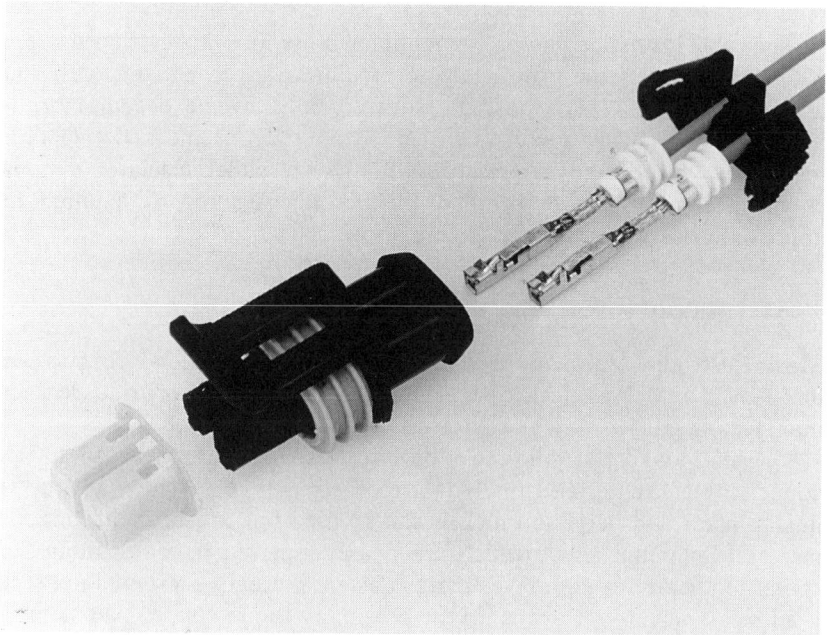

Figure 2.9 Exploded view of the GT Connection System, showing plated terminals, seals and housing components. The connector is designed to prevent the entry of environmental contaminants which can degrade the quality of the interconnection.

which will perform at higher temperatures but, generally, above 125 °C soldering or welding is recommended. At 225 °C soldered crimps are not reliable and only welded crimps are acceptable. Welded crimps are reliable to well over 300 °C.

2.6.3 Housings

Above 150 °C the existing higher cost engineering materials – polyphenylene sulfide (PPS), polyether imide (PEI), diallyl phthalate (DAP) can be used to 200 °C. Housings from liquid crystal polymers (LCPs) can be used to about 235 °C if no flex members are required. Materials are available for use above 235 °C, but they require significant new process technologies which would have a great impact on component design and development. Figure 2.9 shows an exploded view of a typical connector and illustrates the basic features of an automotive grade connector.

2.7 SUMMARY AND CONCLUSIONS

Semiconductor devices and circuits

The materials properties of silicon, the body of research literature on high operating temperature silicon ICs, the recent introduction of trench DMOS power transistors and their potential for low leakage currents at high temperatures, and the low cost of silicon electronics in general lead us to believe that silicon-based circuits and devices will satisfy almost all automotive high temperature electronic requirements for the foreseeable future. Junction operating temperatures in the 200 °C range should be reliably achieved with some modification of existing materials. However, ICs may require redesign to give acceptable performance over the wider operating temperature range. A development program by the supplier industry is required since few high temperature qualified silicon devices and circuits are available today.

Hybrid packaging

Operation to 175 °C is possible using higher temperature plastics for the housing and connector and a terminal material that retains its spring rate at the higher temperature.

Hybrid circuit materials

The thick film inks used in hybrid circuits are stable at around 200 °C, with some change in the parametric values. However, the compatibility of the inks within an ink system must be determined to assure long-term reliability.

Component attachment

To increase the maximum operating temperature, the development of a higher temperature solder system will be necessary. The solder system must be compatible with the thick films and the component metalization, as well as having improved strength and ductility to meet the increased thermal stress requirements. This is a major development project.

Components

There are many types of components used in hybrid circuits and many different manufacturers of each component. Very few of these components are designed to operate above 125 °C. Products that are adapted for the 150–200 °C range require all their components to be appropriately upgraded. Without a complete set of components available, the circuit designs will be limited in scope.

Cables and connectors

The wiring harness and connectors can go to 200 °C ambient temperature with existing higher performance materials. Higher operating temperatures will generally require more expensive insulator, housing and contact materials. For temperatures greater than 200 °C, the development of new materials and processes for housings, platings and insulators is required.

Overall

We conclude that increasing the maximum operating temperature of automotive electronic modules, cables and connectors can for the most part be accomplished through improvements in today's technology, although there are some aspects, such as some components, which require substantial improvement. The desire to increase the operating temperature of automotive electronic circuits has existed for many years. The present desire is to increase operating temperature and long-term reliability without increasing cost. The cost goal is the most difficult to attain, but until the technology questions are addressed, the cost cannot be determined.

An extensive reliability qualification program, using the full battery of reliability tests, is **mandatory** to verify that all the problems standing in the way of increasing the maximum operating temperature of automotive hybrid electronic modules, cables and connectors have been fully resolved.

Although the vehicle manufacturers are vertically integrated as far as designing and manufacturing the powertrain and body-in-white, and designing and assembling the vehicle, they depend on the supplier industry for the majority of the components used in the vehicle. The manufacturers of electronic modules, whether part of a vehicle manufacturing or a supplier company, do the circuit design and manufacturing but purchase most of their components and materials from their supplier industry. Thus, the challenge of increasing the reliable maximum operating temperature of automotive electronics is spread over a large number of supplier companies, some of whom are relatively small and have limited development resources. To realize a higher operating temperature electronic module, the suppliers must upgrade their product. In order to do this effectively and economically, there first must be understanding and agreement about the specifications for the improved product, so that development efforts are directed at the same goals. One of our goals in writing this chapter is to stimulate dialog among the vehicle companies, electronic module and cable and connector manufacturers, and the supplier industry to arrive at a consensus on appropriate high temperature automotive E/E system specifications so that development can proceed.

ACKNOWLEDGMENTS

This work was performed jointly at the GM R&D Center, Delco Electronics Corporation, and Delphi Packard Electric Systems. We thank our many colleagues at the GM R&D Center, Delco Electronics, and Delphi Packard Electric Systems for their inputs on the many facets of increasing the maximum operating temperature of automotive electrical and electronic systems. We thank L. Jacovides, L. Schwartz and A. Lee for their management support.

REFERENCES

1. GM Hughes Electronics Corporation (1994) *1993 Annual Report*, p. 9.
2. Hansen (1995) Underhood electronics packaging – a diverse set of choices. *The Hansen Report On Automotive Electronics*, **8**(5), 1.
3. Erskine, J.C., Carter, R.G., Hearn, J.A., Fields, H.L. and Himelick, J.M. (1994) High temperature automotive electronics: trends and challenges, in *Proceedings of the Second International High Temperature Electronics Conference*, 5 June 1994, p. 13.
4. Society of Automotive Engineers (1978) *Recommended Environmental Practices for Electronic Equipment Design*, Standard J1211.
5. Society of Automotive Engineers (1994) *Joint SAE/TMC Recommended Environmental Practices for Electronic Equipment Design (Heavy Duty Truck)*, Standard J1455.
6. Shoucair, F., Hwang, W. and Jain, P. (1984) Electrical characteristics of large-scale integration (LSI) silicon MOSFETs at very high temperatures. Part II: experiment. *Microelectronics and Reliability*, **24**(3), 487.
7. Shoucair, F., Hwang, W. and Jain, P. (1984) Electrical characteristics of large-scale integration silicon MOSFETs at very high temperatures. Part III:

   modeling and circuit behavior. *IEEE Transactions on Components, Hybrids, and Manufacturing Technology*, **7**(1), 146.

8. Francis, P., Terao, A., Gentinne, B., Flandre, D. and Colinge, J.-P. (1992) SOI technology for high temperature applications. *IEEE Electron Devices Meeting Technical Digest*, p. 353.

9. Shione, N. and Hasimoto, C. (1982) Threshold instability of n-channel MOS-FETs under bias–temperature aging. *IEEE Transactions on Electron Devices*, **29**(3), 361.

10. Haller, G., Knoll, M., Braunig, D., Wulf, F. and Fahrner, W. (1984) Bias–temperature stress on metal–oxide–semiconductor structures as compared to ionizing radiation and tunnel injection. *Journal of Applied Physics*, **56**(4), 1844.

11. Liu, Z.H., Nee, P., Ko, P.K., Hu, C., Sodini, C., Gross, B.J., Ma, T.-P. and Cheng, Y.C. (1992) Field and temperature acceleration of time-dependent dielectric breakdown for reoxidized–nitrided and fluorinated oxides. *IEEE Electron Device Letters*, **13**(1), 41.

12. Farahani, M.M., Turner, T.E. and Barnes, J.J. (1989) Electrical and metallurgical characterization of niobium as a diffusion barrier between aluminum and silicon for integrated circuit devices. *Journal of the Electrochemical Society*, **136**(5), 1484.

13. Khajezdeh, H. and Rose, A. (1975) Reliability evaluation of hermetic integrated circuit chips in plastic packages, in *Proceedings of the IEEE International Reliability Physics Symposium*, p. 87.

14. Palmer, D. and Heckman, R. (1978) Extreme temperature range microelectronics. *IEEE Transactions on Components, Hybrids, and Manufacturing Technology*, **1**(4), 333.

15. Sabnis, A.G. (1990) *VLSI Reliability*, AT&T and Academic Press, New York, Ch. 4.

16. Bulucea, C. and Rossen, R. (1991) Trench DMOS transistor technology for high-current (100 A range) switching. *Solid State Electronics*, **34**(5), 493.

17. Oaks, J.A. (1981) A pressure sensor for automotive application. Paper C181/81 presented at IEE/IMECHE Third International Conference On Automotive Electronics.

18. Putty, M.W. and Najafi, K. (1994) A micromachined vibrating ring gyroscope. *Solid-State Sensor and Actuator Workshop 1994 Technical Digest*, p. 213.

19. Dreike, P.L., Fleetwood, D.M., King, D.B., Sprauer, D.C. and Zipperian, T.E. (1994) An overview of high temperature electronic device technologies and potential applications. *IEEE Transactions on Components, Hybrids, and Manufacturing Technology, Part A*, **12**(4), 594–609.

20. Bhatnagar, M. and Baliga, B.J. (1993) Comparison of 6H–SiC and Si for power devices. *IEEE Transactions on Electron Devices*, **40**(3), 645.

21. Harper, C.A. (ed.) (1974) *Handbook of Thick Film Hybrid Microelectronics*, McGraw-Hill, New York.

22. Tummala, R.R. (1989) Ceramic packaging, in *Microelectronics Packaging Handbook* (eds R.R. Tummala and E.J. Rymaszewski), Van Nostrand Reinhold, New York, pp. 455–522.

Oil-well applications: instrumentation of deep hot holes

3

T. Fallet, G. Forre and J. Gakkestad

3.1 BACKGROUND OF OIL-WELL INSTRUMENTATION

The technology of using electrical measurements in oil prospecting started as early as 1927, when the French Schlumberger brothers first applied their methods of 'electrical prospecting' to boreholes drilled for oil. The results of these first tests were outstanding and led to a new industry which today employs more than 100 000 workers and has gross revenues of over $30 billions.

The boreholes or oil wells extend 6000 m into the ground and have horizontal sections up to 8 km. The temperature gradient of the earth's crust is proportional to the specific thermal resistance of the rock and the local thickness of the crust. Practically speaking, the operating environmental temperature of the equipment used is normally 100–150 °C. Hot oil/gas wells go up to 200 °C, special wells (mostly geothermal) may operate near 350 °C.

From the first simple measurements of resistivity and electrical potential applied in 1927, a large family of measurement equipment has evolved. It includes a multitude of methods for measuring properties of the rock needed for oil prospecting, but also instrumentation for safe and efficient drilling of wells and for optimum drainage of the oil reservoirs.

A major service company, operating in all areas of oil-well logging, will offer more than 100 different services and 'logging tools'. The technologies used range from simple spinner type flowmeters to devices based on nuclear

High Temperature Electronics. Edited by M. Willander and H.L. Hartnagel. Published in 1997 by Chapman & Hall, London. ISBN 0 412 62510 5.

Figure 3.1 California 1932: a Schlumberger truck with its characteristic wireline winch. (Courtesy A.G. Schlumberger.)

magnetic resonance, for element characterization, and inertial guidance systems for drill navigation.

The well, one of the most demanding environments subject to instrumentation, is the most important common factor for these instruments or 'tools' of the industry. In the well, the instruments have to cope with very high pressures, up to 1500 bar (20 000 psi, 150 MPa), as well as very high temperatures. Very rough handling is typical in the oil field. The tools also need to be very compact. The diameter of the reservoir sections of the wells are normally 2–10 in. in diameter range and the tools should move easily inside without obstructing the flow. In addition, some tools are directly connected to the drill bit and subject to sustained severe vibrations.

We will discuss the electronics required for some of these tools plus traditional and new ways of design to obtain the necessary tool strength. But note that the majority of downhole tools are developed, built and used by the service companies of the oil industry themselves. Most of the detailed information concerning the different tools is therefore of proprietary nature. The information presented here is based on open sources and is therefore not very detailed. The idea is to provide an overall picture of the need for high temperature electronics in this industry. Section 3.5 discusses the

HTASIC technique introduced by SINTEF Instrumentation in 1993 as a step towards meeting this need (HTASIC is a registered trademark of SINTEF Instrumentation).

3.1.1 A string of tools characterizing the well profile

Most of the measurements taken in an oil well are normally performed as 'wireline services'. This means that a collection of instruments (tools) are connected together in series forming a 'tool string', up to 30 m long, and lowered into the borehole using a 'wireline winch' (Fig. 3.1).

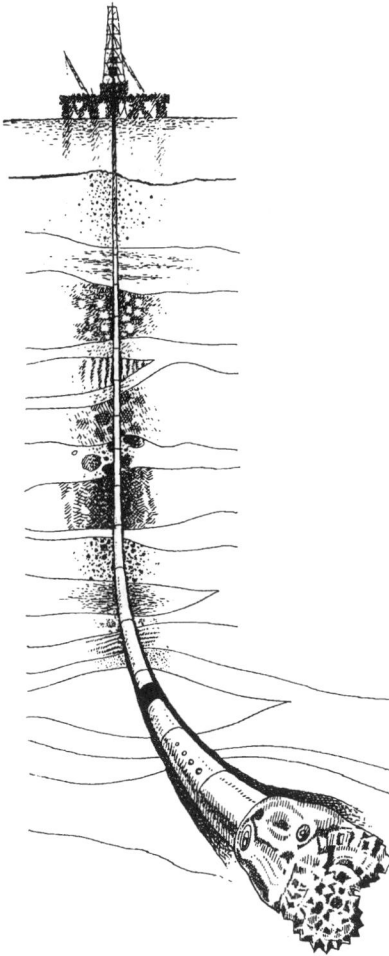

Figure 3.2 The characteristic of the well is measured while drilling.

All measurements are usually taken while the tool string is later hoisted up, thus all measurements, including depth, appear as a function of time. The measurements are later referred to depth. This measurement document, traditionally a long roll of paper, is called the 'log' and the activity is called 'well logging'.

The 'wireline' itself normally consists of seven single conductors packed together in steel armor strong enough (5 tons!) to carry the tool string and itself into the well. The cable length might be up to 10 km or even longer. Communication with topside (above ground) takes place via the seven conductors which also supplies power from topside to the instruments. Traditionally, one conductor was used for each signal, now multiplexed data communication is used, offering up to $100 \, \text{kbits s}^{-1}$ in spite of the very poor characteristics of the wireline cable.

Production instrumentation normally uses a slim 'monocable' with one conductor only inside the steel armor. For high data rates, Schlumberger usually use a coaxial cable with separate screen and armor. The signals and power are in both cases multiplexed on a single conductor.

To do the wireline logging, the drill string has to be removed from the hole first. This is a costly procedure taking up to 24 h, and much effort has been used to overcome this complication.

A method called 'measurement while drilling' (MWD) was introduced around 1980. It uses a measurement section built into the lower part of the drill string. The measurements were at first concentrated on the navigation of the drill bit, ensuring that the hole was drilled in the right direction.

Later came 'logging while drilling' (LWD). Perhaps the most important technical advantage of the technology is that the measurements are made as soon as the formation is penetrated. Otherwise, the drilling mud, which balances the downhole formation pressure, will eventually penetrate the formation and push away the formation fluids. Therefore LWD gives more accurate measurements than conventional well logging. Today LWD includes the most common logging services in the drilling equipment. This technology offers many advantages compared to wireline services but has one primary disadvantage: very poor communication to topside. Cables cannot easily be included in the drill string and acoustic communication is nowadays normal. The mud flow inside the drill string is modulated with pressure pulses. This reduces the data rate available to $<10 \, \text{bits s}^{-1}$ and stresses the need for data compression downhole. Storage of data downhole is also getting increasingly popular. The next section classifies tools using the phases of the oil production process. Some of the tools are going out of use and new ones are introduced continuously, but as an introduction to the high temperature electronics needed in this industry, we believe the following section can provide an overview.

3.2 INSTRUMENT CATEGORIES: FROM PROSPECTING TO PRODUCTION

3.2.1 Open hole logging

Well logging started with measurements of 'spontaneous potential' (SP) and 'unfocused resistivity,' also called 'electric survey,' in newly drilled holes. These two electrical measurements have later developed into four main instrument categories for such 'open hole logging.'

Electromagnetic tools

Electromagnetic tools are mainly developments of the resistivity tool. They provide measurements of the rock resistivity at various depths and with higher or lower vertical resolution. Many of the tools are based upon the induction of eddy currents in the formation and measurements of the signal propagation through the rock. The electronics involved will therefore have to generate and measure AC signals. Frequencies from the kilohertz region to microwaves have been used, and the electronics performs signal processing like phase correlation and amplitude measurements.

Acoustic tools

Acoustic tools are typically used to investigate the porosity of the rock by measuring the interval transit time of a compressional sound wave through the rock itself. The electronics needed are typically acoustic pulse transmitters and receivers together with time measurement devices. Another group of acoustic tools (VSP) is used to correlate traditional seismic data from surface measurements with seismic data from the borehole and also to derive rock mechanical properties.

Nuclear tools

Nuclear tools use neutron beams and gamma rays to investigate the rock. The simple gamma log is merely measuring the natural gamma radiation intensity of the rock. This is used for identification of the layers of the sediments as well as a depth reference for later measurements. But gamma rays are also used for measuring other physical properties of the rock, like density and to some degree composition, normally through backscatter measurements. Neutron measurements are used for water investigation and element analysis (neutron activation). Scintillating crystals, typically NaI, combined with photomultiplier (PM) tubes are normally used for the nuclear measurements. The electronics associated are high stability voltage supplies for the PM tubes as well as pulse amplifiers and counters for the output signals. In later years gamma spectrometers with bulk germanium detectors and up to 4000 channels have been applied. To make such measurements meaningful, a feedback loop is required to stabilize the absolute energy levels of the spectrometer.

Formation and hole geometry

The 'dipmeter' is used to find the geographical orientation of the sediment layers of the reservoir stone. This is done by correlating fine resolution resistivity measurements normally at four points, 90° apart, on the borehole periphery. By combining this data with magnetic directions taken from three orthogonal magnetometers, the orientation of the layers can be found.

The 'caliper' log measures the diameter of the well, normally based on two orthogonal sets of arms which are pressed against the borehole walls. The dimensional measurements are done in several ways, inductive differential transformers (LVDT) are quite normal. The associated electronics are AC generators and phase-sensitive demodulators for measurements.

Acoustic calipers are also employed, using the reflections from the walls to measure the dimensions through transit time measurements.

Directional logs are used to find the trajectory of the well and to orient the drill. Magnetometers together with inclination meters are commonly used, but gyroscopes, mechanical or fiber-optic, have also been employed. The necessary electronics are more or less the same as for traditional applications, disregarding the more stringent temperature requirements.

These measurements are all available as wireline services for low and medium well temperatures and most of them also as MWD and LWD services. MWD tools are built as a part of the drill string. The measurement function and the electronic functional requirements are the same as for 'wireline tools.' However, MWD and LWD instruments need extremely robust electronics. First of all, they are mechanically connected to the drill bit and therefore subject to very hard vibration. Different levels have been quoted, but $30g$ RMS over a wide frequency spectrum may be typical. And, as the drilling goes on for a long time, reliability requirements are much stronger than for wireline services. Due to the fact that mud from topside is continuously pumped through the drilling equipment, the temperatures will normally not reach the very highest levels.

MWD operations have also opened for the following special services related to drilling efficiency and safety:

Borehole pressure: important for safety reasons.

Torque and weight on bit: used for drilling control, values measured at the bit can be quite different from the corresponding values measured at topside.

Temperature and vibration: temperature and vibration of the drill bit inform the driller of potential problems.

The MWD tools are normally powered using a mud turbine and the data is normally compressed and multiplexed before transmission on mud pulses to topside.

3.2.2 Completion tools

Before a borehole can be used as an oil well it needs a completion. This includes the setting of steel casings in cement to prevent the hole from collapsing and to isolate the different reservoir zones, as well as insertion of oil tubing and packers to lead the crude oil from the reservoir areas up to topside. Certain special measurement tools are required.

Cement bond evaluation

Cement bond evaluation is performed using an acoustical instrument with a pulse transmitter and two receivers. The quality of the bond is evaluated from the amplitude of the received pulses versus time. It uses electronics that is very similar to the sonic tools for open hole logging. Ultrasonic instruments with arrays of transducers (CET) are used to evaluate the cement bond.

Pulsed neutron logs

Pulsed neutron logs are normally used for evaluation of water in the geological formation but may also be used as an indicator of water flowing behind the casing. The electronics consists of a pulsed neutron source and either a gamma or a ^{3}H detector for thermal/epithermal neutrons.

Radial differential temperature

Radial differential temperature was developed specifically for the detection and remedial treatment of channels in the cement behind the casing. The basis for this instrument is the fact that the fluids flowing in a channel behind the casing tends to be slightly warmer or colder than the surroundings. Thus a measurement of temperature gradients across the well may reveal such flows. The temperature difference between two opposites sides of the cased hole is measured in practice, with great sensitivity. The temperature differences caused by such flows are in the range 0.003–0.03 °C whereas the absolute temperature may be > 100 °C.

Another tool used for the same purpose is the 'noise log', actually a microphone. This was used with some success during Saga Petroleum's famous blowout in the North Sea.

3.2.3 Production logging

During the production phase, including the production planning, instruments are needed to guide in the optimal drainage of the well. The oil company would generally like to know where in the well the water, oil and gas are produced and their relative rates. This way, water-producing areas may be shut off and the wanted goods may flow freely. The measurement signals are typically transferred to frequencies which are time multiplexed and sent to topside for presentation and storage, together with depth references. Sand control is also an issue for certain wells. Several tools are typically employed in this phase of oil production.

Temperature sensors

Temperature sensors are mainly used for the identification of injection or production zones and are for this purpose one of the 'workhorses' of production logging. Several different principles of sensors are used, ranging from standard platinum-100 elements to quartz-based systems where the oscillation frequency of a temperature-sensitive quartz crystal is the direct measure of the temperature. Extreme accuracy and resolution is of great value for these sensors as we are primarily looking for very small deviations on top of an average well temperature, typically around 90 °C in the North Sea. The electronics needed will differ a lot from analog measurements based on a resistance bridge with platinum-100 elements to the oscillator needed to obtain a very stable frequency from a quartz crystal. For quartz crystals, short-term stability of about 0.01 K is available, and 0.001 K is a fair goal for new designs. Several versions of temperature sensors are employed for different measurement purposes. Time and position responses are important.

Pressure sensors

Next to temperature sensors, pressure sensors are some of the most important tools of the production logging business. The pressures are indirectly used for estimating flow conditions and fluid categories. As for temperature measurements, the absolute pressure is not as important as pressure gradients. Several sensor types are in use. Traditional types based on strain gauges or capacitive detectors are these days being replaced by units based on quartz crystals with superior sensitivity and stability. State of the art for these units is today around 10 mbar (1 kPa) resolution for a rated pressure of 1000 bar (0.1 GPa). The quality of measurement mainly depends upon the quartz crystal, but special precautions are also required for the electronics to obtain the 1 kPa resolution. Typical electronics for these tools are high precision oscillators and frequency converters.

Flow sensors

Flow sensors are of primary interest for the production logger. With a single-phase flow, an accurate flow log will tell where fluids are gained or lost, information of great significance to the reservoir engineers. Several different measurement principles have been tried throughout the years for both single-phase and multiphase flows. Multiphase flows, especially three-phase flows with oil, water and gas, are very difficult to measure exactly, even in the controlled environment topside. Downhole, no really good sensors are yet available. But several systems are operated on a regular basis. The most common instrument today is the 'spinner' flowmeter. It consists of a vane-like propeller with a rotating speed in principle proportional to the flow rate. The number of revolutions are typically counted using a magnetic or eddy current pickup. Quite simple reliable electronics are needed to transmit the counts to topside and the accuracy of the tool is mainly restricted by mechanical performance like friction. But interpretation of the spinner readings is very difficult in a multiphase flow, where the different phases may have completely different superficial speeds.

Other types of sensors have been tried and continue to be used, e.g. hot wire anemometers, differential pressure, rotameters and acoustic sensors measuring transit times in opposite directions or Doppler frequency shifts. The electronics needed are just as diversified as the measurement principles involved. Research and development work is investigating acoustic sensors to obtain the full flow profile over the pipe diameter.

3.2.4   Composition sensors (water cut meters)

Most petroleum wells produce a mixture of oil, gas and water, the proportions differ from one well to another and they change throughout the life of the reservoir. Typically the water is below the oil and, as the oil is produced, increasing amounts of water flow from the lower parts of the well. Instruments to detect the ratio of water to oil, 'water cut meters,' are valuable for determining which part of the well should be sealed off. This is specially true for horizontal wells which may produce water and oil from many different places along the horizontal section. Many different types of instrumentation have been used with limited success. Some of the principles employed are electrical conductivity (much higher for water than oil), capacitance (relative permittivity of water ≈ 80) and density. Within this last group, two different principles are most common: radioactive absorption and pressure drop measurement.

The common gamma densitometer is electronically close to the radio-active gauges already mentioned, no special precautions are taken over the electronics. Extreme accuracy is needed to obtain reasonable accuracy with a pressure-based density gauge. To obtain 10% accuracy in water cut over a 2 ft (610 mm) vertical distance, the pressure difference must be known to 0.017 psi (1.19 mbar, 119 Pa). With a typical operating range of 1000 bar (0.1 GPa) this asks for a pressure gauge with accuracy and resolution at the top of the line. The alternative is a differential pressure gauge normally called a 'gradiomanometer.' The pressure will typically be converted to a force measurement, either using a position sensor combined with a spring or an electrical force feedback system. This is usually achieved with a phase-sensitive detector system to go with an LVDT and the necessary feedback compensation elements and analog-to-digital converters.

The capacitive gauges will function well for low water cuts and a homogeneous mix of oil and water flow. But this is seldom the case for very long. All devices experience severe sampling problems and refined statistical data enhancement will be applied as soon as high quality electronic modules are available. In addition to these items, several of the tools mentioned earlier are also used during the production phase.

3.3 EXAMPLES OF OIL-WELL ELECTRONIC MODULES

As may already be clear, most kinds of electronics are used in downhole equipment for moderate temperatures. For the hot wells, the available tools are typically built with simple electronics or cooling is applied for short-term wireline operations. However, more and more tools need to be qualified for long-term operation at temperatures up to 200 °C. The next few sections discuss block diagrams of a few typical electronic 'building blocks' together with specifications and some circuit details. The examples are picked from tools designed and built by SINTEF Instrumentation; some of them will later be used to illustrate applications of HTASIC, SINTEF's new technique.

3.3.1   Data communication on wireline

A system commonly used for communication from the well to topside in production logging, is based on the simple digital time-multiplexed scheme (Fig. 3.3). A string of tools are connected in series and the 'monocable,' a single steel wire with an insulated center conductor is their only connection to topside. This cable supplies all the instruments with power and simultaneously communicates with topside. A DC level of typically 50 V downhole is used for primary power, the signaling is done by superposing 10 V pulses.

Each instrument sends 1 bit of information in each time frame. With 12 instruments and a time frame of 1 ms, every instrument can send a total of

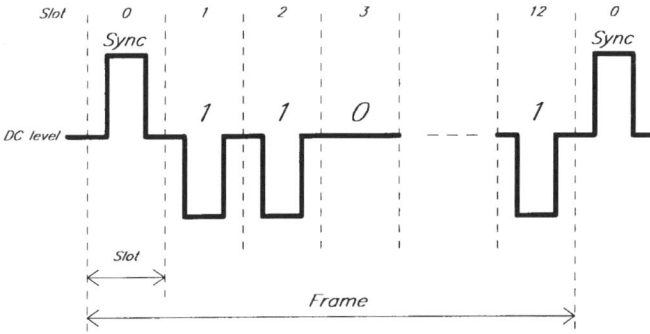

Figure 3.3 Data frame used for time-multiplexed communication.

1000 bits to topside every second. In its simplest form this can be used for the transmission of 12 'analog' frequencies in the range 0–1000 Hz with a resolution of 1 Hz. The only electronics needed downhole is an individual delay circuit in each tool triggered by the sync pulse and making sure the bit is sent in the correct time slot.

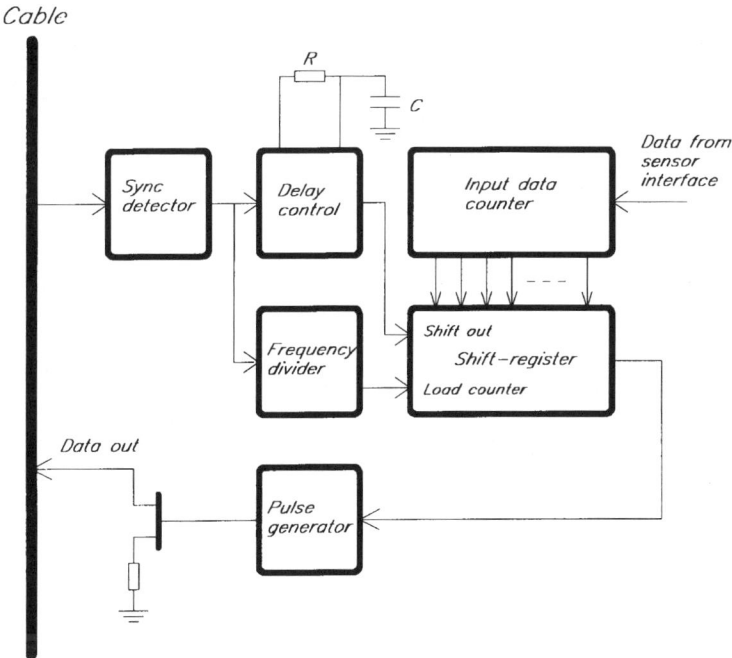

Figure 3.4 A telemetry unit for time-multiplexed digital communication.

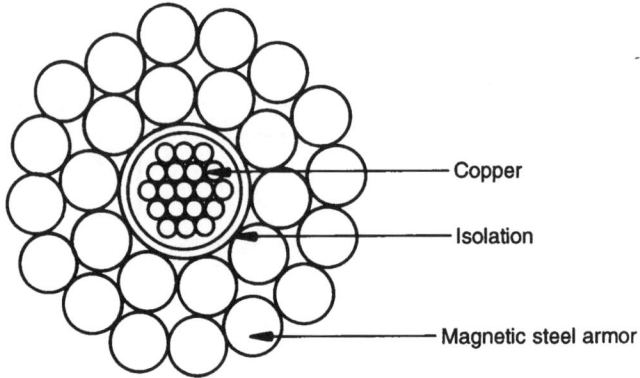

Figure 3.5 A cut-through of a typical 7/32 in. (5.6 mm) monocable.

This communication scheme is very robust but equally inefficient with regard to the data rate available. By doing a small modification (Fig. 3.4), a general data word of perhaps 10 bits can be sent to topside 100 times per second from each tool. Instead of using 1 s to obtain an accuracy of 1‰, this is now done in 10 ms.

The maximum bit rate is limited by the monocable, which has rather poor communication performance but excellent mechanical strength and wear properties in a downhole environment. A cut-through of a 7/32 in. (5.61 mm) monocable is shown in Fig. 3.5. The return lead is the magnetic steel armor itself which, together with the skin effect of the center conductor, presents an attenuation of more than 20 dB at 20 kHz increasing to almost 40 dB at 40 kHz for a cable length of 5 km.

3.3.2 Power supply modules

A typical tool requires a regulated supply voltage of 5 V and/or 12 V with an available current of 5–50 mA. The voltage will have to be generated locally from the unregulated line voltage on the monocable. The combination of power transmission and signaling on the monocable puts some special requirements on the power supply modules:

- good isolation between line and output voltage;
- no influence on the signals on the monocable from the current drawn by the tools;
- good power conversion efficiency;
- small size is crucial – some tools have 1 in. outside diameter only;

- a minimum of large capacitors should be needed as reliable electrolytics for 200°C operation are difficult to obtain.

In principle, three types of conversion circuits exist, each with their pros and cons:

Switch mode regulators have a high potential conversion efficiency. Their main problems are requirements for good filters to reduce influence on the signals. Higher complexity and need for relatively large capacitors are also important cons.

Series regulators are simple and offer good isolation but conversion efficiency is reduced when the voltage difference between line and output is increasing. Some means of avoiding load currents influencing the signals must be found.

Parallel regulators normally have poor conversion efficiency and poor isolation; they are, however, easily designed to overcome the line signal distortion problem.

A composite regulator used for this application is shown in Fig. 3.6. The advantage of a composite regulator is that it combines the good properties of series and parallel regulators. A certain system redundancy is also obtained as the maximum current drawn is limited. A short circuit in one tool will therefore be of limited consequence. Good isolation can be obtained without the use of large capacitors or inductors, but the obvious disadvantage is poor conversion efficiency when the line voltage gets high.

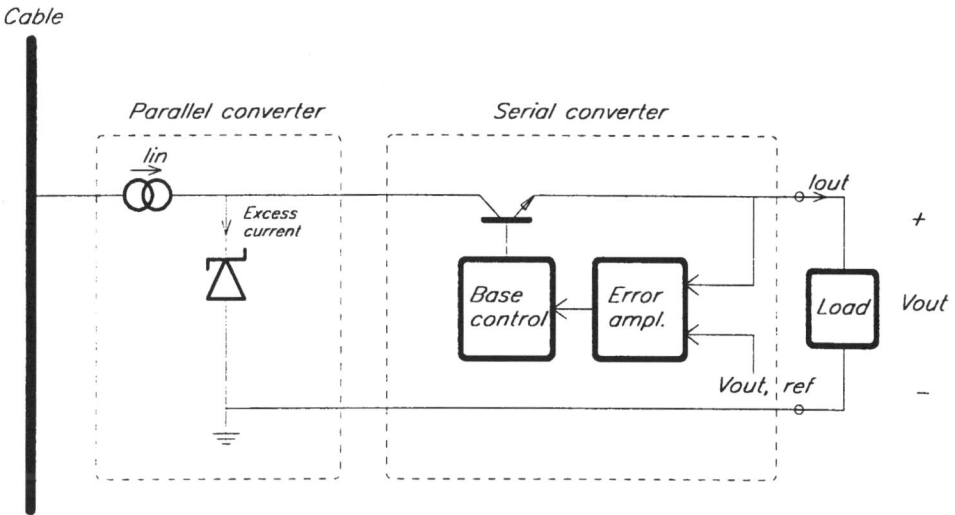

Figure 3.6 Two-stage parallel–series regulator block diagram.

3.3.3 Scintillation counter

Scintillation counters are used in almost all tools based on nuclear detection downhole; this is because a scintillating crystal coupled to a photomultiplier is by far the most common detector. Downhole scintillation counters require some electronics, but this is very similar to the electronics used for scintillation counters at normal temperature ranges. A typical electronic module for a scintillation counter is shown in Fig. 3.7. Certain special tricks may be needed when the environmental temperature approaches 200 °C.

The intensity of the light pulses from the scintillating crystal vary in proportion to the energy of the received gamma quanta. These are then linearly amplified in the photomultiplier tube, its gain strongly dependent upon the high tension (HT) supply. The gain also depends upon the environmental temperature. Pulses are normally counted only within certain energy ranges. For special applications the complete energy spectrum is interesting. Both cases require some method of amplification stabilization. This is normally done by introducing a reference line in the spectrum using a built-in isotope like cesium-137, then automatically controlling the HT supply to keep the reference line at a certain output pulse amplitude. This technique is of special importance in oil-well applications as the environmental temperature may change from 0 °C during calibration at topside on a wintry night to 190 °C when reaching the bottom of a hot well.

To obtain a continuous energy spectrum, an analog-to-digital converter (ADC) may be used to sort the pulses into their appropriate channels according to pulse height. To give the ADC necessary time to do the conversion, a peak-hold circuit is needed (Fig. 3.8). However, to make a hold circuit without drift from 0–200 °C, and in the presence of unavoidable leakage current, may illustrate some of the challenges for this module.

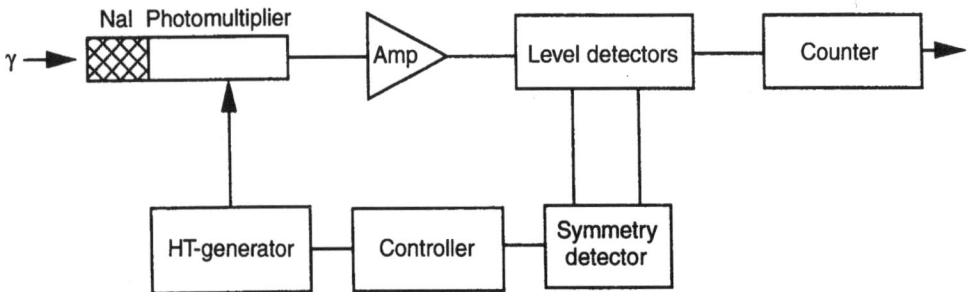

Figure 3.7 Typical electronic module for scintillation counter.

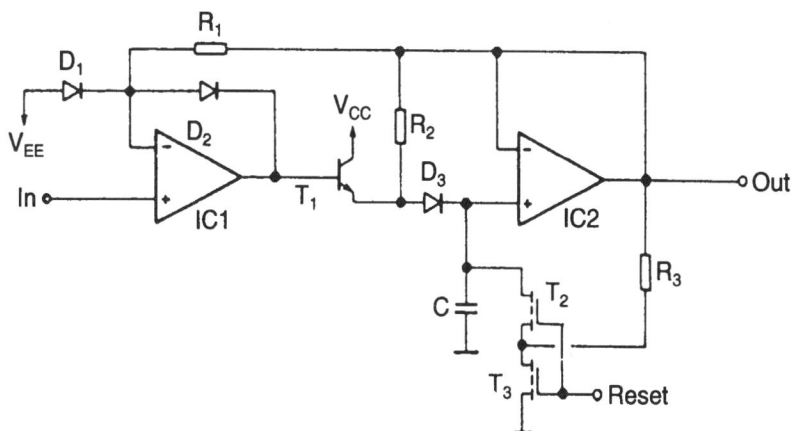

Figure 3.8 A hold circuit with reduced leakage currents.

To keep the voltage over the capacitor 'C' unaffected by the normal leakage, the voltage follower IC2 is used to make the voltage across D3 and T2 zero. With no voltage there is no leakage current. In the amplifier IC1 the leakage in the diode limiter D2 is compensated with the same current in the matched diode D1.

3.3.4 Quartz-based pressure and temperature sensor

Accurate measurements of pressure and temperature in the wells are crucial in oil production and many different types are in use. A measurement principle based on pressure- and temperature-sensitive quartz crystals have gained special interest in later years. It combines a large dynamic range with very high resolution and accuracy. The pressure and temperature determines the frequency of crystal oscillators and the frequency shift is used as a measure for the physical quantity. Typical resonance frequencies used are in the 7–8 MHz range and full-scale deviation is around 30 kHz. Obtainable frequency stability is about 0.01 Hz for accuracy in the region of 10^{-6}. Thus pressure changes of a few millibars (a few hundreds of pascals) can be detected in an environment of 1000 bar (0.1 GPa) static pressure! Corresponding temperature deviations down to 0.001 K are detectable. A typical block diagram for this kind of sensor is shown in Fig. 3.9.

A reference crystal oscillator is used to reduce the need for a good communication line to topside; only the frequency differences are transmitted to topside. An integrated version of this electronics has been designed at SINTEF Instrumentation; it is discussed in a later section.

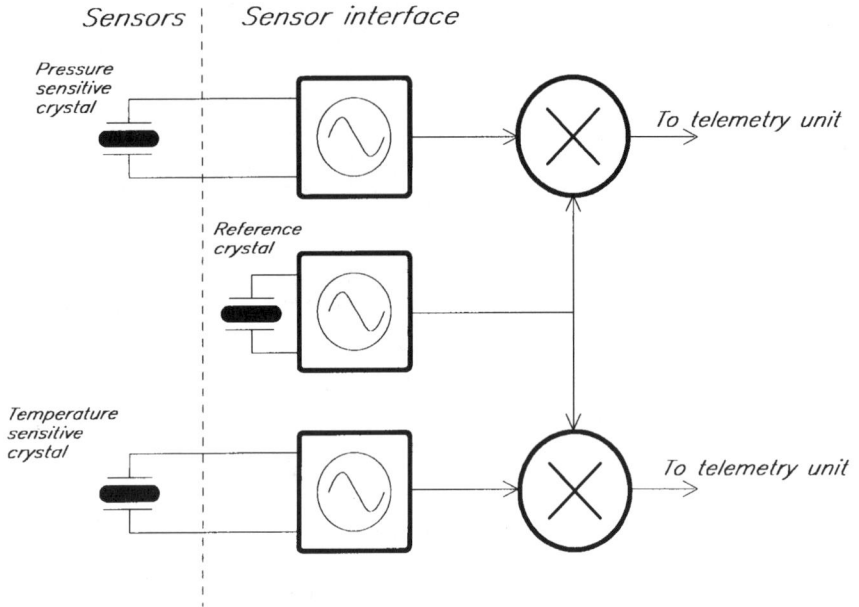

Figure 3.9 Quartz-based pressure and temperature sensors.

3.4 TRADITIONAL TECHNOLOGIES

Downhole electronics (Fig. 3.10) has traditionally been made in the same way as electronics for normal environments, but taking into consideration the higher temperature rating during design and production. Printed circuit

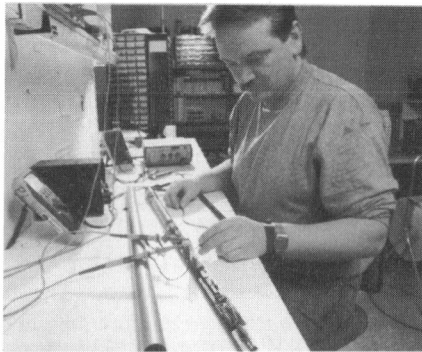

Figure 3.10 Typical electronics module used in well logging.

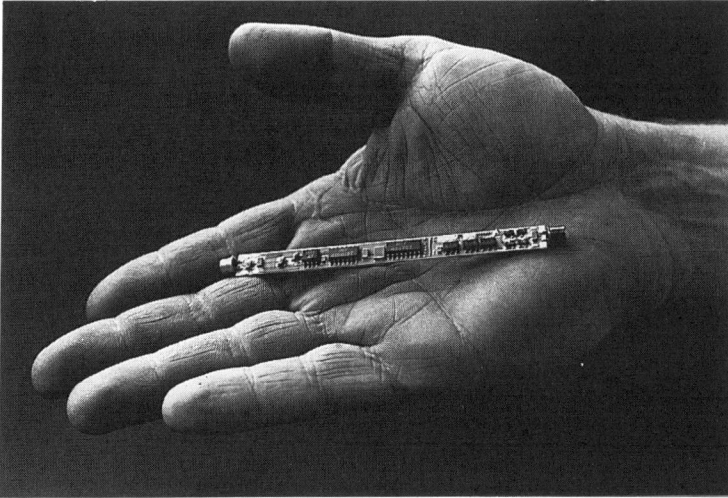

Figure 3.11 Hybrid circuit used for strain gauge measurement and multiplexed data transmission to topside.

board technology based on high temperature laminates and high temperature tin solder has been commonly used for more than 20 years. But the equipment produced was seldom designed to operate at over 150 °C. In hotter wells, a combination of Dewar flasks and melting materials, kept the inside temperature of the tools below 150 °C for long enough to perform the required measurements. Around 1980 some companies started to use alumina substrate and thick film technology for downhole tools (Fig. 3.11).

The electronics were partly based on discrete components and partly on the few available ICs with an adequate temperature specification. Operational amplifiers and ADCs were available for analog design. And CMOS demonstrated adequate performance for digital design, when delivered in hermetic ceramic or metal cans. Designs were largely based on 'clever' solutions to take account of the nonideal performance of the individual components.

This gave acceptable performance for short-term logging operations, even for fairly complex electronics. However, experience also indicated that the reliability necessary for long-term operation (several years) in hot wells was difficult to obtain. One of the reasons was the long-term degradation of interconnections between active and passive parts of the circuits.

3.5 HTASIC–A NEW TECHNIQUE FOR BETTER RELIABILITY

During the late 1980s some researchers began looking for a new method to design practical and economic high temperature electronics. A small team was formed at the SI Center for Industrial Research in Norway. At the start of 1993 SI merged with SINTEF and the high temperature electronics group became part of SINTEF Instrumentation. The technique to be developed needed to be applicable to both analog and digital design, but with emphasis on high reliability analog design for sensor interface. SINTEF's previous experience showed that these were by far the most complicated circuits to obtain. The group had participants from solid-state physics as well as people with experience in oil-well electronics and design of application-specific integrated circuits (ASICs). The conclusion was that CMOS-based full-custom ASIC processes showed interesting possibilities. Some performance data were simulated with our CAD tools and laboratory tests of typical CMOS characteristics, such as mobility, threshold voltages and drain currents, showed good correspondence with the simulated performance (Fig. 3.12). Based on this we concluded that, with some caution, the simulator results could be relied upon up to 200 °C.

Generally the advantages of using ASICs are obvious. First of all, the number of interconnections on the printed circuit board (PCB) or hybrid substrate may be dramatically reduced as more functions are integrated on the chip. Taking full advantage of IC design, the number of external components may also be reduced. Since the internal capacitances of an ASIC are much smaller than for discrete electronics, the power consumption may be reduced substantially. Apart from the obvious system desirability, this would also reduce internal overtemperatures on the chip. Using full custom, any power-consuming element on the chip may be spread out physically to avoid any local hot spots. Thus, higher reliability should be almost guaranteed.

Several methods are available for ASIC design such as 'gate arrays', 'standard cell' or 'full custom.' The two first methods are by far the most economic and easy to use; however, to be able to optimize the circuit design for high temperature, full custom has to be used. In full custom, we can control the dimensions of every transistor and arrange the layout to minimize diffusion areas; we can also control the widths of metal lines on the chip in order to achieve low current densities.

Very important is the flexibility to mix analog and digital functions on the same chip. For borehole applications, the typical situation is analog sensor signals which are converted to a digital word to be transmitted to topside. SINTEF wanted, in principle, to achieve this on one chip.

Full-custom designs of ASICs are expensive even at standard temperatures. With all the additional considerations necessary to obtain high performance at extreme temperatures, designs are very expensive. Based on

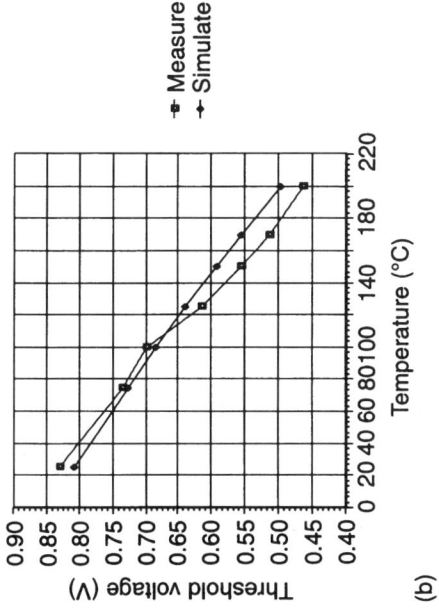

(a)

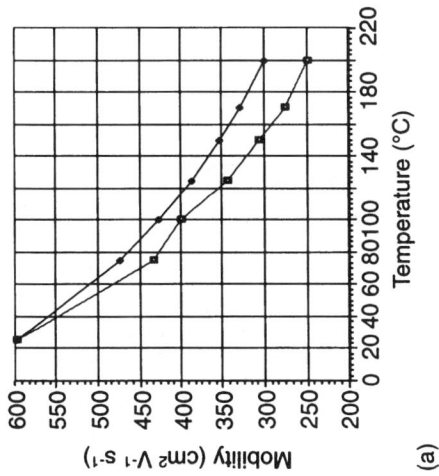

(b)

Figure 3.12 (a) Mobility and (b) threshold voltage of an NMOS transistor as a function of temperature: (□) measured and (◆) simulated.

SINTEF's experience to date, circuits of medium complexity, as illustrated later in this text, will have a design cost of \$70 000–300 000. The time needed to go from a specification to first samples delivered is 6–12 months. On the other hand, high temperature electronics of traditional design is also expensive and has long development times. After finished design, ASIC production incurs costs which are comparable with traditional circuit board production.

Since leakage currents cause major problems at high temperature, the most suitable technologies are silicon-on-sapphire (SOS) or silicon-on-insulator (SOI). These technologies have traditionally been used in military and aerospace projects to obtain good radiation hardness. The major disadvantage is that few silicon foundries are supporting these technologies and production costs are high. A BiCMOS technology with buried layers may also be used to minimize leakage currents and thus offer some advantages over conventional CMOS. BiCMOS also has the advantage of mixing bipolar transistors with CMOS. This is an important feature when designing high performance input stages and high voltage output stages. The HTASIC technique is based on a BiCMOS process.

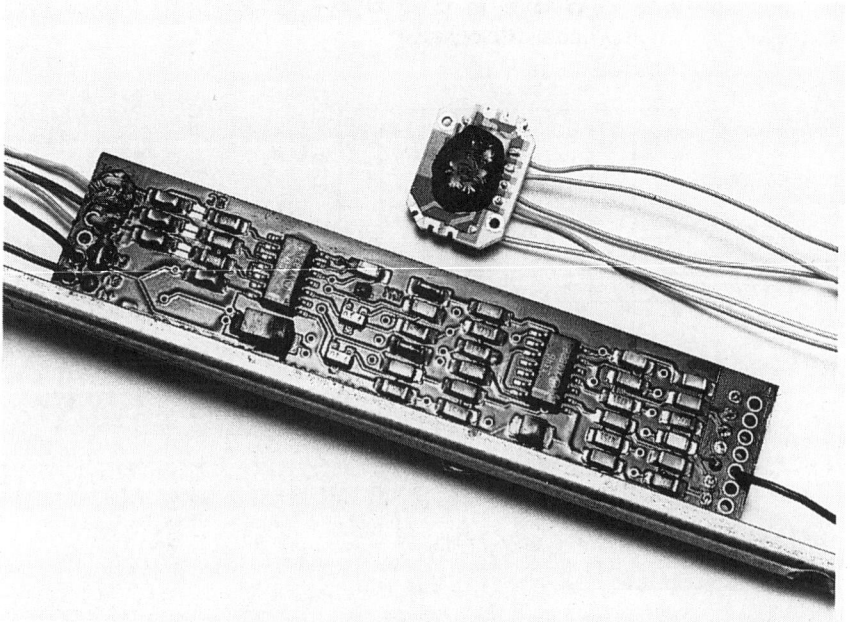

Figure 3.13 Two generations of modern electronics for downhole tools.

Physical size is another very important feature of using ASIC design in the downhole tools (Fig. 3.13). Substituting a large printed circuit board with an ASIC directly mounted to a ceramic hybrid could easily reduce the overall size of the electronics compartments by a factor of 100. This will reduce the overall length of the tool string, important from an operational viewpoint, but crucially allowing more tools on each tool string and the possibility of building advanced tools into narrowhole drilling tools. Such possibilities have a high economic impact in the oil industry. The ELS-001, pressure gauge HTASIC circuit, is mounted to a ceramic substrate for the quartz crystal and output connections. The new hybrid uses 3 mm of tool length whereas the PCB based on surface mount technology uses 85 mm more!

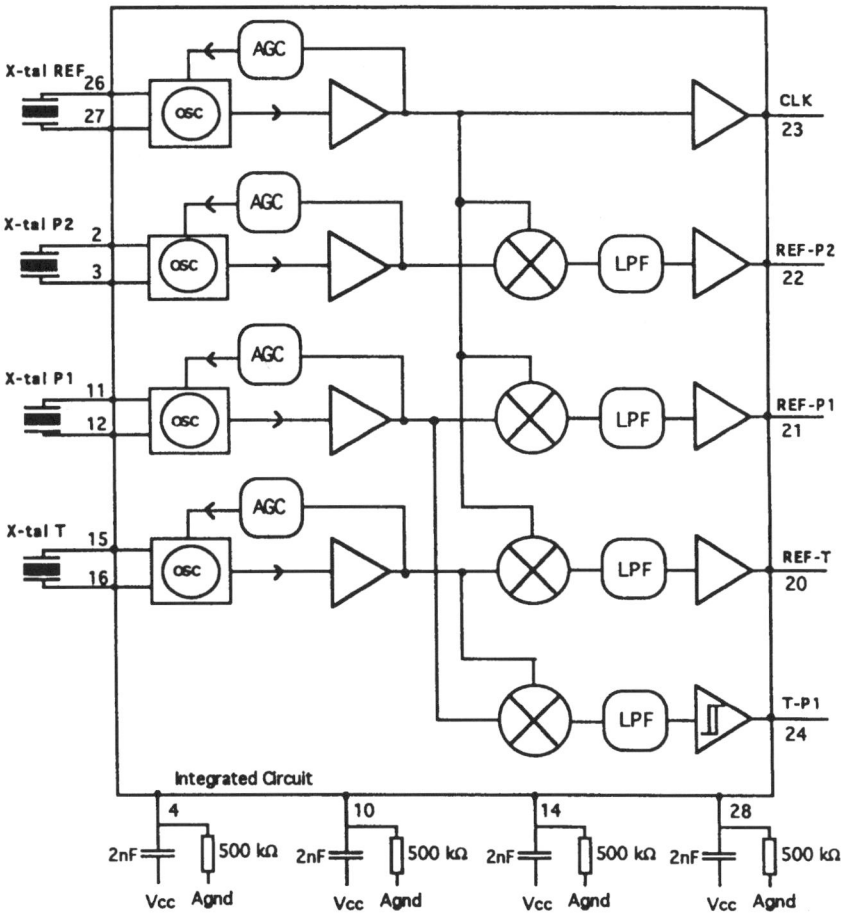

Figure 3.14 Block diagram of the ELS-001.

3.5.1   Quartz-based pressure sensor electronics

Designed and produced using HTASIC, ELS-001 provides all the electronics for a dual pressure and temperature gauge based on pressure and temperature-sensitive quartz crystals. Marketed by Maritime Well Service, Forus, Norway, the ELS-001 has four identical Pierce oscillators with automatic gain control (AGC) four mixers and five outputs. The circuit is connected to one reference crystal, one or two pressure crystals and a temperature-sensitive crystal (Fig. 3.14).

The circuit needs a single 5 V supply; the specified temperature range is 0–200 °C, but it has been operated successfully up to 270 °C. The circuit will work all the way from 4.5 V to 6 V, but reliability is probably reduced at high voltages. The current consumption versus temperature is shown in Fig. 3.15; the current consumption also increases with increasing power voltage.

Frequency stability is a primary performance indicator of this circuit as this is directly related to the system performance. The frequency stability versus temperature (of the circuit itself) is shown in Fig. 3.16. The stability of frequency is better than 0.02 Hz over the entire V_{cc} voltage range. One of the key elements to obtaining this kind of performance is the use of AGC

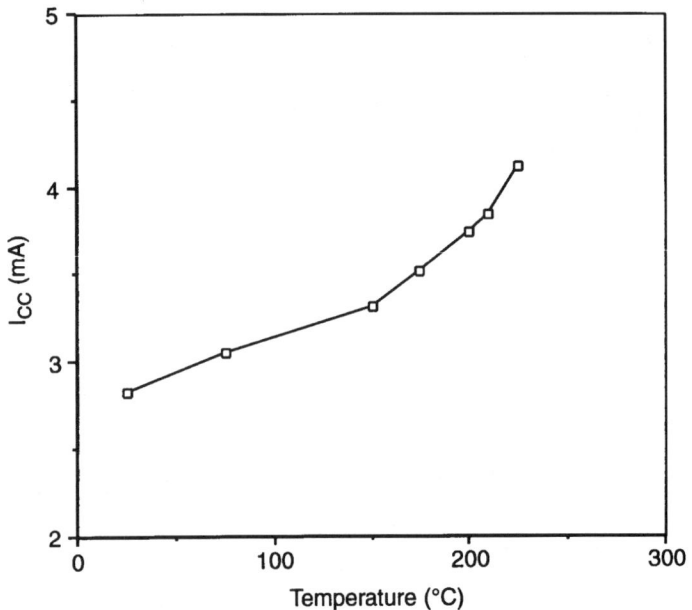

Figure 3.15 Current consumption of the ELS-001.

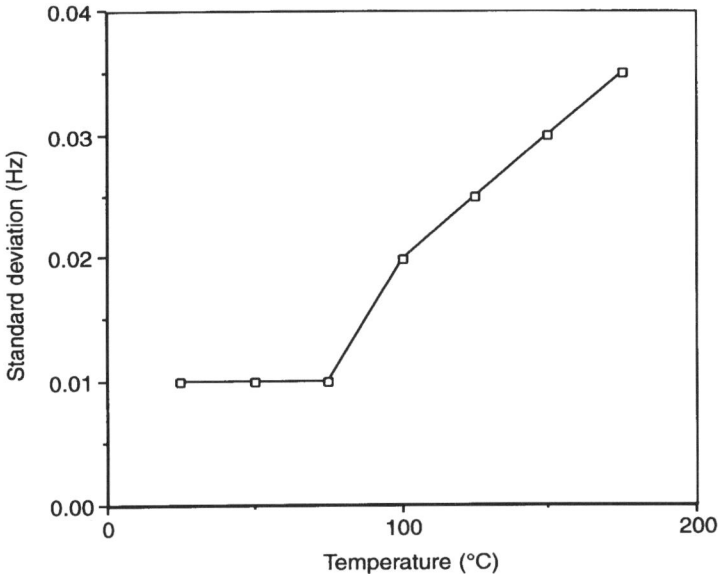

Figure 3.16 Frequency stability versus temperature for the ELS-001 (averaging time 1 s).

circuits to stabilize the operating point of the oscillators over a wide temperature range.

Increased reliability was the primary goal of the HTASIC technique, but reliability figures are difficult parameters to establish. Four units of the ELS-001 taken from the engineering sample production have been tested for 5000 h continuously at 230 °C. No temperature-induced failures have been detected inside the ASICs. Based on a conventional exponential derating with temperature, this indicates a lifetime in excess of 15 years at 175 °C. If this proves to be true, it opens the way for a new generation of reliable downhole electronics. Sensors may then be installed on completion of a newly drilled well and operate for its entire lifetime. The amount of electronics needed in the oil industry may therefore rise considerably.

3.6 PACKAGING AND COOLING

Electronics for oil-well instrumentation is subject to many hazards besides high temperature. The fluid pressures in a deep well may go up to 1400 bar (20 000 psi, 140 MPa) and oil-field operation in general requires great robustness. Some equipment must be placed near a zone to be perforated. Here a series of explosive charges are fired against the steel casing, penetrating it and creating the communication for fluids from the reservoir to the inside of the well. The shock impact on nearby tools can be quite severe.

Equipment which is going to be mounted together with the drill bit, typically MWD/LWD equipment, will experience very severe vibration during its whole life span as the drill bit is rotating and crushing the rock formations on its way. Thus well equipment should be designed to tolerate repeated shocks in the 100g range and continuous vibration up to 30g over a broad frequency spectrum for MWD operations.

Such an environment calls for good packaging. However, efficient vibration and shock isolation may easily lead to good heat isolation as well, which is quite detrimental to high temperature performance and reliability. The thermal resistance of a normal IC package may introduce a temperature increase of 14 K at 100 mW dissipated. If little care is taken in the thermal design, this number can easily double before the environmental temperature is reached. Having 180 °C in the well is therefore no guarantee that the hot spots of the chip will be below 200 °C!

Heat expansion is another interesting point. The design may indicate materials of different thermal expansion, so the mechanical stress levels due to thermal effects can be tremendous. Be careful when using alumina substrates. They are very strong but brittle materials which can easily be damaged due to thermal expansion if no forgiving soft materials are present. Thus, the best of high temperature electronics may experience low reliability due to inadequate packaging. Most downhole electronics will be made with low power dissipation and offer no great challenge with regard to thermal design. However, for special circuits like power supply units, highly efficient cooling designs may be appropriate. Examples are liquid and evaporation cooling; using these techniques, thermal resistances of 12 K W^{-1} are available for devices of 1 cm^2. This is 100 times less than typical values for still air.

Last but not least, oxidation will be detrimental to electronics when operated for long periods at the highest temperatures. Packaging techniques which allow complete removal of oxygen from the electronics will considerably increase the lifetime of downhole equipment.

3.7 ALTERNATIVES TO ELECTRONICS

Instrumentation for very hot logging environments (> 200 °C) has traditionally been designed using fully mechanical or mechanical sensors with simple electrical readouts such as LVDTs. No electronics or signal conditioning were placed downhole. However, these instruments generally have an inadequate sensitivity for today's requirements. They have also been limited

Table 3.1 Main specifications of FOWM

Pressure range	0–1000 bar (0–0.1 GPa)
Resolution	<0.01 bar (<1 kPa)
Accuracy	<0.4 bar (<40 kPa)
Temperature range	20–200 °C
Time response	$1\,\text{bar}\,\text{s}^{-1}$ ($0.1\,\text{MPa}\,\text{s}^{-1}$)
Vibration	$20g$, 30–100 Hz
Shock	$190g$, 9 ms
Useful life	$\geqslant 5$ years 90% probability

to simple gauges such as calipers, pressure sensors and temperature sensors. Some of these sensors have included mechanical recording without an electrical connection to topside. The operation was performed using 'slick line,' a simple steel wire (piano wire) lowered into the well with the tool string connected to it.

In later years another alternative to electronics has been developed: fully optical devices. An optical pressure sensor has been developed in Norway and Great Britain for a Norwegian branch of Alcatel and is briefly described below.

3.7.1 Optical pressure transducer for permanent well installation

Instead of electronics, optical devices may be used for sensing and signal transmission to topside. Fiber-optic well monitoring has been developed by Alcatel Kabel Norge and listed under the trade name FOWM. A fiber-optic cable provides the link between reservoir and topside. The fibers are partly used for excitation of the optical sensor and partly used to transmit the sensor signal to topside. The primary pressure-sensing element is a vibrating beam of silicon which is tensioned by the pressure applied. Excitation of the beam is assisted by the optical power. The primary readout signal is a frequency proportional to the pressure. Temperature is measured using similar principles. The pressure response is typically $45\,\text{Hz}\,\text{bar}^{-1}$ ($450\,\text{Hz}$ MPa^{-1}) and the temperature response is $15\,\text{Hz}\,°\text{C}^{-1}$. The main specifications are given in Table 3.1.

Two cables are used, each containing a single-mode fiber encapsulated in tubing of 1.8 mm diameter made of stainless steel (316); better materials, such as Inconel 625, may be used for certain applications. The cable is flexible, like a slick line, so the gauge can be used for wireline services but is mainly intended for permanent installations in oil wells. Fiber-optic data transmission is also used for some electronic devices with very high data rates used in well monitoring. Video cameras are typical examples.

3.8 FUTURE OF OIL-WELL ELECTRONICS

Demand is rising for high temperature electronics in oil-well applications with regard to temperature range, advanced applications and reliability. Future trends of development are difficult to foresee, but guestimates based on two different scenarios may shed some light onto what may happen.

3.8.1 High temperature electronics for the automotive industry is booming

There is now an increasing interest all over the world for high temperature electronics, but mostly in academic research and for very high temperatures. The main mass market is the automotive industry. Here the combined pressure for high reliability and low prices together with few but very strong users, will most probably lead to the use of custom-designed application-specific integrated circuits (ASICs). New processes may also be developed that combine relatively high voltage ratings and robustness against electrostatic discharge (ESD) with good high temperature performance. Low prices for typical high temperature automotive modules may then be expected and probably a certain range of standard ICs such as op-amps, microcontrollers and memory.

For simple oil-well tools, the available circuits will be adequate and they will probably be used with thick film hybrids as carriers. HTASIC technology may be used for highly competitive new designs. The main reasons are reliability and packaging density. The fact that development prices will be somewhat higher is probably of minor interest as the complete tools are already expensive devices and the environment they must withstand are even worse than usual. The available production processes for oil well specific circuits will improve considerably together with connection technologies.

3.8.2 No high volume demand for high temperature electronics

If the automotive industry continues to use standard electronics, the interest for high temperature electronics from the major production houses will be meager; few new processes will be developed and put into production. There is likely to be a slow evolution of today's situation with very few producers of standard components. A few will probably concentrate on typical standard circuits needed for sensors like op-amps and microcontrollers with limited memory. The volumes needed in the oil industry have never been large enough to pay for a major IC process development alone.

Competitive new designs may be based on HTASIC or other competing technologies for custom-designed ASICs. With a more limited supply of good standard circuits, the drive for custom-designed ICs will probably be even higher. Increased compactness is easy to obtain and the need for reliability will favor the reduced number of interconnections typical for ASICs.

In all cases there appear to be obvious advantages of using ceramic substrates for comounting ICs and the unavoidable surrounding components. Efficient cooling of the devices to avoid local hot spots will be an important element. Liquid cooling is a promising solution.

PART TWO

Silicon

High temperature operation of silicon MOS transistors

4

R. B. Brown and K. Wu

4.1 INTRODUCTION

The applications discussed in Part One call for higher temperature operation than is available in conventional electronic parts, which are specified to function properly to only 125 °C. A survey of researchers and potential users of high temperature electronics [1] shows, however, that the majority of applications for high temperature electronics are below 300 °C: 87% of automotive, 68% of aircraft and 98% of well-logging applications. According to the survey results, even in space and military applications a large share of needs could be met below 300 °C: 56% for military and 33% for space applications.

The effects of temperature on materials and devices have been of great interest throughout the history of semiconductor research [2,3]. As the electronics industry developed, silicon became the dominant semiconductor material. A number of higher bandgap semiconductor systems (GaAs, SiC, diamond and nitrides) now offer alternatives to silicon for high temperature applications. Silicon bipolar circuits have also been used in high temperature applications [4]; they require special care in circuit design to avoid second breakdown (thermal runaway). The most common and cost-effective integrated circuit technology is now silicon CMOS. Though CMOS circuits are typically specified to operate only as high as 125 °C, there is no fundamental reason why they cannot operate reliably at temperatures considerably higher than that, covering many of the high temperature electronics applications. Conventional CMOS circuitry does have one major reliability concern, latchup, which is strongly dependent on temperature.

High Temperature Electronics. Edited by M. Willander and H.L. Hartnagel. Published in 1997 by Chapman & Hall, London. ISBN 0 412 62510 5.

The first objective, then, in high temperature CMOS work must be to evaluate and reduce susceptibility to latchup.

As CMOS operation is extended to its temperature limits, the validity of assumptions commonly made to simplify the device equations for moderate temperature operation is compromised. To successfully design high temperature CMOS circuits, one must have a thorough understanding of the implications of high temperature and accurate models of semiconductor behavior at elevated temperatures. This section begins by briefly describing the effects of high temperatures on basic semiconductor physical properties (energy bandgap, intrinsic carrier density, impurity carrier concentration, Fermi potential and carrier mobility) and on MOS transistor parameters (leakage current, latchup, threshold voltage, subthreshold current and surface mobility). Several general approaches are then described for increasing the temperature limits of silicon: modifying the process technology, changing physical layout design rules and using special high temperature circuit methods [5]. Each of these strategies offers a range of possible approaches with corresponding costs and benefits. They should all be evaluated in the development of any high temperature integrated circuit; in most cases, a combination of these approaches will yield the most cost-effective solution.

4.2 PHYSICAL PROPERTIES AT HIGH TEMPERATURES

4.2.1 Energy bandgap

Energy bandgap, the minimum energy separation between conduction and valence bands, is the most fundamental semiconductor property. At room temperature and atmospheric pressure, the bandgaps of high purity Si, Ge and GaAs are 1.12 eV, 0.66 eV and 1.424 eV, respectively. Increasing temperatures decrease the bandgaps of most semiconductors. This effect can be ascribed to two factors. The first is a volume change with temperature. The other exists for fixed volume and is related to electron–phonon scattering. Therefore, the change in an energy state with temperature is given by

$$\frac{\partial E}{\partial T} = \left(\frac{\partial E}{\partial T}\right)_{\text{exp}} + \left(\frac{\partial E}{\partial T}\right)_{\text{ph}} \tag{4.1}$$

where the first term on the right-hand side of equation (4.1) is derived from expansion of the lattice and the second term accounts for the constant-volume electron–phonon scattering. For most semiconductors, the contribution of these two effects to bandgap change with temperature is comparable.

Table 4.1 Parameter values in Thurmond's energy bandgap model for Si, Ge and GaAs (After Thurmond [6])

Material	E_g at 0 K (eV)	α (eV K^{-1})	β (K)
Si	1.170	4.73×10^{-4}	636
Ge	0.7437	4.774×10^{-4}	235
GaAs	1.519	5.405×10^{-4}	204

Thurmond provides a universal function for the temperature dependence of energy bandgaps for Si, Ge and GaAs [6]:

$$E_g(T) = E_g(0 \text{ K}) - \frac{\alpha T^2}{T + \beta} \tag{4.2}$$

The values of model parameters $E_g(0 \text{ K})$, α and β in Thurmond's model are listed in Table 4.1. Figure 4.1 shows the decrease in bandgaps with

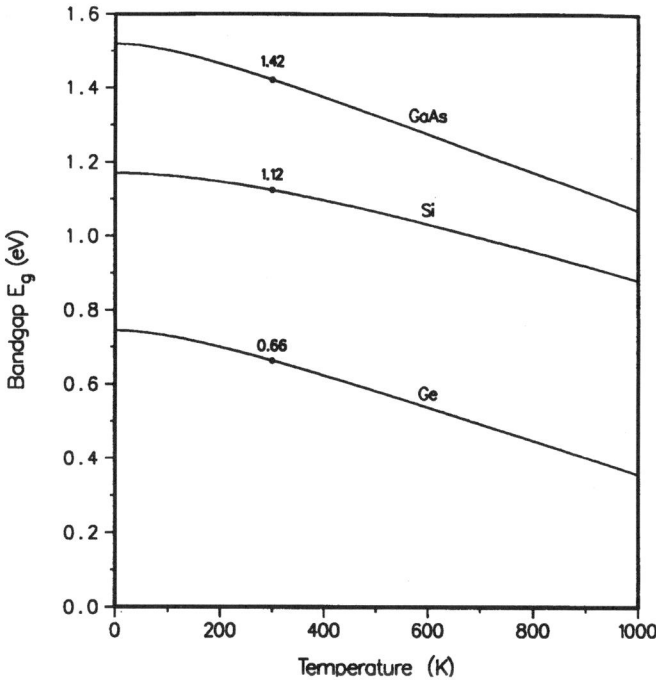

Figure 4.1 Energy bandgaps of Ge, Si and GaAs as functions of temperature. (After Thurmond [6])

increasing temperature for Si, Ge and GaAs [7]. The second term in equation (4.2) is proportional to T at high temperatures and proportional to T^2 at low temperatures.

Smaller bandgaps cause larger intrinsic carrier densities, larger leakage currents in p–n junctions, poorer junction rectification in power switching devices, and poorer device isolation by reverse-biased junctions.

4.2.2 Intrinsic carrier density

The equilibrium concentrations of electrons and holes in semiconductors are given by

$$n = N_C \exp\left(-\frac{E_C - E_F}{kT}\right) \tag{4.3}$$

and

$$p = N_V \exp\left(-\frac{E_F - E_V}{kT}\right) \tag{4.4}$$

where k is Boltzmann's constant, T the absolute temperature, E_F the Fermi energy, E_C and E_V the energies at the conduction and valence band edges, and N_C and N_V the effective densities of states in the conduction and valence bands, respectively. The above equations are obtained from Boltzmann statistics in nondegenerate semiconductors (the Fermi level is at least several kT away from the band edges). The effective densities of states, N_C and N_V, are given by

$$N_C = 2 \left(\frac{2\pi m_{de} kT}{h^2}\right)^{3/2} M_C \tag{4.5}$$

and

$$N_V = 2 \left(\frac{2\pi m_{dh} kT}{h^2}\right)^{3/2} \tag{4.6}$$

where h is Planck's constant, m_{de} and m_{dh} the density-of-state effective masses for electrons and holes, respectively, and M_C the number of equivalent minima in the conduction band. M_C has a value of 6 for Si, 4 for Ge and 1 for GaAs.

The thermal agitation in a semiconductor crystal at temperatures above absolute zero excites electrons from the valence band to the conduction band, leaving behind holes. Equilibrium is maintained by recombination of electrons in the conduction band with holes in the valence band. In undoped (intrinsic) semiconductors, the densities of free electrons n and holes p are equal and known as the intrinsic carrier density (n_i). The product of electron and hole densities is obtained from equations (4.3) and (4.4):

$$np = n_i^2 = N_C N_V \exp\left(\frac{-E_g}{kT}\right) \tag{4.7}$$

This relationship, $np = n_i^2$, is called the mass action law for semiconductors.
The intrinsic carrier density can also be expressed as

$$n_i = \sqrt{N_C N_V} \exp\left(\frac{-E_g}{2kT}\right)$$

$$= 2\left(\frac{2\pi m_0 k}{h^2}\right)^{3/2} \left(\frac{m_{de} m_{dh}}{m_0^2}\right)^{3/4} M_C^{1/2} T^{3/2} \exp\left(\frac{-E_g}{2kT}\right)$$

$$= 4.9 \times 10^{15} \left(\frac{\langle m \rangle}{m_0^2}\right)^{3/2} M_C^{1/2} T^{3/2} \exp\left(\frac{-E_g}{2kT}\right) \tag{4.8}$$

where m_0 is the free-electron mass and $\langle m \rangle$ is the density-of-state average
effective mass, which is the geometric mean of the electron and hole
density-of-state effective masses:

$$\langle m \rangle = (m_{de} m_{dh})^{1/2} \tag{4.9}$$

Intrinsic carrier density can be obtained from the experimental results of
Hall measurements. The intrinsic carrier densities for Si, Ge and GaAs as
functions of temperature and reciprocal temperature are shown in Fig. 4.2
[7]. At room temperature, the intrinsic carrier density has a value of
$1.45 \times 10^{10}\,\text{cm}^{-3}$ for Si, $2.4 \times 10^{13}\,\text{cm}^{-3}$ for Ge and $1.79 \times 10^6\,\text{cm}^{-3}$ for
GaAs. As expected, at a given temperature, n_i is larger in smaller bandgap
materials. The average effective mass, $\langle m \rangle$, has been shown to have only a
weak temperature dependence [8], which can usually be neglected in
calculations of n_i. At room temperature, n_i is small compared to device
doping levels, but it increases rapidly with temperature, doubling every
11 °C for silicon.

From equation (4.8) the temperature dependence of n_i is

$$n_i \propto \langle m \rangle^{3/2} T^{3/2} \exp\left(\frac{-E_g}{2kT}\right) \tag{4.10}$$

or

$$\log n_i \propto \frac{3}{2}\log\langle m \rangle + \frac{3}{2}\log T - \frac{E_g \log e}{2kT} \tag{4.11}$$

For silicon at temperatures below 300 °C, n_i is approximately an exponential
function of reciprocal temperature, that is, a straight line for $\log n_i$ versus
$1/T$. The slope of this straight line is approximately equal to $-E_g \log e/2k$,
demonstrating that the $1/T$ term inside the exponential bracket of equation

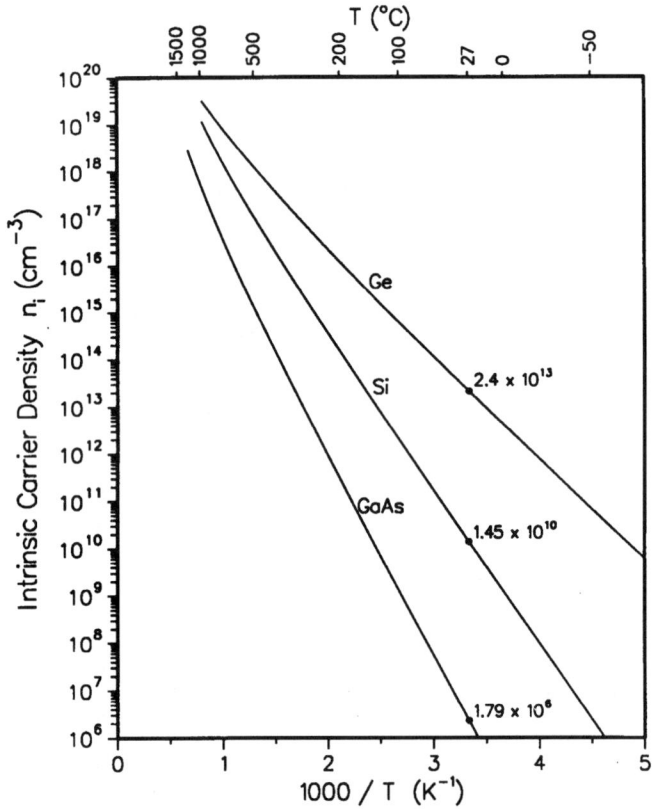

Figure 4.2 Intrinsic carrier densities in Si, Ge and GaAs as functions of reciprocal temperature. (After Thurmond [6])

(4.10) is by far the dominant temperature factor, compared to the $T^{3/2}$ term outside the exponential bracket. At temperatures higher than 500 °C, the curve of $\log n_i$ versus $1/T$ departs from a straight line due to the increasing importance of $T^{3/2}$, and to a decrease of E_g with temperature.

4.2.3 Carrier concentrations

The carrier concentrations can be obtained from solving the charge neutrality condition

$$n + N_A^- = p + N_D^+ \tag{4.12}$$

where n and p are given by equations (4.3) and (4.4) and N_D^+ and N_A^- are respectively the numbers of ionized donors and acceptors, which according

to Sze [7], are given by

$$N_D^+ = \frac{N_D}{1 + 2\exp(E_F - E_D/kT)} \tag{4.13}$$

and

$$N_A^- = \frac{N_A}{1 + 4\exp(E_A - E_F/kT)} \tag{4.14}$$

where E_D and E_A are the donor impurity level and acceptor impurity level, respectively.

Figure 4.3 is a plot of electron density in lightly doped silicon as a function of reciprocal temperature [9]. Three ranges can be identified for the electron density: an intrinsic range at high temperatures ($T > 500\,\mathrm{K}$), a saturation range at moderate temperatures ($100\,\mathrm{K} < T < 400\,\mathrm{K}$) and a freezeout range at low temperatures ($T < 75\,\mathrm{K}$). In the intrinsic range, the intrinsic carrier density by thermal generation is larger than the background doping, so $n_i \approx n \approx p \gg N_D$. The slope of $\log n_i$ versus $1/T$ in this range is

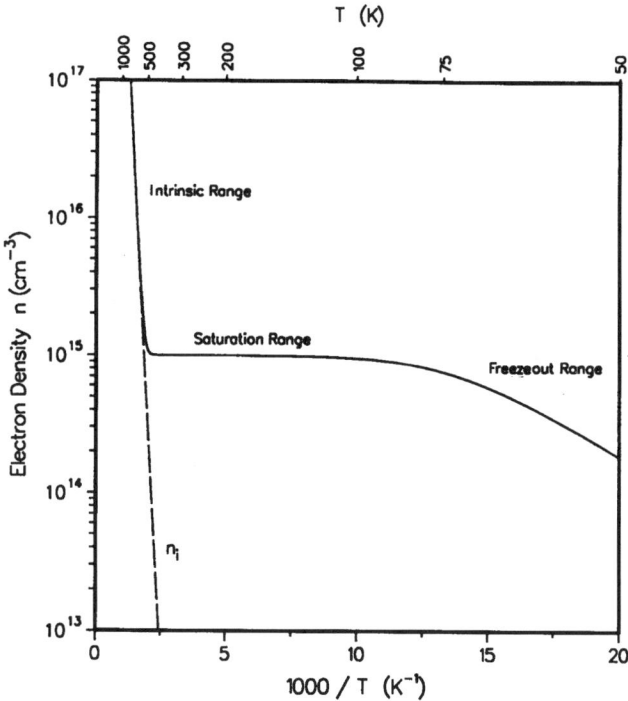

Figure 4.3 Electron density in silicon as a function of reciprocal temperature with a donor impurity concentration of $10^{15}\,\mathrm{cm}^{-3}$ and donor level at $0.044\,\mathrm{eV}$ below the conduction band. (After Smith [9])

given by $-E_g \log e/2k$. At very low temperatures, thermal energy is not sufficient to supply the donor ionization energy, so most carriers are frozen out. The slope in the freezeout range is given by $-(E_C - E_D) \log e/2k$. The electron density remains essentially constant at $n \approx N_D$ over a wide range of temperatures, 100–400 K. Most semiconductor devices are designed to operate in this temperature range, so that majority carrier concentration is almost independent of temperature.

Figure 4.4 shows the electron density as a function of temperature for Si, Ge and GaAs with a doping level of $10^{16} \, cm^{-3}$. The donor level is assumed to be at 0.045, 0.012 and 0.058 eV below the conduction band for the three semiconductors, respectively. As seen in this figure, a larger energy bandgap material, such as GaAs, has a wider carrier saturation range than a smaller bandgap material with the same doping level, and would therefore be expected to have a wider operational temperature range. The saturation range in Fig. 4.4 has a width of about 200, 300 and 400 K for Ge, Si and GaAs, respectively.

At high temperatures, thermal generation becomes the dominant process of carrier generation. As shown in Fig. 4.4 for a doping level of $10^{16} \, cm^{-3}$,

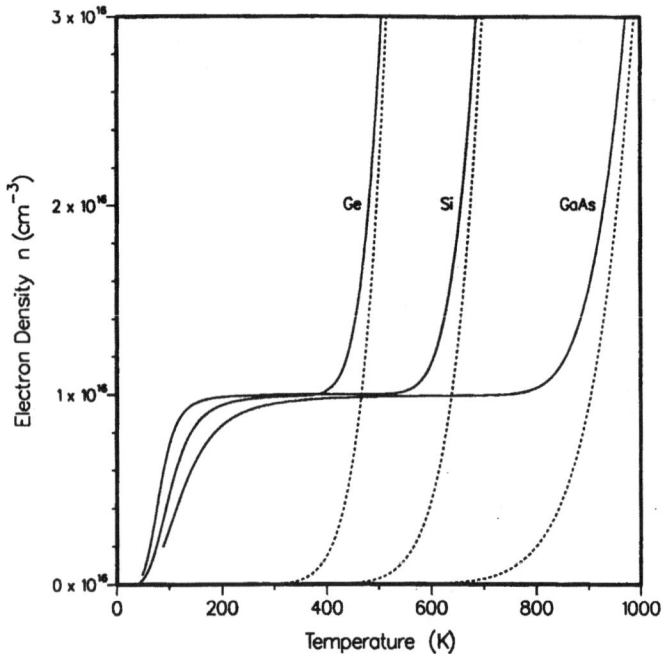

Figure 4.4 Electron density versus temperature for Si, Ge and GaAs with the donor impurity concentration of $10^{16} \, cm^{-3}$, where the donor level is 0.045, 0.012 and 0.058 eV below the conduction band for the three semiconductors, respectively: (—) n and (---) n_i.

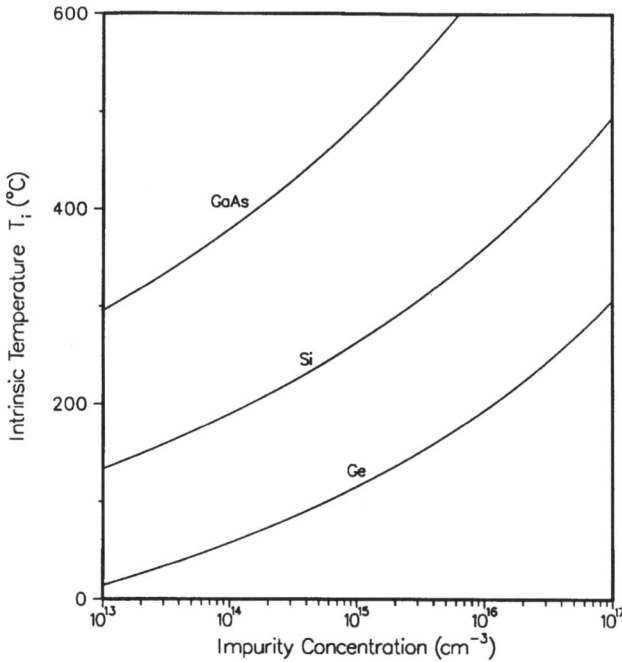

Figure 4.5 Intrinsic temperature as a function of background concentration for Ge, Si and GaAs. (After Sze [7])

the intrinsic carrier density dominates the electron density at temperatures of about 450, 600 and 900 K for Ge, Si and GaAs, respectively. The intrinsic temperature, T_i, is defined as the temperature at which the intrinsic carrier density becomes equal to the background impurity concentration. Figure 4.5 plots intrinsic temperatures of Ge, Si and GaAs as functions of impurity concentration. As expected, higher doping levels and larger bandgaps result in higher T_is. Therefore, semiconductors for high temperature applications should have wide bandgaps or high doping levels to obtain high intrinsic temperatures and wide operating temperature ranges. Moderately doped silicon having an impurity concentration of 10^{16} cm^{-3} becomes intrinsic at a temperature near 350 °C.

At normally specified operating temperatures (above 100 K for silicon), most donors and acceptors are ionized ($N_D^+ \approx N_D$ and $N_A^- \approx N_A$), so the neutrality condition of equation (4.12) can be written

$$n + N_A = p + N_D \tag{4.15}$$

The electron and hole densities at temperatures above the freezeout range can be obtained by solving the charge/neutrality condition of equation

(4.15) and the mass action law of equation (4.7). Electron density in n-type semiconductors is then given by

$$n_n = \frac{(N_D - N_A) + \sqrt{(N_D - N_A)^2 + 4n_i^2}}{2}$$ (4.16)

and hole density in p-type semiconductors is given by

$$p_p = \frac{(N_A - N_D) + \sqrt{(N_A - N_D)^2 + 4n_i^2}}{2}$$ (4.17)

The majority carrier concentration in silicon is shown in Fig. 4.6 as a function of temperature with background doping as the parameter. At lower temperatures, n_i is negligible compared to the background doping, i.e. $n_i \ll n_n$ or $n_i \ll p_p$. The flat regions in the figure are the carrier saturation ranges for each doping level, in which the majority carrier concentrations are relatively temperature independent. At the intrinsic carrier temperatures, n_i becomes comparable to the background doping, and majority carrier concentration begins to follow intrinsic carrier density. At temperatures much above T_i, majority carrier concentration follows n_i as it rises exponentially with temperature. Though they become very close, the majority carrier concentration remains greater than n_i.

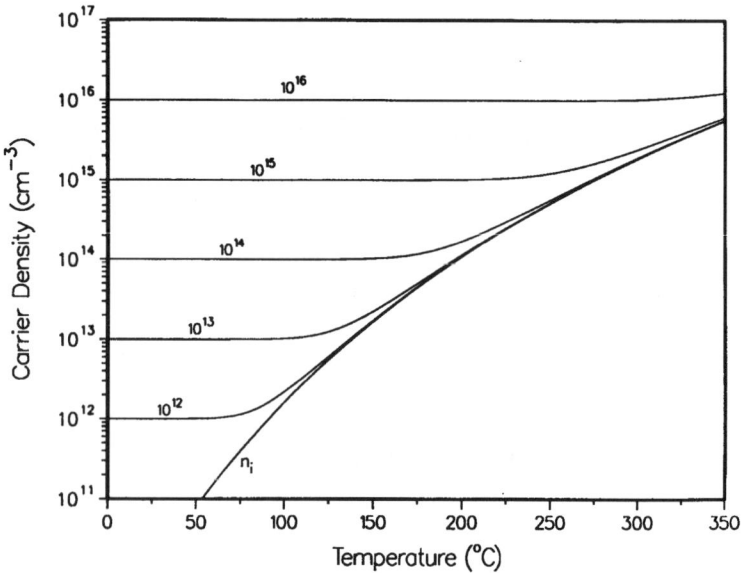

Figure 4.6 Majority carrier density as a function of temperature with background doping as the parameter.

The hole density in an n-type semiconductor is given by

$$p_n = n_i^2/n_n \tag{4.18}$$

and the electron density in a p-type semiconductor by

$$n_p = n_i^2/p_p \tag{4.19}$$

In the carrier saturation range, majority carrier concentrations have the value of background doping concentration, so minority carrier concentrations are

$$p_n \approx n_i^2/N_D \tag{4.20}$$

and

$$n_p \approx n_i^2/N_A \tag{4.21}$$

Note that, since the np product is always equal to n_i^2, and since in the saturation range majority carrier concentration is relatively independent of temperature, the minority carrier concentration in the saturation range has the same temperature dependence as n_i^2. This means that minority carrier concentration increases much faster with temperature than majority carrier concentration at normal operating temperatures. In the intrinsic range, both majority and minority carrier concentrations approach the intrinsic carrier concentration, and have the same temperature dependence as n_i.

4.2.4 Fermi potential

The Fermi energy E_F, the average energy of a free carrier in a semiconductor, is used as an indicator of how far from intrinsic the material appears to be, or in other words, how heavily doped. Intrinsic silicon has a Fermi energy that is close to the midgap at room temperature (7.3 meV below), so for convenience, the intrinsic Fermi energy E_i is usually assumed to be at midgap.

Following Sze [7], the electron and hole densities, equations (4.3) and (4.4), can then be written in terms of intrinsic carrier concentrations and intrinsic Fermi level:

$$n = n_i \exp\left(\frac{E_F - E_i}{kT}\right) \tag{4.22}$$

and

$$p = n_i \exp\left(\frac{E_i - E_F}{kT}\right) \tag{4.23}$$

The difference in average carrier energies can also be described as a difference in potentials, ϕ_f, by dividing the average energy of a carrier by the electronic charge. Fermi potential ϕ_f is relative to the intrinsic level

$$\phi_f = \frac{E_F - E_i}{q} \tag{4.24}$$

Carrier concentrations can then be written as

$$n = n_i \exp\left(\frac{q\phi_f}{kT}\right) \tag{4.25}$$

and

$$p = n_i \exp\left(-\frac{q\phi_f}{kT}\right) \tag{4.26}$$

and one can solve for the Fermi potential ϕ_f for n-type semiconductors as

$$\phi_n = \frac{kT}{q} \ln\left(\frac{n}{n_i}\right) \tag{4.27}$$

and for p-type semiconductors

$$\phi_p = -\frac{kT}{q} \ln\left(\frac{p}{n_i}\right) \tag{4.28}$$

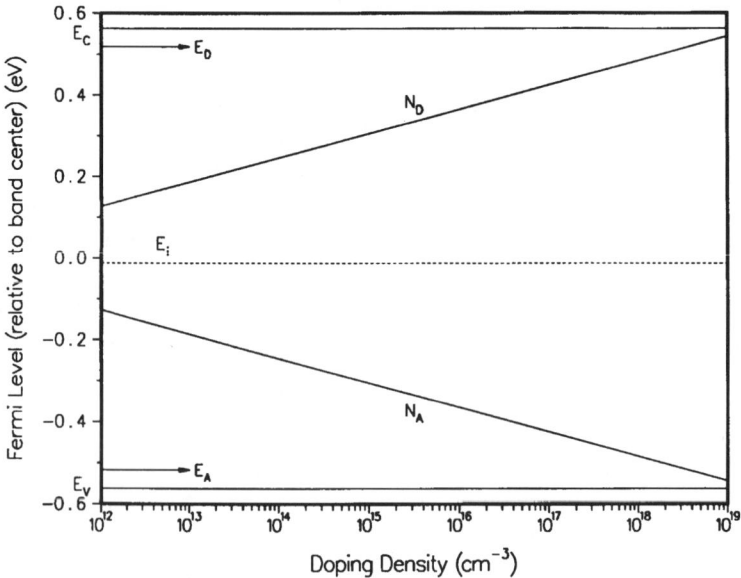

Figure 4.7 Fermi level versus doping concentration for silicon at room temperature. The donor level is 0.044 eV below the conduction band and the acceptor level is 0.045 eV above the valence band.

Figure 4.7 shows Fermi levels as functions of doping density in silicon at room temperature. The Fermi level is above E_i for n-type semiconductors and below E_i for p-type semiconductors.

At temperatures above the carrier freezeout range, Fermi potentials can be obtained by substituting equations (4.16) and (4.17) into equations (4.27) and (4.28):

$$\phi_n = \frac{kT}{q} \ln\left(\frac{N_D + \sqrt{N_D^2 + 4n_i^2}}{2n_i}\right) \tag{4.29}$$

and

$$\phi_p = -\frac{kT}{q} \ln\left(\frac{N_A + \sqrt{N_A^2 + 4n_i^2}}{2n_i}\right) \tag{4.30}$$

Figure 4.8 shows the temperature dependence of electron Fermi potentials in silicon, with impurity doping concentration as the parameter. As temperature increases, the electron Fermi potential at first decreases linearly with temperature, then approaches zero at temperatures above T_i, meaning that extrinsic semiconductors will become intrinsic at sufficiently high temperatures. Semiconductors having higher doping levels maintain extrinsic behavior to higher temperatures.

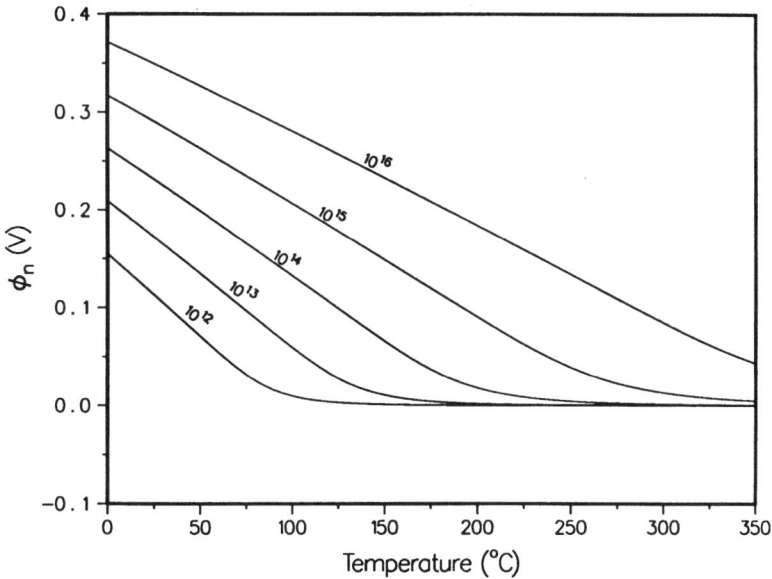

Figure 4.8 Electron Fermi potential versus temperature for silicon with donor impurity concentration as the parameter.

At moderate temperatures, majority carrier concentration is approximately equal to doping level, so the Fermi potentials can be simplified to

$$\phi_n \approx \frac{kT}{q} \ln\left(\frac{N_D}{n_i}\right) \tag{4.31}$$

and

$$\phi_p \approx -\frac{kT}{q} \ln\left(\frac{N_A}{n_i}\right) \tag{4.32}$$

These approximations are valid only when the intrinsic carrier density is negligible compared to the impurity concentration. When n_i exceeds the impurity concentration, the fractional terms inside the bracket of the natural logarithm in equations (4.31) and (4.32) become less than one, making the natural logarithm negative and incorrectly indicating a reversal of the semiconductor type. Unfortunately, equations (4.31) and (4.32) are used in popular circuit simulators instead of equations (4.29) and (4.30), prohibiting proper high temperature simulation.

The effects of higher temperature on Fermi potential can be summarized as follows: the bandgap becomes narrower, the intrinsic Fermi level drifts slowly away from the bandgap center, and the Fermi levels for n- and p-type materials approach the intrinsic Fermi level. These effects are more gradual in heavily doped silicon.

4.2.5 Carrier mobilities

Semiconductor carrier mobilities, as well as concentrations, as discussed above, are affected by temperature. Solid-state physics theory states that a perfectly periodic lattice will not scatter free carriers, or in other words, the carriers will not interchange energy with a stationary periodic lattice. However, at any temperature above absolute zero, lattice vibrations disturb the periodicity, allowing energy to be transferred between the carriers and the lattice. The interactions with lattice vibrations can be viewed as collisions between free carriers and phonons. Lattice scattering is more effective at higher temperatures, as seen in the following relationship from Sze [7]:

$$\mu_L \sim (m^*)^{-5/2} \, T^{-3/2} \tag{4.33}$$

where μ_L is the mobility from lattice scattering and m^* is the conductivity effective mass.

Dopant impurities also cause local distortions in the lattice and they scatter free carriers. Impurity scattering becomes less significant at higher temperatures because the carriers are moving faster, so they remain near the impurity atom for a shorter time and are less affected by the interaction.

According to Sze [7], the mobility from ionized impurities, μ_I, can be written

$$\mu_I \sim (m^*)^{-1/2} N_I^{-1} T^{3/2} \tag{4.34}$$

where N_I is the ionized impurity density. For nonpolar semiconductors, such as silicon, lattice and impurity scattering are the two dominant factors in determining mobility. The combined mobility from these mechanisms is given by

$$\mu = \left(\frac{1}{\mu_L} + \frac{1}{\mu_I} \right)^{-1} \tag{4.35}$$

The resultant mobility is therefore smaller than that determined by any of the individual scattering mechanisms.

Figures 4.9 and 4.10 show carrier mobilities in silicon as functions of temperature for electrons and holes, respectively. At lower temperatures the mobility increases as temperature rises since impurity scattering dominates, whereas at higher temperatures the mobility decreases with temperature because lattice (acoustic phonon) scattering dominates. These competing temperature variations lead to characteristic maxima in the mobility versus temperature relation (Figs 4.9 and 4.10). Note that the temperature at which carrier mobility reaches its maximum is concentration dependent, increasing as doping concentration increases. Since impurity scattering dominates at

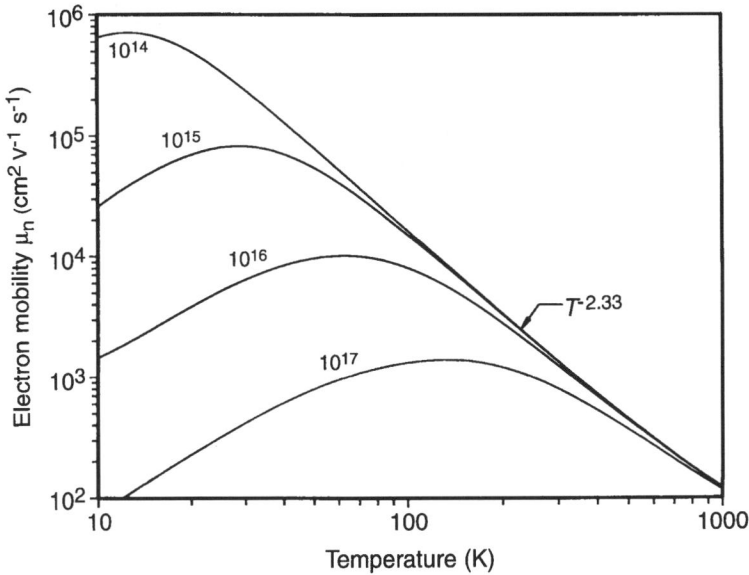

Figure 4.9 Low field electron mobility in silicon as a function of temperature for different impurity concentrations (calculated).

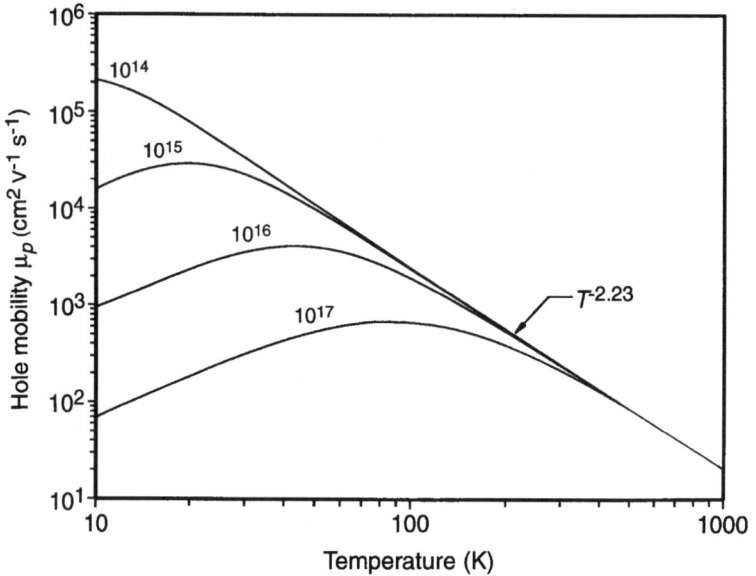

Figure 4.10 Low field hole mobility in silicon as a function of temperature for different impurity concentrations (calculated).

lower temperatures, the resultant mobility is doping concentration dependent; see equation (4.34). At higher temperatures, mobility is relatively insensitive to the impurity concentration because lattice scattering dominates.

The lattice-scattering mobility, found from experimental data of lightly doped silicon in the temperature range 150–500 K [10], is well represented by the following expressions for electrons:

$$\mu_{Ln} = 8.56 \times 10^8 T^{-2.33} \tag{4.36}$$

and for holes

$$\mu_{Lp} = 1.58 \times 10^8 T^{-2.23} \tag{4.37}$$

The temperature dependence of carrier mobility in silicon over 0–300 °C is shown in Figs 4.11 and 4.12. At higher dopant concentrations, mobility decreases more slowly with increasing temperature, and at sufficiently high doping ($\geqslant 10^{19} \, cm^{-3}$ for donors and $\geqslant 10^{20} \, cm^{-3}$ for acceptors), the mobility remains relatively independent of temperature. Under these conditions the two scattering mechanisms, which have opposite temperature dependences, are balanced, leaving mobility nearly insensitive to temperature.

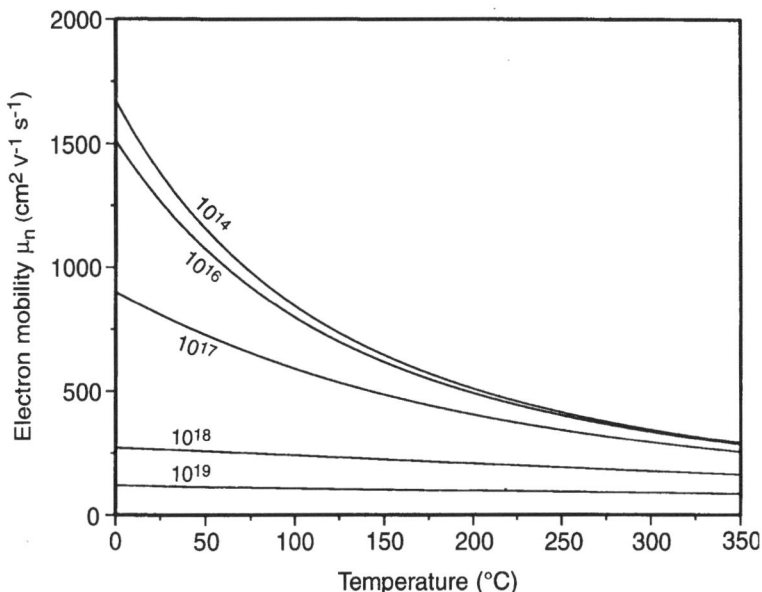

Figure 4.11 Temperature dependence of electron mobility in silicon for sample dopings ranging from 10^{14} cm^{-3} to 10^{19} cm^{-3}. (After Arora *et al.* [10])

4.2.6 Depletion-approximation limitations

Junction theory is fundamental to understanding high temperature effects in semiconductor devices. Many of the basic equations used in semiconductor device work are based on p–n junction depletion layer width. Examples are: depletion layer capacitance and recombination–generation current in junction diodes; current density, current gain and frequency response in bipolar transistors; and threshold voltage, channel capacitance and short-channel effects in field effect transistors (FETs).

Depletion layer widths are commonly obtained by using the complete depletion approximation, which is based on two assumptions: (1) complete depletion of mobile charges (majority and minority carriers) in the depletion region and (2) an abrupt transition between the depletion region and neutral region.

Although the depletion approximation is very useful in the normal operating temperature range, it is inaccurate at high temperatures. Minority carrier concentration increases rapidly with temperature, in the depletion region as well as in neutral material. At the intrinsic temperature, the minority carrier concentration is comparable to the background doping level, and certainly cannot be neglected in depletion width calculations.

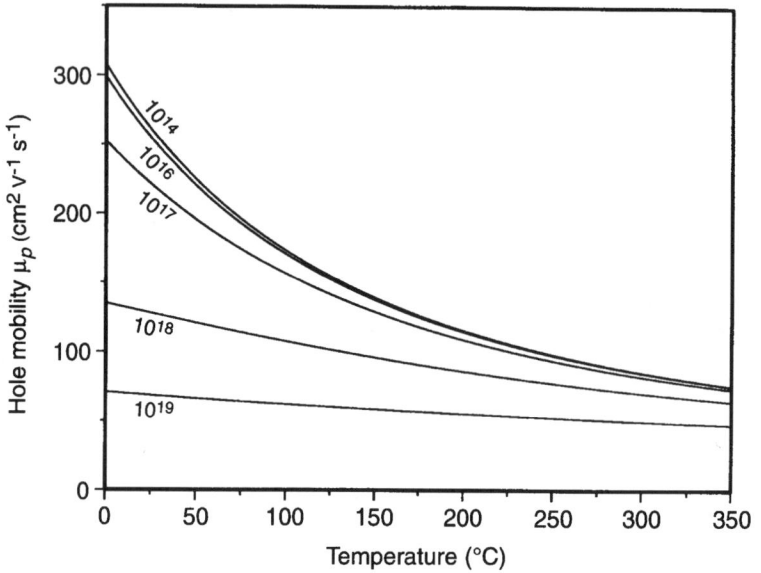

Figure 4.12 Temperature dependence of hole mobility in silicon for sample dopings ranging from 10^{14} cm^{-3} to 10^{19} cm^{-3}. (After Arora *et al.* [10])

A sharp transition between the depletion region and the neutral region exists only at very low temperatures. At higher temperatures there is a gradual transition between these regions, which becomes smoother and wider as temperature increases.

The correct space-charge distributions and current flows in p–n junctions can be found [11, 12] by solving concurrently the Poisson equation

$$\nabla^2 \psi = -\frac{q}{\varepsilon_s}(p - n + N_D^+ - N_A^-) \tag{4.38}$$

and the two-carrier current–continuity equations

$$\frac{\partial n}{\partial t} = G_n - U_n + \frac{1}{q}\nabla \cdot \boldsymbol{J}_n$$

$$\frac{\partial p}{\partial t} = G_p - U_p - \frac{1}{q}\nabla \cdot \boldsymbol{J}_p \tag{4.39}$$

where ψ $(= -E_i/q)$ is the intrinsic Fermi potential, G_n and G_p are the generation rates, U_n and U_p are the recombination rates, and $\boldsymbol{J}_n$ and $\boldsymbol{J}_p$ are the current densities, for electrons and holes, respectively. Solving these coupled equations, one can find the actual charge distribution in p–n junctions at any temperature. The deviation of results from those of the depletion approximation are greatest for junctions that are lightly doped,

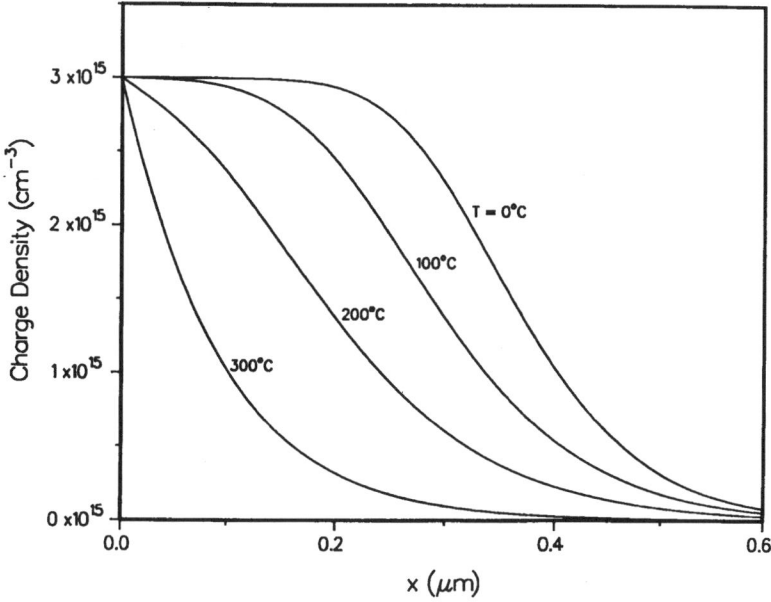

Figure 4.13 Space-charge density versus distance on the n-type side of symmetrically doped junction with $N_A = N_D = 3 \times 10^{15} \, \text{cm}^{-3}$ and $V_a = 0 \, \text{V}$.

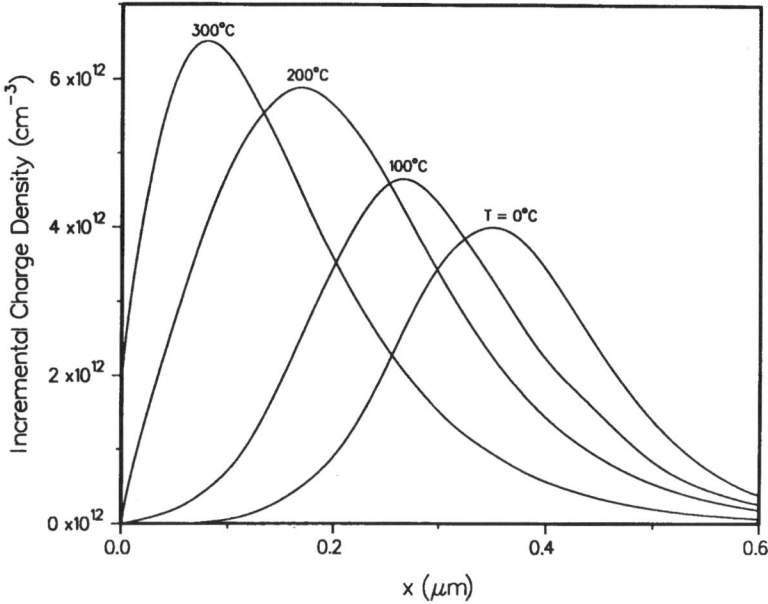

Figure 4.14 Incremental space-charge density versus distance on the n-type side of a symmetrically doped junction with $N_A = N_D = 3 \times 10^{15} \, \text{cm}^{-3}$, $V_a = 0 \, \text{V}$ and $\Delta V_a = -1 \, \text{mV}$.

nonsymmetrically doped, positively biased and operating at high temperatures.

Figure 4.13 shows the change in space charge with temperature for a lightly doped junction at zero bias. This figure is a rather extreme case, but the effect is present to some degree under any doping and bias conditions. Figure 4.14 is a plot of incremental space-charge density (location and quantity of new charge uncovered as junction bias changes) at four temperatures. Incremental space charge is the parameter of greatest interest in small-signal applications. In both the large-signal (Fig. 4.13) and small-signal (Fig. 4.14) cases, high temperatures obviously affect the junction capacitance. See Chapter 4 of Wu [11] for a macromodel of these capacitances, which is accurate for various doping profiles and concentrations over the range 0–300 °C.

One of the most important implications of this high temperature junction capacitance model is that small-signal capacitance has a higher value at high temperatures than predicted by the complete-depletion model. This effect further depresses the speed of silicon circuits at high temperatures.

4.2.7 Summary of semiconductor parameter effects

The energy bandgaps of most semiconductors decrease with increasing temperature. The intrinsic carrier density increases rapidly with temperature. Semiconductors with higher doping levels or wider bandgaps have higher intrinsic temperatures and therefore wider operating temperature ranges. The magnitude of the Fermi potential decreases approximately linearly with temperature, and approaches zero above the intrinsic temperature. At room temperature and above, the mobility decreases with temperature due to lattice scattering. The temperature effects on MOSFET behavior resulting from changes in these fundamental semiconductor parameters are described in the next section.

4.3 HIGH TEMPERATURE BEHAVIOR OF MOS TRANSISTORS

MOS devices are well suited for microelectronics operating at elevated temperatures because they do not suffer thermal runaway effects, as do bipolar transistors. Increasing temperature causes a decrease in carrier mobilities, which in MOSFETs reduces channel current, thereby reducing power dissipation.

High temperature does have negative effects on the performance of MOS transistors through leakage current, latchup threshold voltage shifts, subthreshold current and surface mobility changes [13]. In this discussion, the transistor current–voltage characteristics in strong inversion are assumed to follow the Shichman–Hodges model [14], and, except as noted, the depletion approximation is used under the channel. As for the space-charge region of p–n junctions, the validity of the depletion approximation for the space-charge region under the MOSFET gate is compromised at high temperatures. The small error in threshold voltage caused by this approximation is discussed in section 4.3.2.

4.3.1 Leakage current and latchup

Leakage currents in MOS transistors are attributed primarily to p–n junction leakage because they show quite similar temperature dependencies. The leakage current of a p–n junction is composed mainly of the diffusion (or Shockley) current and the space-charge generation current. The total leakage current density, J_R, can be expressed as

$$J_R = J_{R,diff} + J_{R,gen} \tag{4.40}$$

where $J_{R,diff}$ is the diffusion current density and $J_{R,gen}$ is the space-charge generation current density. The diffusion current is generated in neutral regions where there is no significant electric field. The carriers move by diffusion from regions of higher concentration to regions of lower concentration. When these carriers reach the edge of a depletion region, they are swept across the junction by the electric field. The generation current is due to electron–hole pairs thermally generated within the depletion region. These carriers are separated and swept across the region by the electric field.

The carrier generation current in the depletion region is

$$J_{R,gen} = \frac{qn_i W}{\tau_e} \tag{4.41}$$

where q is the electronic charge, n_i is the intrinsic carrier concentration, W is the p–n junction depletion-layer width and τ_e is the effective carrier lifetime. The exponential temperature dependence of n_i overshadows the small variations with temperature seen in the other parameters of this equation, making the temperature dependence of the generation current proportional to n_i.

The diffusion current is proportional to the minority carrier density. As discussed in section 4.2, for a wide range of temperatures, the minority carrier density has a temperature dependence of n_i^2. At very high temperatures, when the material becomes intrinsic, the minority carrier density has a temperature dependence of n_i. Based on this simple analysis, one would expect the leakage current in the saturation (extrinsic) range to have a temperature dependence of n_i when generation current dominates and a dependence of n_i^2 when diffusion current is the dominant component.

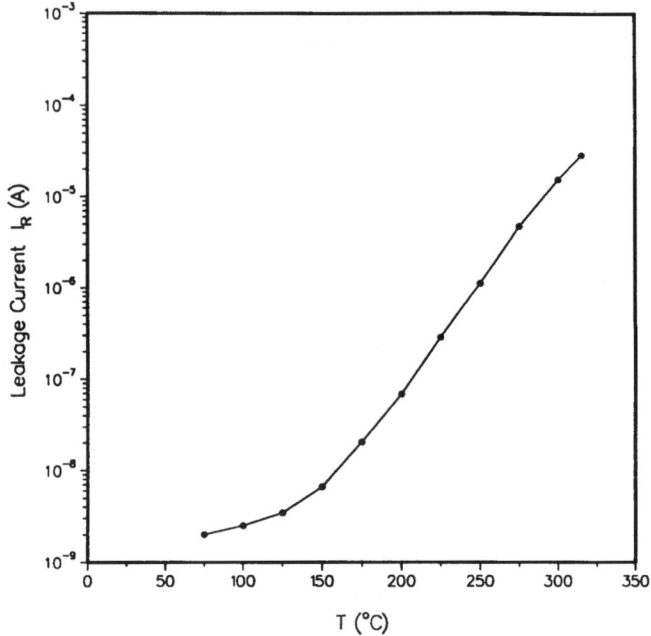

Figure 4.15 Leakage current as a function of temperature for a p-well/n-substrate diode on an epi CMOS wafer. The diode dimensions are $350 \, \mu m$ by $150 \, \mu m$; $V_a = -2 \, V$.

Figure 4.15 is a plot of the leakage current as a function of temperature for a reverse-biased p-well/n-substrate diode on an epi CMOS wafer. Figure 4.16 shows the leakage current as a function of reciprocal temperature for the same diode. The leakage current has a slope corresponding roughly to n_i dependence between 125 °C and 170 °C, and to n_i^2 dependence between 200 °C and 315 °C. This indicates that the generation current component dominates at lower temperatures, and the diffusion current component dominates at higher temperatures. At temperatures below 125 °C, the leakage current has less temperature dependence than expected.

In CMOS integrated circuits, junction leakage at high temperatures can be problematic in several areas: drain-to-source leakage, drain-to-substrate/well leakage, well-to-substrate leakage and input protection circuitry leakage [15–17]. Leakage in the large junctions of input protection circuitry can pin input signals at a voltage between V_{DD} and ground. An even worse consequence of leakage in CMOS circuitry is latchup.

Bulk CMOS latchup stems from the pnpn structure formed by parasitic npn and pnp bipolar transistors in the CMOS process, as seen in Fig. 4.17a. One of these bipolar transistors is a lateral transistor (the pnp, Q1, in this p-well case) formed in the substrate, and the other is a vertical device (Q2)

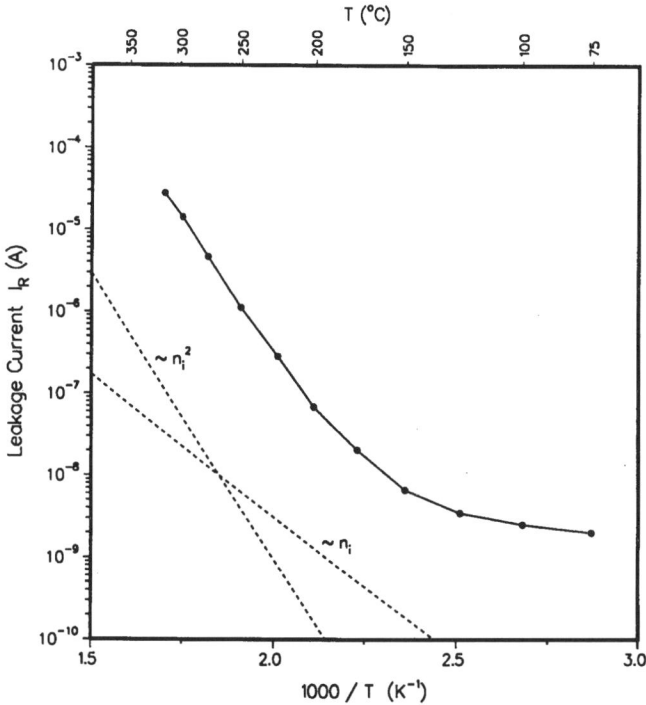

Figure 4.16 Leakage current as a function of reciprocal temperature on the same diode as shown in Fig. 4.15; $V_a = -2\,\text{V}$.

formed in the well. The two are inherently connected as a pair, with the reverse-biased well-to-substrate junction acting as the collector for both transistors. Figure 4.17b is a schematic of the cross-coupled circuit. The larger the parasitic resistance, the more likely a structure is to latch. If enough charge is injected into the n-type substrate (as by junction leakage at high temperatures) and collected by the p-well, the voltage drop across R_S will be of sufficient magnitude to switch on the bipolar transistor Q1, which in turn provides base current to Q2, turning it on, and forcing the pnpn structure into a low resistance mode; positive feedback from each transistor to the other keeps the circuit in latchup until power is removed. This leads to extraordinary static power dissipation in the device which may cause catastrophic failure. These effects are moderated by reducing the gain of the bipolar transistors and by reducing the values of the parasitic resistors R_W and R_S, as will be discussed in section 4.4.

Junction leakage is among the most serious challenges in silicon high temperature electronics. It increases self-heating of the circuit, shifts operating points in analog circuits and can be the source of latchup-triggering current.

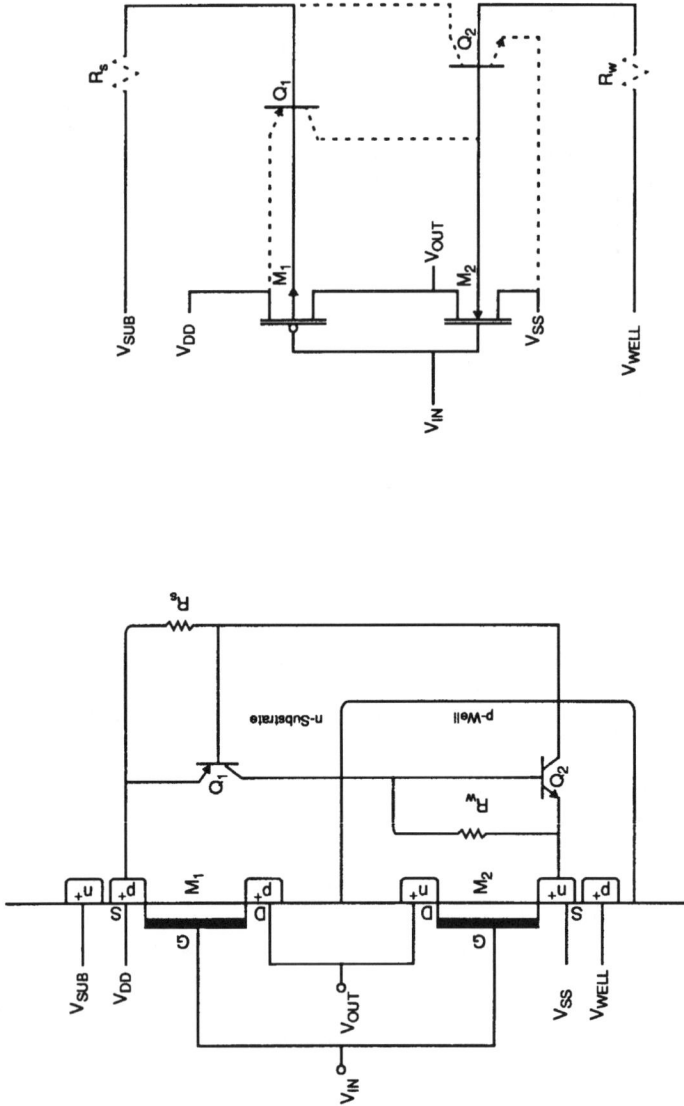

Figure 4.17 Cross section (a) and schematic (b) of a CMOS inverter showing parasitic bipolar transistors which cause latchup.

4.3.2 Threshold voltage

The threshold voltage of an n-channel MOSFET [18] is

$$V_T = V_{FB} + 2|\phi_p| - \frac{Q_d}{C_{ox}} \qquad (4.42)$$

where V_{FB} is the flat band voltage, ϕ_p is the Fermi potential of the p-substrate, Q_d is the charge per unit area contained within the surface depletion region (a negative charge in the n-channel case), and C_{ox} is the oxide capacitance per unit area. The temperature sensitivity of the threshold voltage can be determined by expanding each of these terms and taking the partial derivative. For an n-channel device, the threshold voltage sensitivity to temperature [11] is

$$\frac{\partial V_T}{\partial T} \cong -\frac{1}{2q}\frac{\partial E_g}{\partial T} + \frac{1}{T}\left(|\phi_p| - \frac{E_{g0}}{2q}\right)\left[1 - \frac{Q_d}{C_{ox}(2|\phi_p| - V_{BS})}\right] \qquad (4.43)$$

The threshold voltage of a p-channel MOSFET is given by

$$V_T = V_{FB} - 2\phi_n - \frac{Q_d}{C_{ox}} \qquad (4.44)$$

The temperature sensitivity of V_T for a p-channel MOSFET can be derived similarly as

$$\frac{\partial V_T}{\partial T} \cong -\frac{1}{2q}\frac{\partial E_g}{\partial T} - \frac{1}{T}\left(\phi_n - \frac{E_{g0}}{2q}\right)\left[1 + \frac{Q_d}{C_{ox}(2\phi_n + V_{BS})}\right] \qquad (4.45)$$

It is interesting to note that the threshold voltages of n- and p-channel devices have different variations with temperature. The first term on the right-hand side of equations (4.43) and (4.45), $-(1/2q)(\partial E_g/\partial T)$, is positive, since $\partial E_g/\partial T$ is negative. The terms inside the square brackets on the right-hand side of equations (4.43) and (4.45) are positive; the terms outside the square brackets are negative. Therefore, $\partial V_T/\partial T$ in equation (4.43) for an n-channel device is a small positive number plus a large negative number, whereas for a p-channel device it is a small positive number minus a large negative number. This means that the temperature sensitivity of threshold voltage is negative for n-channel devices and positive for p-channel devices, and the threshold voltage of p-channel devices has a slightly greater temperature dependence than for n-channel devices [19].

The calculated threshold voltage is shown in Fig. 4.18 as a function of temperature with substrate doping as the parameter. This figure is for a zero-biased substrate, oxide thickness of 50 nm and 25 °C threshold voltage of 1.2 V for n-channel devices, − 1.2 V for p-channel devices. The negative change in threshold voltage with increasing temperature for n-channel MOSFETs and the positive change for p-channel MOSFETs are due to the tendency of the Fermi level toward midgap with increasing temperature. For

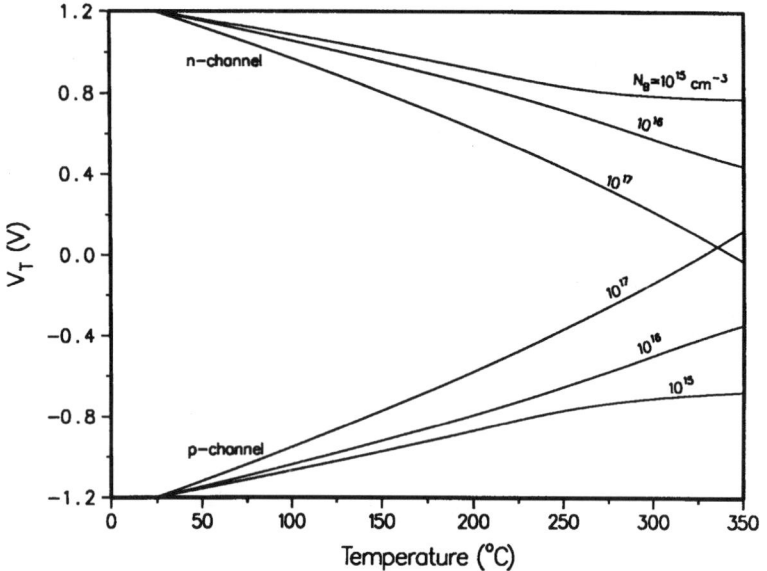

Figure 4.18 Threshold voltages of silicon n- and p-channel MOSFETs as functions of temperature with substrate doping as the parameter; $V_{BS} = 0\,\mathrm{V}$, $t_{ox} = 50\,\mathrm{nm}$ and $V_T|_{T=25\,°C}$ is set at $1.2\,\mathrm{V}$ for n-channel devices and $-1.2\,\mathrm{V}$ for p-channel devices with threshold-setting implants.

enhancement-mode MOSFETs, the threshold voltage decreases in magnitude with rising temperature, and may change sign at a sufficiently high temperature, making enhancement-mode devices into depletion-mode devices. Figure 4.18 shows the slightly greater temperature dependence for p-channel than n-channel transistors, as well as the increased temperature dependence of MOSFETs having higher substrate doping levels. This can be described in terms of the body-effect parameter, γ, which is defined as

$$\gamma \equiv \frac{\sqrt{2q\varepsilon_s N_B}}{C_{ox}} \tag{4.46}$$

The threshold voltage of an n-channel MOSFET with $V_{BS} = 0\,\mathrm{V}$ can be written as

$$V_T = V_{FB} + 2|\phi_p| + \gamma\sqrt{2|\phi_p|} \tag{4.47}$$

A larger value of γ due to higher substrate doping tends to amplify the variation of Fermi potential with respect to temperature. Increasing the substrate doping is an effective way to reduce leakage current; however, as seen in these equations, higher substrate doping will increase the temperature dependence of threshold voltage.

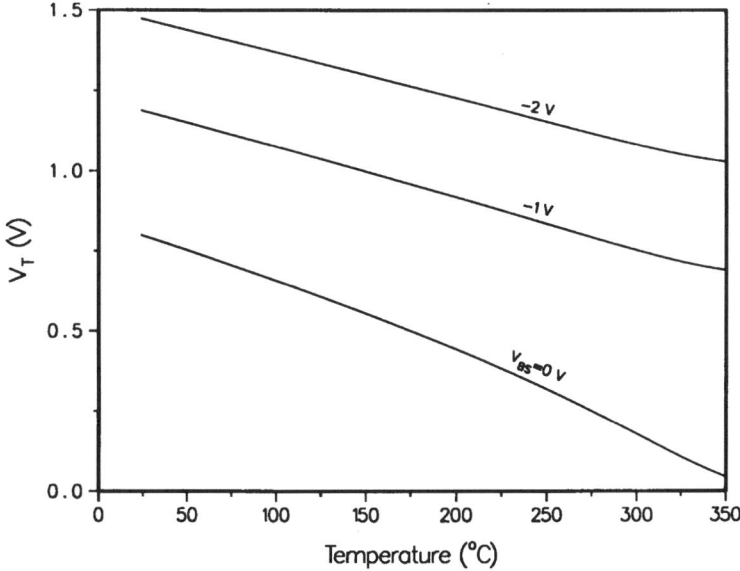

Figure 4.19 Calculated threshold voltage of an n-channel Si MOSFET as a function of temperature with substrate bias as the parameter; $N_B = 10^{16}\,\text{cm}^{-3}$, $t_{ox} = 50\,\text{nm}$ and $V_T|_{T=25\,°C,\, V_{BS}=0\,\text{V}} = 0.8\,\text{V}$.

The threshold voltage associated with low substrate doping (e.g. $N_B = 10^{15}\,\text{cm}^{-3}$ in Fig. 4.18) becomes nearly insensitive to temperatures above 250 °C. This follows from the fact that in lightly doped semiconductors, the magnitude of Fermi potential becomes very small (close to zero) and nearly temperature insensitive at high temperature. Of course, this temperature stability is of no use at these temperatures because the lightly doped semiconductor becomes intrinsic and the channel can no longer be controlled by the gate.

The effect of substrate bias on threshold voltage is shown in Fig. 4.19 as a function of temperature. Here $N_B = 10^{16}\,\text{cm}^{-3}$, $t_{ox} = 50\,\text{nm}$, and $V_T|_{T=25\,°C,\, V_{BS}=0\,\text{V}} = 0.8\,\text{V}$. In addition to the expected threshold voltage increase with a reverse bias applied to the substrate-to-source junction, a change in temperature sensitivity is also associated with backgate bias. The threshold voltage change [18] can be expressed as

$$\Delta V_T = V_T|_{V_{BS}} - V_T|_{V_{BS}=0\,\text{V}}$$

$$= \gamma(\sqrt{2|\phi_p| - V_{BS}} - \sqrt{2|\phi_p|}) \qquad (4.48)$$

Since a large value of $|V_{BS}|$ will overshadow the small variation of $|\phi_p|$ with temperature, the threshold voltage becomes less temperature dependent when a large substrate bias is applied.

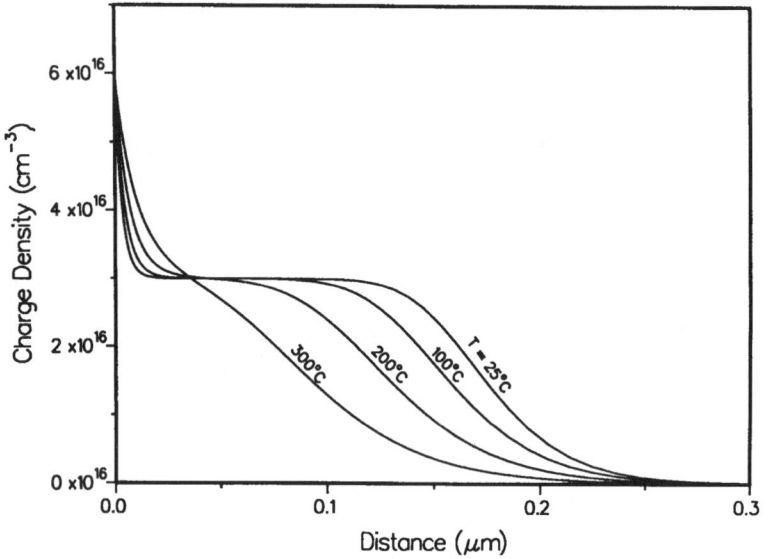

Figure 4.20 Magnitude of charge density versus distance below the surface of an n-channel MOSFET at threshold; $N_B = 3 \times 10^{16}\,\text{cm}^{-3}$, $V_{BS} = 0\,\text{V}$ and $t_{ox} = 25\,\text{nm}$.

The conventional equations for MOS threshold voltages, i.e. equations (4.42) and (4.44), are derived based on the complete-depletion approximation (ignoring the high temperature effects on space charge). Q_d in equation (4.42), the charge per unit area in the surface depletion region, is expressed as

$$Q_d = -\sqrt{2q\varepsilon_s N_B(2|\phi_p| - V_{BS})} \qquad (4.49)$$

which assumes complete depletion of mobile charge in the surface depletion region. A more accurate view of this region can be seen in Fig. 4.20, which shows the magnitude of charge density versus distance below the surface of an n-channel MOSFET, having substrate doping of $3 \times 10^{16}\,\text{cm}^{-3}$ at threshold as simulated with PISCES [20]. The transition region becomes wider and the complete-depletion region becomes narrower at higher temperatures. The complete-depletion region disappears at 300 °C. A negative bulk-to-substrate bias generates more charge under the gate, so the change in charge with temperature is a smaller percentage of Q_d.

The dependence of threshold voltage on temperature and substrate doping is shown in Fig. 4.21. The threshold voltage change $(V_T|_{T=25°C} - V_T)$ of an n-channel MOSFET is plotted as a function of substrate doping in this figure. Results calculated from the conventional equation are shown as dashed lines and PISCES simulation results are shown by dots. At low

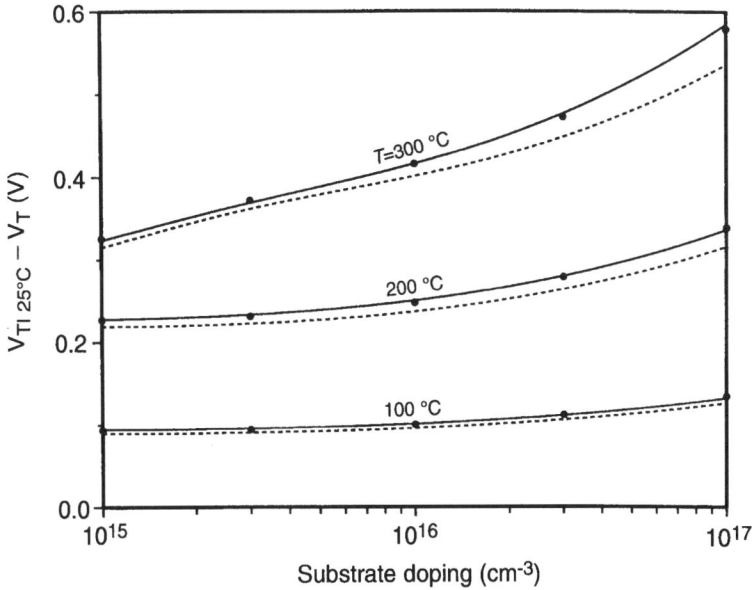

Figure 4.21 Threshold change $(V_T|_{T=25\,°C} - V_T)$ versus substrate doping for n-channel MOSFETs with $t_{ox} = 25\,nm$ and $V_{BS} = 0\,V$. Results from the conventional equation are shown as dashed lines, and simulation results are shown as dots.

temperatures the conventional threshold voltages are in good agreement with the more accurate PISCES simulation results, but at high temperatures MOS threshold voltages have a slightly larger temperature sensitivity than predicted by the conventional model.

Measured threshold voltages for n- and p-channel devices fabricated through MOSIS are shown in Fig. 4.22 as functions of temperature. MOSIS is a foundry interface service funded by the US Advanced Research Projects Agency and National Science Foundation. It runs multiproject mask sets on commercial fabrication lines for university and industry researchers. The threshold voltage is approximately a linear function of temperature in the range 50–250 °C for both n- and p-channel devices. However, at temperatures higher than 275 °C, the threshold voltage changes quite dramatically. The p-well doping in this process is about $10^{16}\,cm^{-3}$ and the n-substrate doping is $1.4 \times 10^{15}\,cm^{-3}$. The plot of threshold voltage for the p-channel device versus temperature would be expected to have a quasi-flat region (Fig. 4.18) instead of a dramatic change at temperatures above 300 °C. This discrepancy may arise from (1) inaccuracy of measurements due to increased leakage current (worse in p-channel than n-channel devices because of lower doping level) and (2) modification of the effective substrate doping by the channel implant.

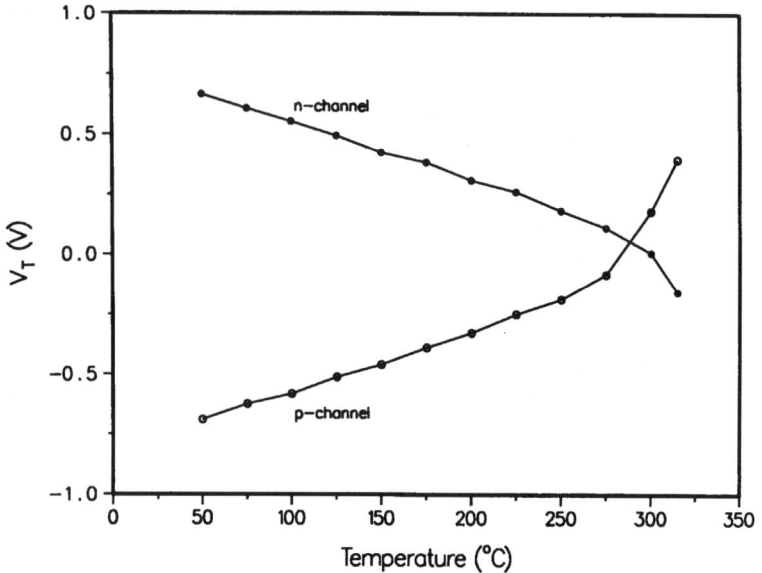

Figure 4.22 Measured threshold voltage as a function of temperature for MOSIS-fabricated n- and p-channel devices with $W/L = 4.5\,\mu\mathrm{m}/3\,\mu\mathrm{m}$.

Threshold voltage change with temperature is one of the most important effects that a circuit designer must address in high temperature applications. As temperature increases, the threshold voltage of n-channel transistors becomes more negative, and that of p-channel transistors becomes more positive. At sufficiently high temperatures, enhancement-mode transistors of either p- or n-type may become depletion devices, in which the drain current cannot be turned off with $V_{GS} = 0$. High temperature effects on the space-charge region cause the threshold voltage to have a slightly larger temperature dependency than predicted by the conventional model.

4.3.3 Subthreshold current

When gate voltage is below the threshold voltage and the semiconductor surface is in weak inversion, the corresponding drain current, called the subthreshold current, is dominated by diffusion [21]. The subthreshold behavior of a MOSFET is strongly dependent on temperature.

The subthreshold current for a long-channel NMOS device [22] is given by

$$I_D = \frac{qWD_n n_i L_{Di}}{L\sqrt{u_S + 1}} e^{-(3u_F/2)} e^{u_S}(1 - e^{-(qV_D/kT)}) \qquad (4.50)$$

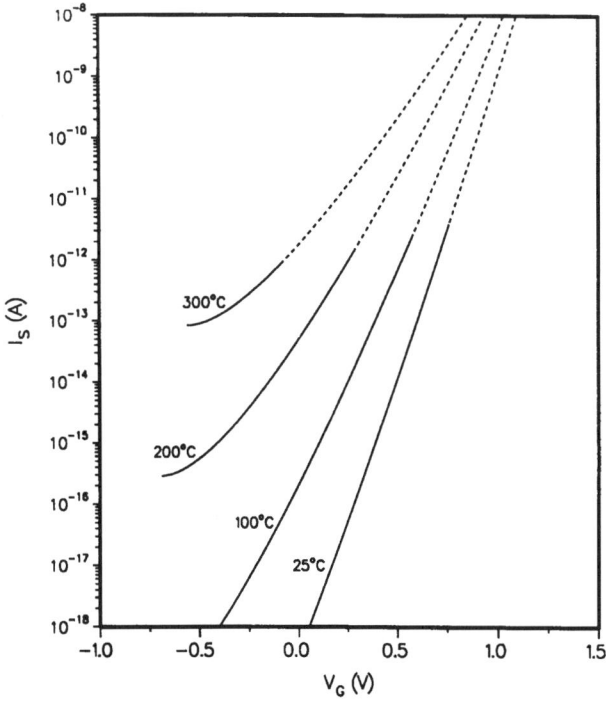

Figure 4.23 Calculated subthreshold current of a long-channel NMOS device having $N_B = 10^{16}\,cm^{-3}$, $W/L = 1$, $t_{ox} = 50\,nm$, $Q_{ss} = 10^{10}\,cm^{-2}$, $V_D \gg 3kT/q$.

where L_{Di} is the intrinsic Debye length, u_F is the normalized equilibrium Fermi potential of the substrate and u_S is the normalized surface potential.

When the transistor is in the weak inversion regime, channel current is directly proportional to carrier density in the channel. The carrier density is roughly exponentially proportional to u_S, and u_S has been shown in this region to be strongly proportional to the gate voltage [22]. Therefore, the subthreshold current is expected to be an exponential function of gate voltage. For a long-channel device, the subthreshold drain current is independent of drain voltage when $V_D > 3kT/q$; see equation (4.50).

Figure 4.23 shows the subthreshold characteristics of a long-channel n-type device having $N_A = 10^{16}\,cm^{-3}$ at different temperatures as functions of V_G. The plot is calculated from equation (4.50) representing channel current in a MOSFET without the inclusion of junction leakage (which dominates at high temperatures). (Source current is used as the dependent variable of the y-axis instead of drain current because junction leakage is not included in the figure.) The following assumptions were made in generating this figure: the ratio of channel width to channel length, W/L, is

1; the drain voltage is much greater than $3kT/q$; the p-type substrate is uniformly doped; the gate is made of n^+-polysilicon with a work function of $\phi_M = \chi$; the oxide thickness is 50 nm; the interface state density is $10^{10}\,cm^{-2}$; and the charge density in the gate oxide is zero. The solid lines correspond to weak inversion currents ($u_S \leqslant 2u_F$) and the dashed lines correspond to strong inversion diffusion currents ($u_S \geqslant 2u_F$). As V_G increases toward threshold, the ability of V_G to control the surface potential, u_S, decreases dramatically. When the device is in the strong inversion regime, drift current dominates and drain current departs from the exponential dependence on V_G, showing instead a linear or square dependence. The gate voltages at points where solid and dashed lines meet are threshold voltages. Again, the threshold voltage is shown to decrease with increasing temperature. It is interesting to note that, at $V_G = V_T$ over this 275 °C temperature range, the source current changes less than an order of magnitude.

Figure 4.24 is a measured subthreshold current plot for a MOSIS-fabricated n-channel device. The slope of $\log I_D$ versus V_G in the weak inversion regime decreases with increasing temperature, as expected from semiconductor theory. In the high temperature range, the drain current is a mixture of subthreshold and leakage currents, the latter dominating in magnitude. For $T > 200$ °C, the drain–substrate leakage currents are so large as to make the subthreshold region in the plot difficult to identify. For

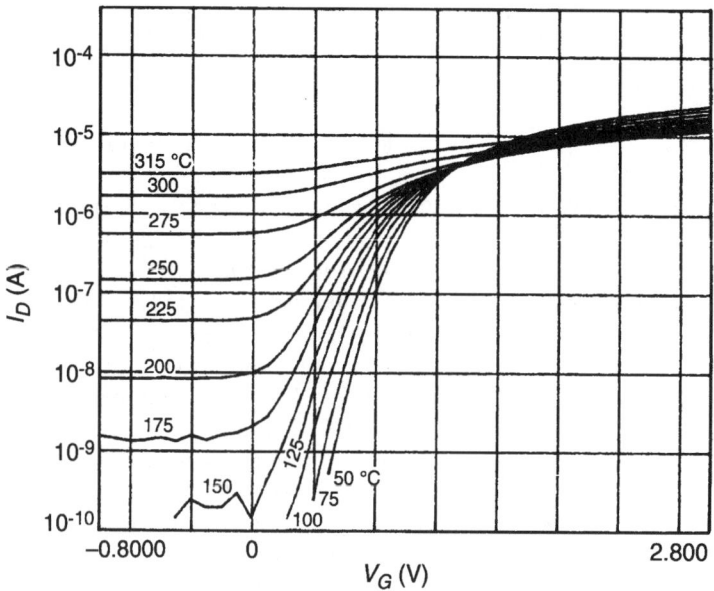

Figure 4.24 Measured subthreshold current in an n-channel MOSFET; $W/L = 4.5\,\mu m/3.0\,\mu m$ and $V_D = 200\,mV$.

this device, at temperatures above 200 °C, there exists no gate voltage which unambiguously separates on and off. A CMOS inverter at this temperature, for example, will allow a current having a magnitude equal to the leakage current to flow from V_{DD} to V_{SS} regardless of the value of the voltage applied to the gate. Power is thereby dissipated at all times, and the likelihood of latchup is increased. Logic voltage levels and transfer characteristics may become unacceptable under these conditions.

4.3.4 Surface mobilities

The effects of temperature on carrier mobility in bulk semiconductors were discussed in section 4.2.5. In MOSFETs it is mobility in the channel, or surface mobility, which affects device characteristics. Surface mobility is always lower than bulk mobility because of the vertical electric field, which causes carriers to interact with the silicon–silicon dioxide interface, losing energy and speed.

Like bulk mobility, surface mobility is known to decrease with temperature. In MOSFETs, lower mobility degrades device transfer characteristics. The temperature dependence of surface mobility is commonly said to be proportional to T^{-n} where n ranges from 1 to 2.5, depending on the oxide growth condition and temperature range.

The channel mobility can be found by measuring the transconductance, which is defined as the slope of I_D versus V_G at a constant V_D. In the linear region, the transconductance is given by

$$g_m \equiv \frac{\partial I_D}{\partial V_G}\bigg|_{V_D = \text{const.}} = \mu C_{ox} \frac{W}{L} V_D \qquad (4.51)$$

In the saturation region

$$g_m = \frac{\partial I_D}{\partial V_G}\bigg|_{V_D = \text{const.}} = \mu C_{ox} \frac{W}{L} (V_G - V_T) \qquad (4.52)$$

Figure 4.25 is a plot of I_D versus V_G for an n-channel MOSFET. The device is operated in the linear region by applying a small drain-to-source voltage (0.2 V). The experimental data show that I_D is approximately a linear function of V_G. The slope of I_D versus V_G (g_m), decreases with temperature, indicating that surface mobility decreases with temperature. Results are similar for p-channel devices. In real devices of either type, g_m is slightly dependent on V_G because of the vertical field dependency of surface mobility.

Figure 4.26 shows the temperature dependence of electron surface mobility for the device of Fig. 4.25. This transistor exhibits a $T^{-1.9}$ dependence on temperature in the range 50–225 °C. The mobility falls even faster at higher temperatures. Bulk mobility also follows a temperature dependence of T^{-n}, with 2.5 being most commonly used for n. SPICE assumes a

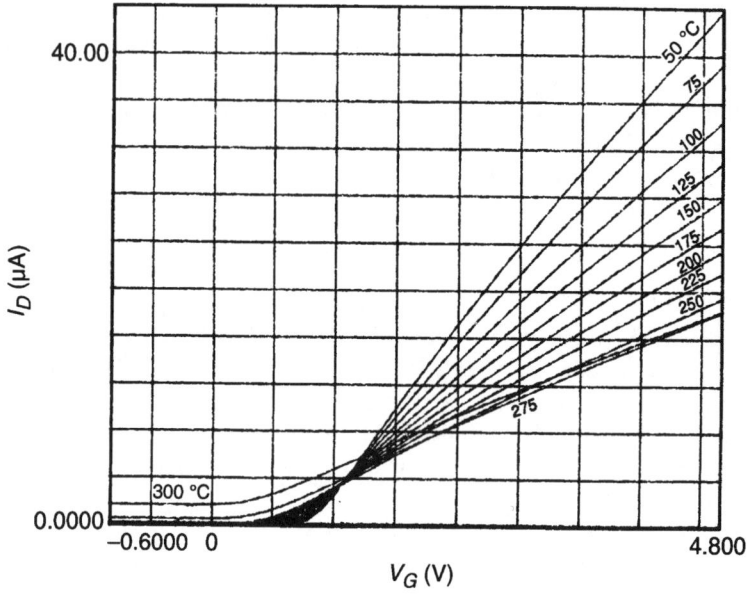

Figure 4.25 I_D versus V_G for a MOSIS-fabricated n-channel device with $W/L = 4.5\,\mu\text{m}/3\,\mu\text{m}$ and $V_D = 0.2\,\text{V}$.

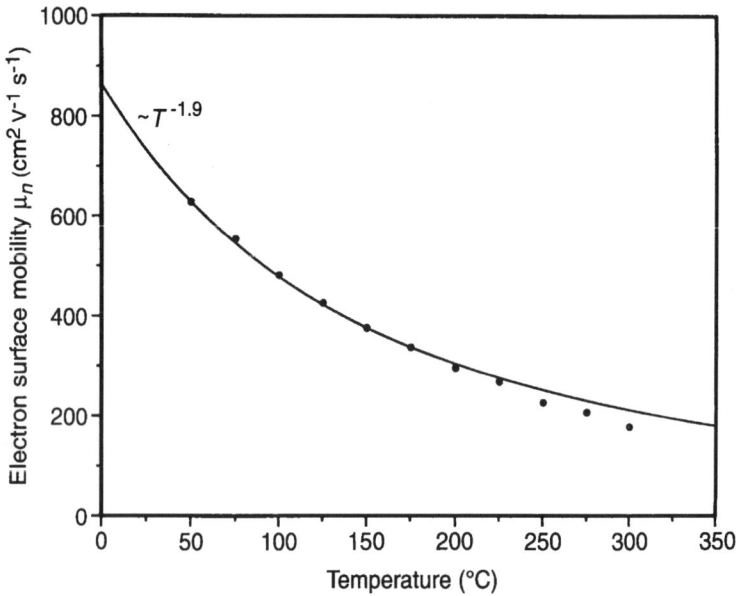

Figure 4.26 Electron surface mobility versus temperature; the data points (dots) are from Fig. 4.25.

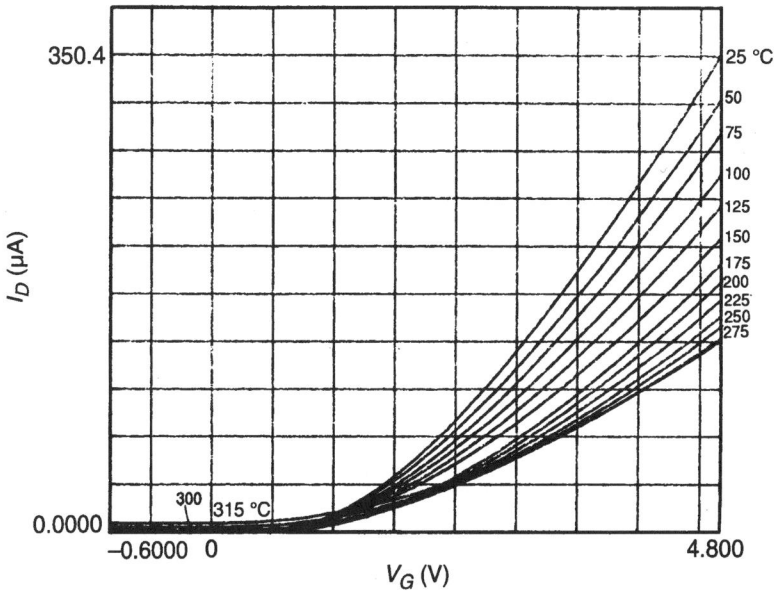

Figure 4.27 I_D versus V_G for an n-channel device in the saturation region with $W/L = 4.5\,\mu m/3\,\mu m$ and $V_D = 8$ V.

temperature dependence of $T^{-1.5}$ for surface mobilities of both n- and p-channel devices [23], which would overestimate the surface mobility of the device of Fig. 4.25 at high temperatures.

Figures 4.27 and 4.28 are plots of I_D versus V_G in the saturation region for n- and p-channel devices, respectively. Transconductance for the n-channel transistor in the saturation region is plotted in Fig. 4.29. In this region, transconductance is slightly dependent on the drain voltage. For a large value of V_G, g_m falls slightly below a linear extrapolation from low field values due to the large vertical field. The decrease in surface mobility with increasing temperature reduces MOSFET transconductance, which reduces the gain of analog circuits and the speed of digital logic gates.

4.3.5 Zero temperature coefficient operation

An important feature of Figs 4.25 and 4.28 is the existence of a zero temperature coefficient (ZTC) bias point, at which the I_D versus V_G characteristics are substantially independent of temperature [15, 24, 25]. Some MOSFETs, such as the one represented in Fig. 4.27, do not present well-defined ZTC points, but instead, a range of gate voltages over which the drain current remains approximately constant with changes in temperature.

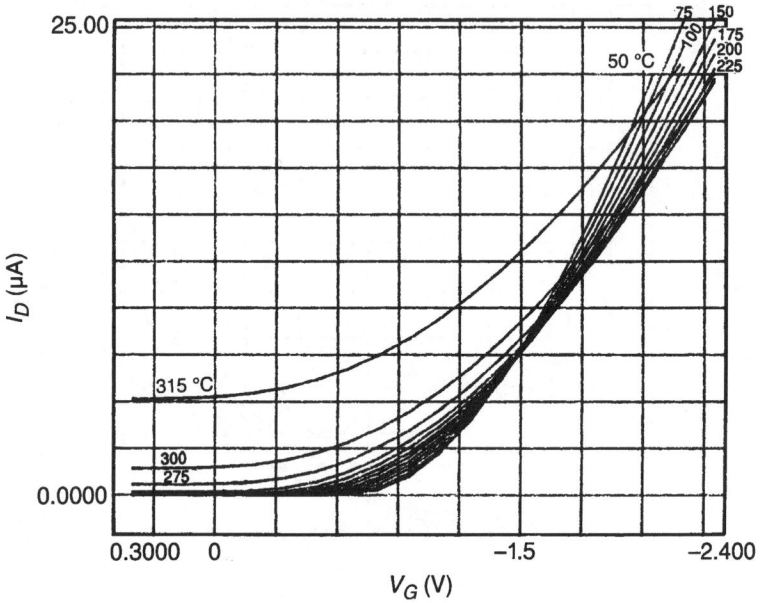

Figure 4.28 I_D versus V_G for a p-channel device in the saturation region with $W/L = 4.5\,\mu m/3\,\mu m$ and $V_D = -8\,V$.

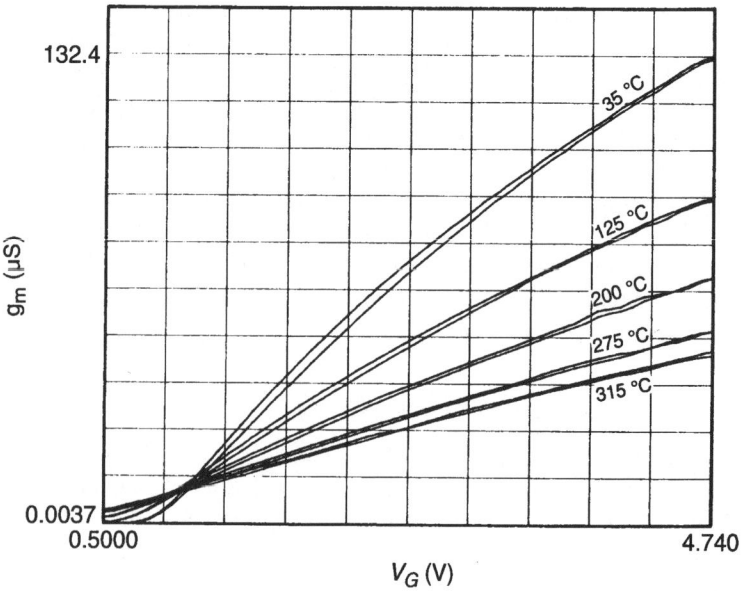

Figure 4.29 Transconductance versus V_G for an n-channel device in the saturation region with $W/L = 9\,\mu m/6\,\mu m$; $V_D = 5\,V$ and $8\,V$ for each temperature test.

The temperature sensitivity of I_D in the linear region is given by

$$\frac{\partial I_D}{\partial T} = \frac{I_D}{\mu}\frac{\partial \mu}{\partial T} - \frac{I_D}{V_G - V_T}\frac{\partial V_T}{\partial T} \qquad (4.53)$$

and in the saturation region by

$$\frac{\partial I_D}{\partial T} = \frac{I_D}{\mu}\frac{\partial \mu}{\partial T} - \frac{2I_D}{V_G - V_T}\frac{\partial V_T}{\partial T} \qquad (4.54)$$

V_D is assumed to be much smaller than $(V_G - V_T)$ in the linear region, and is neglected in the derivation of equation (4.53). The change of I_D with temperature is primarily due to changes in mobility and threshold voltage with temperature. As an example, in an n-channel device, the electron mobility decreases with temperature ($\partial \mu_n/\partial T < 0$) and the threshold voltage also decreases with temperature ($\partial V_T/\partial T < 0$). The decrease of mobility will cause a decrease in I_D; however, the decrease of V_T will cause an increase in I_D. These two effects might be balanced at a specific gate bias, $V_{G,ZTC}$, making $\partial I_D/\partial T$ approximately equal to zero.

Figures 4.30 and 4.31 show $V_{G,ZTC}$ versus substrate doping in the linear and saturation regions – based on equations (4.53) and (4.54) – with the other parameters listed in the figure captions. The $V_{G,ZTC}$ points are defined as the gate voltages at which I_D has the least variation with temperature

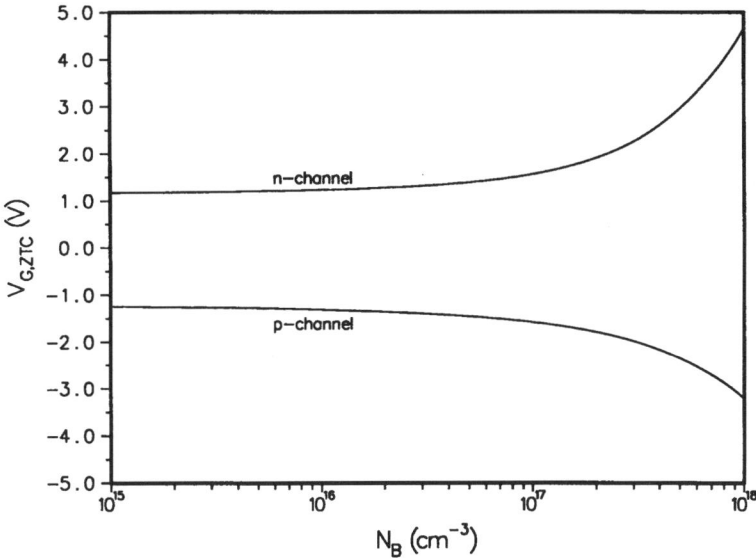

Figure 4.30 $V_{G,ZTC}$ versus N_B in the linear region; $V_T|_{T=25°C} = 1$ V for n-channel devices and -1 V for p-channel devices, $|V_D| = 10$ mV, $t_{ox} = 50$ nm and $Q_{ss} = 10^{10}$ cm^{-2}.

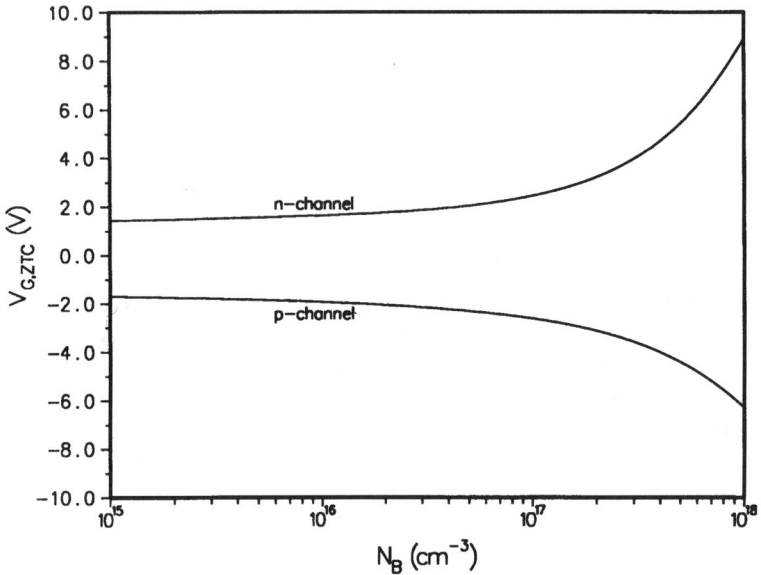

Figure 4.31 $V_{G,ZTC}$ versus N_B in the saturation region; $V_T|_{T=25°C} = 1$ V for n-channel devices and -1 V for p-channel devices, $|V_D| = 10$ mV, $t_{ox} = 50$ nm and $Q_{ss} = 10^{10}$ cm^{-2}.

between 25 °C and 300 °C. In both figures, the threshold voltage at room temperature is normalized to 1 V for n-channel devices and -1 V for p-channel devices. In both the linear and saturation regions, $V_{G,ZTC}$ is almost constant with respect to substrate doping for $N_B < 10^{17}$ cm^{-3}. As N_B rises beyond 10^{17} cm^{-3}, the magnitude of $V_{G,ZTC}$ increases dramatically. Due to the factor of 2 in equation (4.54), $V_{G,ZTC}$ in the saturation region has about twice the variation with N_B as in the linear region.

By definition, when a transistor is biased at its ZTC point, the drain current will not change with temperature. If biased further into inversion than the ZTC point, devices tend to be self-stabilizing – an increase in temperature causes a decrease in current and therefore a decrease in internal power dissipation. For moderate substrate doping (10^{15} cm$^{-3} < N_B < 3 \times 10^{16}$ cm^{-3}), $V_{G,ZTC}$ is only slightly greater than V_T; at $V_{G,ZTC}$ devices are biased just into strong inversion and drain currents are small – too small to be useful for many circuit applications. When the substrate doping is increased, $V_{G,ZTC}$ is a few volts greater than V_T; at this $V_{G,ZTC}$ devices are biased well into the strong inversion region. Drain current at the ZTC point is shown to increase several fold as N_B increases from 10^{15} cm^{-3} to 10^{18} cm^{-3}, in spite of the decrease of mobility due to greater impurity scattering. By careful selection of substrate doping, $V_{G,ZTC}$ and the corresponding I_D can be made appropriate for a particular circuit design. The concept of ZTC operation is especially applicable to analog circuits.

4.3.6 Breakdown voltage and hot carriers

Though most transistor characteristics deteriorate with increasing temperature, breakdown voltage and hot-carrier injection are better at elevated temperatures. The breakdown voltage, as shown in Fig. 4.32 for the $I-V$ characteristics of an n-well to p-substrate junction diode, increases with increasing temperature. Breakdown voltage due to avalanche breakdown has a positive temperature coefficient because hot carriers passing through the depletion layer under a high field lose part of their energy to optical phonons after traveling each electron–phonon mean free path, the length of which decreases with increasing temperature. Therefore, the carriers must pass through a greater potential difference before they can acquire enough energy to generate electron–hole pairs.

The substrate current generated from impact ionization by hot carriers is the most accepted measure of hot carrier injection. The substrate current at fixed drain bias is found to decrease with increasing temperature due to lower carrier mobilities, more phonon scattering, lower localized electric

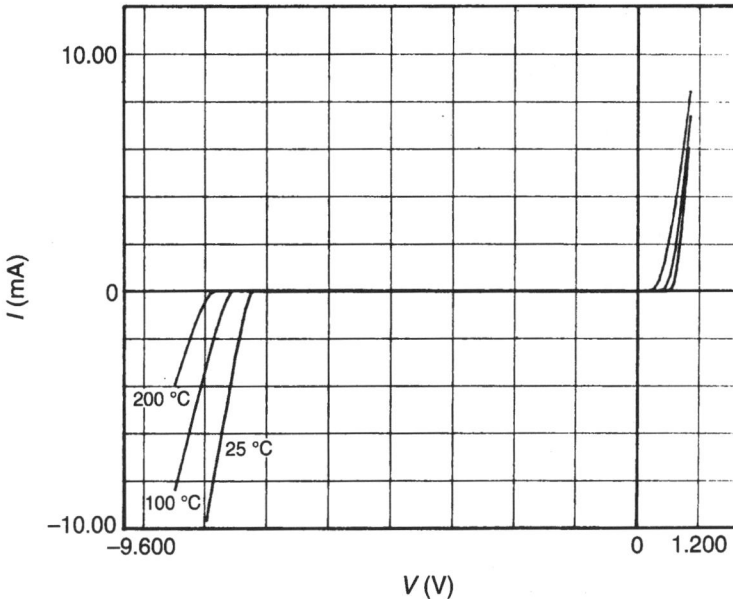

Figure 4.32 $I-V$ characteristics of an n-well to p-substrate junction diode, showing avalanche breakdown.

field and higher breakdown voltage [26]. Data from this study indicates that the useful life of n-channel devices, biased so that hot carriers are generated, is extended by an order of magnitude when the operating temperature is increased from 20 °C to 100 °C.

4.3.7 Summary of MOSFET temperature effects

The performance of MOS transistors has been shown to degrade at high temperatures due to leakage current increase which can cause latchup in CMOS circuits, threshold voltage shift, surface mobility decrease and increased subthreshold currents. These effects combine to make the transistor channels difficult to turn off, and to make the transistors slow. MOSFETs at high temperatures become more like passive resistors and less like active devices. Transistors can sometimes be biased so that the decrease in mobility and threshold voltage have compensating effects, and drain current is independent of temperature. Two transistor characteristics are improved at higher temperatures, breakdown voltage and resistance to hot carrier effects.

4.4 HIGH TEMPERATURE CMOS APPROACHES

Having reviewed the effects of temperature on basic semiconductor properties and on MOSFET parameters, we are prepared to discuss methods for improving the performance of CMOS circuits at high temperatures. Strategies for alleviating the detrimental high temperature effects in microelectronics can be divided into three areas: process technology, physical design and circuit techniques. Each of these areas offers a range of possible approaches with corresponding costs and benefits. The most cost-effective solution is usually a combination of methods from the three areas.

4.4.1 Process technology

Bulk CMOS, the mainstay of the semiconductor industry, is particularly liable to high temperature failure because it is junction isolated and susceptible to latchup. As stated in section 4.3.1, leakage currents through large p–n junctions trigger latchup when the voltage drop through the substrate or well is great enough to turn on the base–emitter junction of one of the coupled parasitic bipolar transistors (Fig. 4.17). Physical layout design rules can be modified, as explained in section 4.4.2, and circuit techniques can be employed, as described in section 4.4.3, to ameliorate the tendency to latchup in bulk CMOS. But when a designer is able to modify a process to improve its high temperature characteristics, or to specify the use of epitaxial wafers (a thin layer of lightly doped silicon on top of a more heavily doped substrate), or better still, a silicon-on-insulator (SOI) process, the high temperature performance of CMOS circuits can be greatly improved.

Silicon-on-insulator

Silicon-on-insulator processes are implemented on wafers having a thin layer of silicon over an insulating layer (typically silicon dioxide). Active devices are formed in the thin layer and isolated from each other with dielectric material. Several techniques to make SOI fabrication more cost-effective have recently been developed; the leading approaches now are separation by ion implantation of oxygen (SIMOX), bond and etchback (BESOI), and zone-melt recrystallization (ZMR) [27]. Though there are few commercial silicon-on-insulator processes, and these are expensive, SOI is the best alternative for high temperature CMOS because its dielectric isolation eliminates latchup (while still allowing dense layouts). Because this technology has no large junctions and usually uses higher doping levels, its power supply leakage is typically several orders of magnitude smaller than for bulk devices [28]. SOI has other benefits as well, including less susceptibility to ionizing radiation and a lower parasitic capacitance, which leads to higher speed operation.

High temperature functionality of SOI has been demonstrated with a variety of circuits: operational amplifiers have been shown to work at 300 °C; CMOS NAND gates have been used at 380 °C; and 4K SRAMs have operated at 300 °C (with doubled access time) [29, 30]. Encouraged by these results, researchers have projected operation of SOI integrated circuits to 400 °C for digital circuits and 325 °C for operational amplifiers [30]. Figure 4.33 is a plot of output voltage versus input voltage for different switching currents in a silicon-on-sapphire (SOS) inverter (an older form of SOI) at temperatures from 30 °C to 315 °C. The magnitude of high temperature leakage current is small compared to the maximum switching current (less than 15% at 315 °C). No significant shift of input switching voltage is seen, and the SOS inverter retains good noise margins at high temperatures.

Though latchup is overcome with SOI technologies, the problems related to mobility degradation, subthreshold leakage and threshold voltage drift are still present in SOI; these problems must be addressed using circuit and geometrical approaches.

Epi CMOS

Despite the benefits of SOI, there remains a defensible argument for using more conventional CMOS processes for high temperature circuits. By using the most common and cost-effective processes, one is able to exploit ongoing developments which are made through the enormous research investments

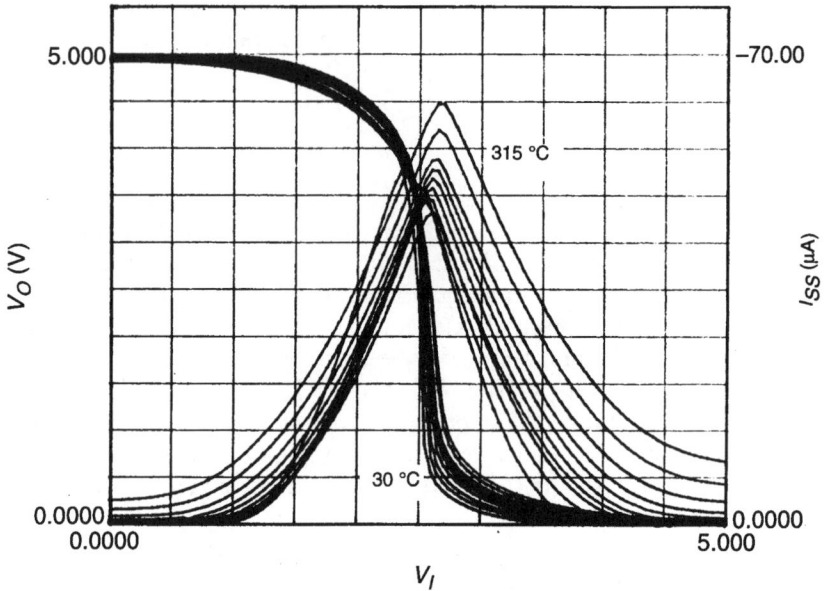

Figure 4.33 The transfer characteristics and switching currents of an SOS inverter.

being made in mainstream silicon technology. An example of such a spin-off is epi CMOS. This variation of bulk CMOS was developed primarily to prevent latchup resulting from the unrelenting scaling of feature sizes in digital circuits. The resulting improvement in latchup robustness translates to functionality at higher temperatures for features of a given spacing. Epi CMOS, therefore, represents an intermediate improvement in high temperature performance, which is realized at a modest increase in cost.

A comparison of latchup susceptibility was made for three $1.2\,\mu m$ CMOS processes [31], one on bulk wafers and two on epi substrates, to quantify the benefits of epitaxial layers of different thicknesses and the effects of different doping levels. This work was done by the authors in conjunction with Mario Ghezzo, Dale M. Brown, Evan Downey and Dave Hanchar of General Electric Corporate Research and Development Center. The process used as a starting point for this work [32, 33] was a 5 V digital and analog CMOS process with a minimum feature size of $1.2\,\mu m$. The basic process included lightly doped drains, self-aligned polysilicon gates, a thin $(3.5\,\mu m)$ n on n+ epitaxial layer and retrograde p-wells. The retrograde wells (implanted so they are more heavily doped at the bottom) improve latchup performance by reducing lateral voltage gradients in the wells and by reducing the gain of the parasitic vertical bipolar transistors. The metallization was composed of molybdenum/titanium–tungsten first-level metal, a molybdenum capacitor–electrode layer and aluminum top-layer metal.

The bulk wafers were processed identically to the standard epi process. The second epi process was modified in several ways to achieve better high-temperature performance. The first of these modifications addressed the electromigration problem by replacing the standard top-level metal (aluminum) with molybdenum. Typically, power rails are routed in top-level metal, and the electromigration problem is aggravated both by the additional thermal energy available at high temperatures, and by additional current density in power traces due to junction leakage. Because the resistance of sputtered molybdenum is about twice that of aluminum interconnect alloys, traces having high current densities must be made wider to avoid resistively coupled noise. On the other hand, the electromigration activation energy for molybdenum is so high as to effectively eliminate electromigration as a reliability concern [34].

Other modifications made in the high temperature CMOS process included thinning the epitaxial layer from $3.5\,\mu m$ to $3.0\,\mu m$ and doubling the well implant to 2×10^{13} ions cm^{-2}. The thinner epi causes even more of the trigger current to be shunted harmlessly into the substrate. Higher doping levels reduce parasitic resistances, lower junction leakage current and raise intrinsic temperatures; they also cause lower junction breakdown voltages, in this case reducing drain breakdown by 30% to a still acceptable level of $9\,V$.

All of the wafers were processed with the same parameter extraction mask set. Data were taken on minimum-size inverters configured as two-terminal devices with cathodes tied to p-well contacts and n-transistor sources; anodes tied to n-substrates and p-transistor sources; gates connected to the anodes; and outputs open. Figure 4.34 shows the well-known deterioration of latchup performance with temperature through holding currents and holding voltages for the bulk, epi and the special high temperature epi (HT-epi) processes. The special high temperature process improves both holding current and holding voltage at 300 °C by more than a factor of 2 over the already excellent performance of the standard $1.2\,\mu m$ process. Holding voltage at 300 °C is four times better in the high temperature proces than in bulk CMOS, and holding current is 30 times better than in the bulk process.

Bulk CMOS

Increasing the doping concentration of the substrate improves latchup robustness of even conventional bulk CMOS. Higher substrate doping levels not only reduce leakage in MOS devices, but also increase intrinsic temperatures, ensure enhancement-mode operation to higher temperatures,

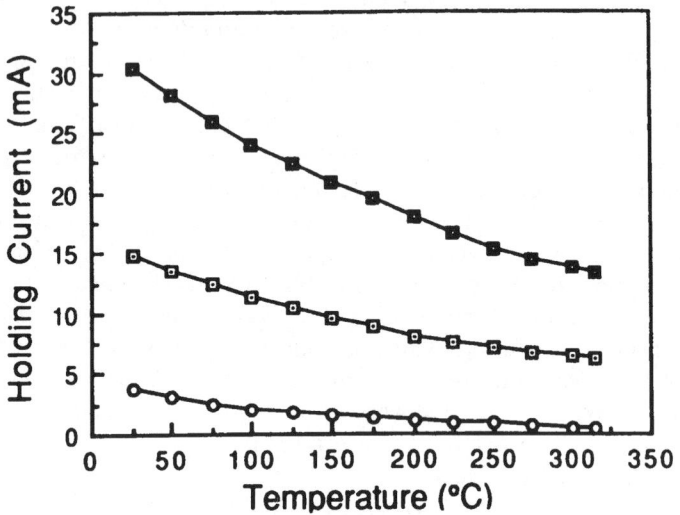

(a)

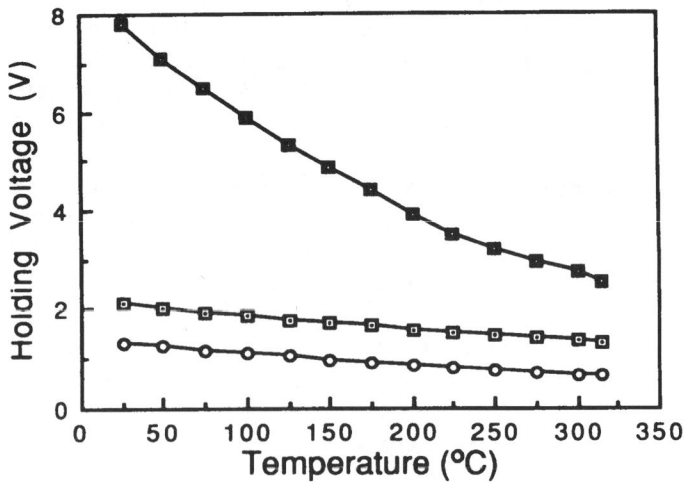

(b)

Figure 4.34 Comparison of (a) holding current and (b) holding voltage for (○) bulk CMOS, (▫) epi CMOS and (■) special high temperature epi CMOS inverters. (After Brown *et al.* [31] © 1989 IEEE)

reduce temperature dependence of carrier mobilities, raise $V_{G,ZTC}$ for ZTC operation and reduce parasitic resistances. However, higher doping levels will cause lower breakdown voltage (which is inversely proportional to the doping level), larger temperature sensitivity of threshold voltage and lower surface mobilities and transconductance.

Long-term reliability of MOS devices may also be compromised by operating at high temperatures. Of particular concern are damage to the gate insulator and the silicon–insulator interface [35]. Reliability research done in standard processes may also be applicable to high temperature technologies. Since bias temperature stress, ionizing radiation and hot carrier injection all generate similar oxide and interface damage [36], processing schemes which harden a technology to radiation or hot carriers may also make it high-temperature hard from the viewpoint of the gate insulator.

Summary

With high temperature modifications, junction-isolated CMOS processes can produce circuits that operate reliably at temperatures above the traditional limit of 125 °C. Epitaxial CMOS processes have much better latchup hardness than bulk processes. And submicron processes, which are made on epitaxial material, have thinner gate oxides, producing higher transconductance for a device of given dimensions, an important feature for high temperature operation, in which transconductance is depressed by loss of surface mobility.

The dielectric isolation of a silicon-on-insulator process eliminates the problem of latchup. But even when latchup is avoided, the other high temperature effects, discussed in sections 4.3.1 and 4.3.2, must still be addressed through physical design and circuit techniques.

4.4.2 Physical design

Latchup

Well-documented physical design methods can be used to improve the latchup characteristics of any bulk or epitaxial CMOS process [37, 38]. These methods include providing adequate spacing between p- and n-active areas; including majority and minority carrier guard structures; using multiple, butted, source-to-well and source-to-substrate contacts; and positioning transistors so that p- and n-devices are separated from each other and aligned at their sources with other devices of the same type. To achieve high temperature latchup immunity, the core of a circuit can employ the more stringent latchup rules normally required only in input/output circuits.

Designing geometrically strong structures from a latchup point of view requires one to compromise circuit density.

Using only layout techniques, researchers have demonstrated functionality of small conventional bulk CMOS circuits at up to 300 °C [11]. This is not to imply that reliable latchup immunity was achieved, but only that geometric methods can significantly raise the temperature at which reliable operation is possible. Costs involved in these layout solutions to the latchup problem derive from the additional chip area which these methods require. The initial gains in temperature performance from design-rule techniques (implemented with the least area-demanding approaches) are very cost-effective; using only layout rules to achieve 300 °C operation with bulk CMOS, however, is not cost-effective. Combined with an epi CMOS process, though, design rule techniques for improving latchup performance are very effective.

This point was made using the baseline epitaxial process described above, with inverters of three sizes, one implemented in the standard 1.2 μm design rules and the others having geometries uniformly scaled to 75% and 50% of those dimensions. Figure 4.35 shows latchup characteristics over temperature for the epi and identically processed bulk devices. As expected, the holding current is significantly greater for the epitaxial process across the temperature range, and both bulk and epi processes have better latchup characteristics with larger geometries. It is noteworthy that for both holding current and holding voltage (not shown) the bulk curves for the different

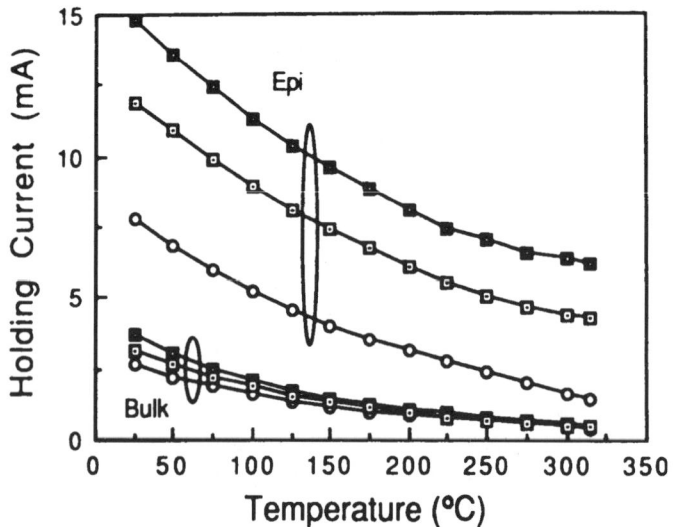

Figure 4.35 Comparison of holding currents for bulk and epi CMOS inverters having channel lengths of 1.2, 0.9 and 0.6 μm: 1.25 μm rules (■) normal size, (□) 25% undersize, (○) 50% undersize. (After Brown *et al.* [31] © 1989 IEEE)

size structures are close together, whereas epi curves are spaced apart. These effects can be explained by the fact that most of the triggering current forced into the test-structure anode on an epi wafer is injected directly into the low resistivity substrate, making it unavailable for forward biasing the lateral pnp transistor. As the anode and cathode are spaced further apart, the percentage of lateral current in the epitaxial layer decreases rapidly due to the epi–substrate resistivity differential. In bulk wafers, the uniform substrate resistance spreads the current flow more evenly. As a result, scaling up of design rules is a very efficient method for improving latchup performance in epitaxial CMOS. This advantage is even more pronounced at higher temperatures.

Latchup holding voltage is generally considered a meaningful measure of latchup immunity: a CMOS structure is latchup-free if its holding voltage is greater than the power supply voltage [38–42]. A conductivity modulation model for holding voltage in junction-isolated processes [43] predicts, based upon extracted model constants for a given process, the required anode–cathode spacing to achieve latchup immunity at a desired temperature. Using this model with the epi and high temperature epi CMOS processes discussed above, Fig. 4.36 shows the anode–cathode spacing required to

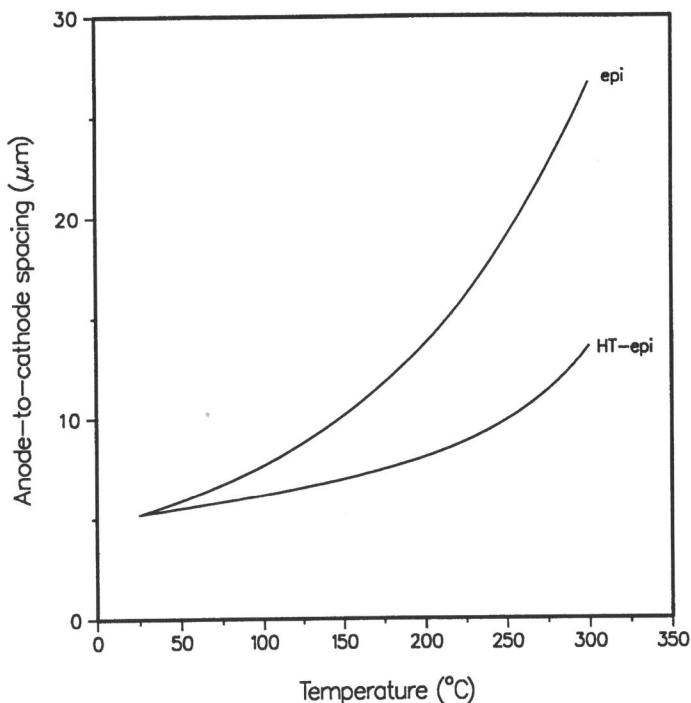

Figure 4.36 Anode-to-cathode spacing versus temperature to maintain constant holding voltage for CMOS inverters in standard epi and HT-epi CMOS processes. (After Wu [11])

maintain room temperature holding voltages as temperature increases for CMOS inverters. The HT-epi process not only has the best holding voltage at room temperature (Fig. 4.35), but is also superior in that smaller increases in anode–cathode spacing are required to achieve a given improvement in latchup hardness. The anode–cathode spacing increases 2.6 times for HT-epi and 5.13 times for the standard epi process to maintain their room temperature holding voltages at 300 °C.

Transconductance

In addition to preventing latchup, physical layout adjustments can be employed to compensate for the loss of surface mobility which comes with high temperature. Figure 4.37 is a plot of the resulting reduction in transconductance with temperature from a SPICE simulation, using model parameters for n- and p-transistors from a conventional bulk CMOS process. The curves are normalized to the transconductance at 75 °C, the upper temperature limit for commercial integrated circuits. This transconductance degradation affects all MOSFETs (SOI devices included), slowing the speed of digital circuits and decreasing gains in analog circuits. A circuit fails if it cannot operate at its specified speed, even if it is functional at lower speeds. Because g_m in both the linear and saturation regions is a function of $\mu(W/L)$, the mobility and transistor width-to-length ratio, CMOS circuits

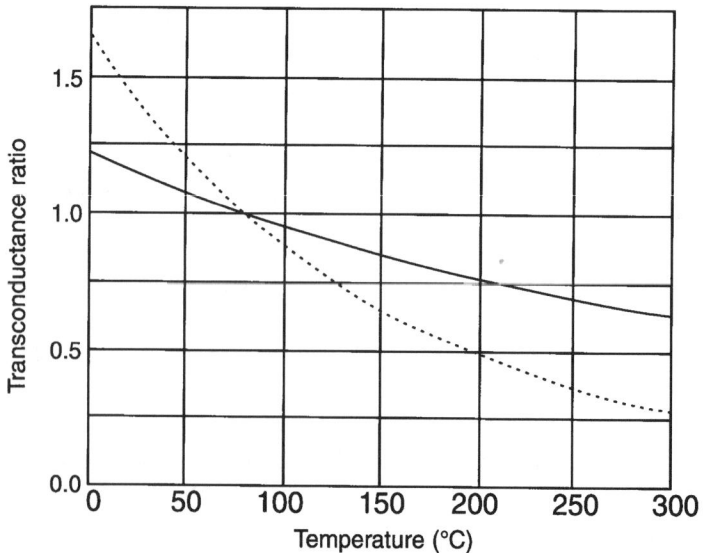

Figure 4.37 Transconductance variation with temperature for bulk CMOS transistors: (—) NMOS and (---) PMOS. (After Brown *et al.* [5] © 1987 IEEE)

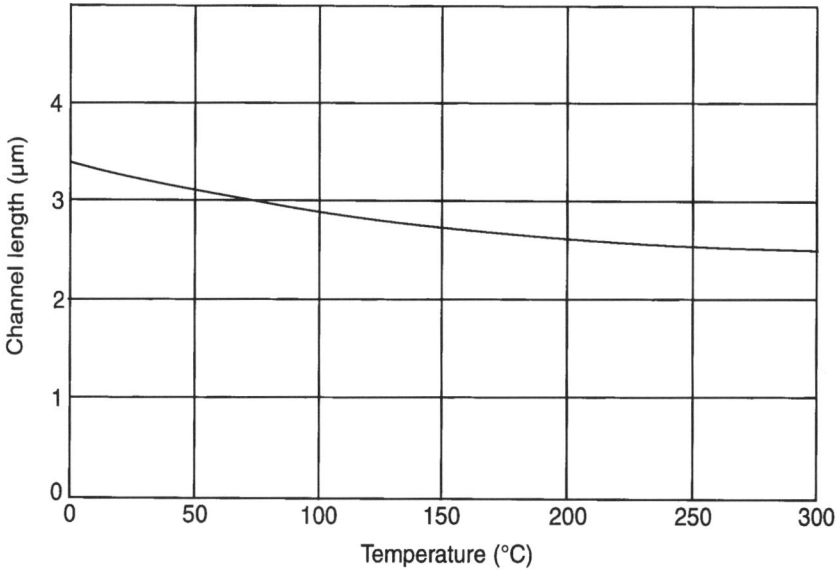

Figure 4.38 Channel length scaling required to maintain digital circuit speed for an inverter (from SPICE). (After Brown *et al.* [5] © 1987 IEEE)

can be made to function at a given speed or gain at an elevated temperature by increasing the width-to-length ratio of their transistors to compensate for the transconductance loss. If constant gain over temperature is required, it can be achieved by using negative feedback techniques.

Figure 4.38 is a plot showing the change in channel length (assuming p- and n-transistors are scaled in length by the same factor to adjust W/L) required to maintain operating speed over temperature. In this case, the speed improves in part because of lower gate capacitance as the gates are made shorter. The simulation assumes a fanout to three gates identical to the driving gate, plus parasitic (interconnect) capacitance of 100 fF. A comparison of Figs 4.37 and 4.38 shows that the method is efficient.

Physical design parameters can be modified to improve latchup immunity in junction-isolated CMOS processes or to assure adequate transconductance for proper circuit operation at elevated temperatures. In addition, a variety of circuit design methods, typically implemented at modest expense, can be used to improve high temperature performance of microelectronics.

4.4.3 Circuit techniques

Among the easiest circuit variations for improving high temperature performance of bulk CMOS digital circuits is reducing the power supply voltage. Although this unfortunately reduces speed and noise margins, it

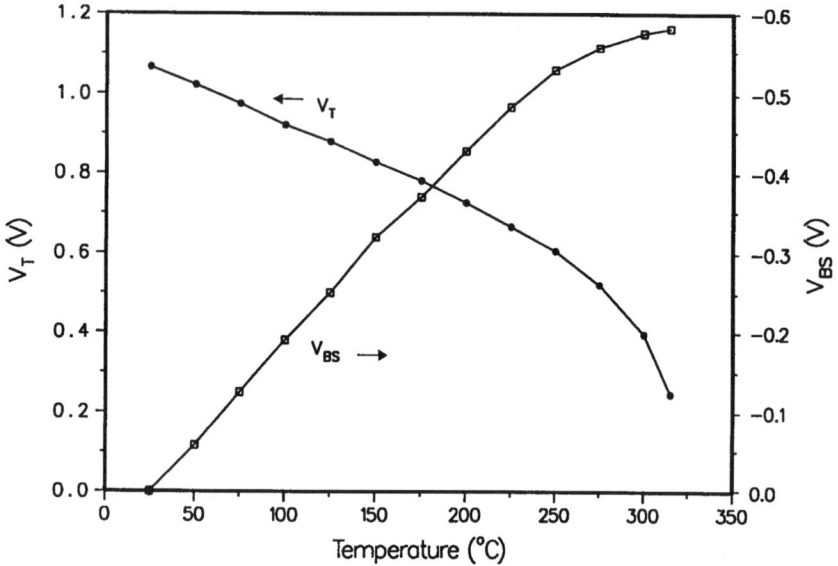

Figure 4.39 Uncompensated threshold voltage versus temperature for an n-channel device, and the substrate bias required to compensate the threshold voltage change with temperature, maintaining the room temperature threshold.

significantly reduces leakage currents for the circuit, thereby raising the temperature-related latchup point and mitigating electromigration effects.

A somewhat more involved approach for improving high temperature performance is based on section 4.3.2. Equation (4.48) shows that threshold voltage is a function of substrate bias. This presents an opportunity for stabilizing transistor characteristics. A reverse source-to-backgate bias increases the magnitude of threshold voltage, maintaining enhancement-mode operation at higher temperatures. Feedback can be used to adjust the backgate voltage so as to maintain a constant threshold voltage over temperature. For example, Fig. 4.39 shows uncompensated n-transistor threshold voltage and the substrate bias required to stabilize the threshold voltage with temperature, maintaining the room temperature threshold. This approach was used with on-chip charge pumps to stabilize thresholds in NMOS memories [44]. In CMOS, charge pump circuits act as minority carrier injectors, which can trigger latchup [45], so integrating the charge pump is probably not advisable. On-chip regulation of external supply voltages to stabilize threshold voltage is practical if the well and substrate voltages can be established before the source voltages.

In addition to stabilizing threshold voltage, the backgate bias significantly improves subthreshold and latchup characteristics at high temperatures. With only 0.25 V bias, the onset of temperature-induced latchup in bulk

CMOS has been raised by 50 °C [5]. Unfortunately, reverse-biasing junctions that would have otherwise had no potential across them aggravate junction leakage, which could cause other problems, especially in analog circuits. Driving all of the wells and substrate with a separate bias requires routing twice as many power supply rails on the chip, which might be unacceptable for larger circuits.

Analog circuits, in particular, are needed not only to function, but to have stable characteristics over temperature. A well-publicized approach [46, 47] for designing such circuits exploits the zero temperature coefficient (ZTC) point of MOSFETs introduced in section 4.3.5 (Figs 4.27 and 4.30) by biasing critical transistors near this point. To make best use of this approach, a designer needs to have the freedom to modify the process so that the ZTC occurs at a current level in the desired range. A high temperature CMOS op-amp which uses ZTC biasing is described in the literature [48].

4.4.4 Conclusions

Process technology, physical design and circuit techniques can be utilized to raise the operating temperatures of CMOS integrated electronics well above the standard commercial and military temperature ranges. Combinations of these techniques provide a continuum of solutions (and costs) to meet the temperature and reliability specifications of a particular application. Employing only circuit and layout techniques, one can raise the ambient temperature range of conventional bulk CMOS to 200 °C. A combination of epi CMOS, conservative layout rules, supply voltage reduction and scaling of transistor width-to-length ratio provides a cost-effective option which can achieve good performance and reliable operation to 250 °C. Use of a silicon-on-insulator process, with appropriate high temperature circuit and layout design methods, can yield silicon circuits that operate reliably at 300 °C.

REFERENCES

1. Thome, F.V. and King, D.B. (1988) *High Temperature Electronics Workshop and Survey Results: Recommendations for Research*, Sandia National Labs, Albuquerque NM.
2. Cobbold, R.S.C. (1966) Temperature effects in MOS transistors. *IEEE Electron. Device Lett.*, **2**(6), 190–191.
3. Shoucair, F.S., Hwang, W. and Jain, P. (1984) Electrical characteristics of large scale integration (LSI) MOSFETs at very high temperatures. Part II: experiment. *Microelectron. Reliab.*, **24**(3), 487–510.
4. Draper, B.L. (1979) Behavior of silicon devices at high temperatures. Sandia National Labs *SAND-80-0834C*. High Temperature Electronics and Instrumentation Seminar, pp. 111–114.

5. Brown, R.B., Terry, F.L. and Wu, K. (1987) High temperature microelectronics – expanding the applications for smart sensors. *IEDM Tech. Digest*, pp. 274–277.
6. Thurmond, C.D. (1975) The standard thermodynamic function of the formation of electrons and holes in Ge, Si, GaAs and GaP. *J. Electrochem. Soc.*, **122**, 1133.
7. Sze, S.M. (1981) *Physics of Semiconductor Devices*, Wiley, New York.
8. Barber, H.D. (1967) Effective mass and intrinsic concentration in silicon. *Solid-State Electronics*, **10**, 1039.
9. Smith, R.A. (1979) *Semiconductors*, 2nd edn, Cambridge University Press, Cambridge.
10. Arora, N.D., Hauser, J.R. and Roulston, D.J. (1982) Electron and hole mobilities in silicon as a function of concentration and temperature. *IEEE Trans. Electron Devices*, **29**(3), 292–295.
11. Wu, K. (1990) Modeling CMOS for high-temperature applications. PhD Dissertation, Electrical Engineering, University of Michigan.
12. Seitchik, J.A., Chern, J.-H. and Yang, P. (1987) Latchup modeling using PISCES. *IEEE Circuits and Devices*, **3**(6), 35–38.
13. Vadasz, L. and Grove, A.S. (1966) Temperature dependence of MOS transistor characteristics below saturation. *IEEE Trans. Electron. Devices*, **13**(12), 863–866.
14. Shichman, H. and Hodges, D.A. (1968) Modeling and simulation of insulated gate field-effect transistor switching circuits. *IEEE J. Solid-State Circuits*, **3**(5), 285–289.
15. Draper, B.L. and Palmer, D.W. (1979) Extension of high-temperature electronics. *IEEE Trans. Components, Hybrids, Manuf. Technol.*, **2**(4), 399–404.
16. McBrayer, J.D. (1979) High-temperature complementary metal oxide semiconductors (CMOS). Sandia National Labs, Oct. 1979, pp. 7–53.
17. Kronberg, J.W. (1981) High-temperature behavior of MOS devices. *IEEE Southeast-con*, 81CH1650-1, pp. 735–739.
18. Muller, R.S. and Kamins, T.I. (1977) *Device Electronics for Integrated Circuits*, Wiley, New York.
19. Palmer, D.W., Draper, B.L., McBrayer, J.D. and White, K.R. (1978) Active devices for high temperature microcircuitry. Sandia National Labs, Feb. 1978, pp. 3–114.
20. Technology Modeling Associates (1989) TMA PISCES-2B two-dimensional device analysis program, version 8908.
21. Taylor, G.W. (1978) Subthreshold conduction in MOSFETs, *IEEE Trans. Electron Devices*, **25**(3), 337–350.
22. Barron, M.B. (1972) Low level currents in insulated gate field effect transistors. *Solid-State Electronics*, **15**, 293–302.
23. Vladimirescu, A. and Liu, S. (1980) The simulation of MOS integrated circuits using SPICE2. *ERL-M80/7*, Electronics Research Lab., University of California, Berkeley.
24. Heiman, F.P. and Miller, H.S. (1965) Temperature dependence of n-type MOS transistors. *IEEE Trans. Electron Devices*, **12**(3), 142–148.
25. Osman, A.A., Osman, M.A., Dogan, N.S. and Imam, M.A. (1995) Zero-temperature-coefficient biasing point of partially depleted SOI MOSFETs. *IEEE Trans. Electron Devices*, **42**(9), 1709–1711.
26. Hsu, F.C. and Chiu, K.Y. (1984) Temperature dependence of hot-electron-induced degradation in MOSFETs. *IEEE Electron Device Lett.*, **5**(5), 148–150.
27. Jastrzebski, L. (1984) Silicon CVD for SOI: principles and possible applications. *Solid State Tech.*, **Sep.**, 239–243.
28. McKitterick, J.B. (1988) Silicon-on-insulator for 300 °C CMOS applications. *High Temperature Electronics Workshop*, Albuquerque NM, April 12–14.
29. Krull, W. and Lee, J. (1988) Characterization of CMOS/SIMOX SRAMs at high temperatures. *High Temperature Electronics Workshop*, Albuquerque NM, April 12–14.
30. Beasom, J.D. (1988) High temperature dielectrically isolated ICs: status and projection, in *Proceedings of the High Temperature Electronics Workshop*, Albuquerque NM, April 12–14.
31. Brown, R.B., Wu, K., Ghezzo, M., Brown, D.M., Downey, E. and Hanchar, D. (1989) Junction-isolated CMOS for high-temperature microelectronics. *IEEE Trans. Electron Devices*, **39**(9), 1854–1856.
32. Jerdonek, R., Ghezzo, M., Weaver, J. and Combs, S. (1982) Reduced geometry CMOS technology. *IEDM Tech. Digest*, **Dec.**, 450–453.

33. Brown, D.M., Chu, S., Kim, M., Gorowitz, B., Milkovich, M., Nakagawa, T. and Vogelsong, T. (1985) Advanced analog CMOS technology. *IEDM Tech. Digest*, pp. 260–263.

34. Kim, M.J., Skelly, D.W. and Brown, D.M. (1987) Electromigration of bias-sputtered Al and comparison with others. *Proceedings of 1987 Reliability Physics Conference*, April 1987, pp. 126–129.

35. Shiau, W.-T. and Terry, F.L. Jr (1989) Bias–temperature stability of nitrided oxides and reoxidized nitrided oxides. *Journal of Electronic Materials*, **18**(6), 767–773.

36. Haller, G., Knoll, M., Braunig, D., Wulf, F. and Fahrner, W.R. (1984) Bias–temperature stress on metal–oxide–semiconductor structures as compared to ionizing irradiation and tunnel injection. *J. Appl. Phys.*, **56**, 1844.

37. Troutman, R.R. (1986) *Latchup in CMOS Technology*, Kluwer, Hingham MA.

38. Sangiorgi, E., Johnston, R.L., Pinto, M.R., Bechtold, P.F. and Fichtner, W. (1986) Temperature dependence of latch-up phenomena in scaled CMOS structures. *IEEE Electron Device Lett.*, **7**(1), 28–31.

39. Hu, G.J. and Bruce, R.H. (1984) A CMOS structure with high latchup holding voltage. *IEEE Electron Device Lett.*, **5**(6), 211–214.

40. Seitchik, J.A., Chatterjee, A. and Yang, P. (1987) An analytic model of holding voltage for latch-up in epitaxial CMOS. *IEEE Electron Device Lett.*, **8**(4), 157–159.

41. Gupta, R.K., Sakai, I. and Hu, C. (1987) Effects of substrate resistance on CMOS latchup holding voltages. *IEEE Trans. Electron Devices*, **34**(11), 2309–2316.

42. Chatterjee, A., Seitchik, J.A., Chern, J.-H., Yang, P. and Wei, C.-C. (1988) Direct evidence supporting the premises of a two-dimensional diode model for the parasitic thyristor in CMOS circuits built on thin epi. *IEEE Electron Device Lett.*, **9**(10), 509–511.

43. Wu, K. and Brown, R.B. (1991) High-temperature design rules, in *Proceedings of the First International High Temperature Electronics Conference*, June 16–20, 1991, Albuquerque NM, pp. 267–272.

44. Huffman, D., Sergers, D. and Green, B. (1979) Minimizing threshold voltage temperature degradation with a substrate bias generator. *MOSTEK 1979 Memory Data Book and Designer's Guide*, MOSTEK Corp., Carrollton TX.

45. Piro, R.A., Sportck, R.R. and DuPasquier, M.P. (1985) Latchup-free substrate bias generators in CMOS. *IEEE 1985 Custom Integrated Circuits Conference*, pp. 524–527.

46. Shoucair, F.S., Hwang, W. and Jain, P. (1984) Electrical characteristics of large scale integration silicon MOSFETs at very high temperatures. Part III: experiment. *IEEE Trans. Components, Hybrids, Manuf. Technol.*, **7**(1), 146–153.

47. Shoucair, F.S. (1986) Design considerations in high-temperature analog CMOS integrated circuits. *IEEE Trans. Components, Hybrids, Manuf. Technol.* **9**(3), 242–251.

48. Kirschman, R.K. (1986) *High-Temperature Electronics*, IEEE Press, New York.

5

Silicon-on-insulator: CMOS devices and processes for high temperature applications

G. Burbach and R. Werner

5.1 INTRODUCTION

Silicon technologies, particularly CMOS, are well established worldwide. There is evidence of constant progress in the efforts to increase the circuit densities and the complexities of integrated functions.

However, the operation of silicon ICs is restricted to temperatures up to 125 °C. Major concerns are increased leakage currents, raised latchup susceptibility and reduced reliability caused by electromigration, oxide wear-out, etc. It is shown that all these problems may be solved by careful design strategies. In order to guarantee circuit operation above 125 °C, however, will cause penalties in circuit complexity.

The silicon-on-insulator (SOI) technology provides an extension of the possibilities on silicon technologies. By introducing the SOI substrates, the process architecture is maintained and process complexity may be reduced. The devices are fully dielectrically isolated so that no latchup paths exist. This is especially important for high temperature operation as the latchup susceptibility greatly increases with temperature. The area of p–n junctions causing leakage currents is drastically reduced. The temperature shifts of device parameters are less severe for SOI devices in comparison to their bulk counterparts.

High Temperature Electronics. Edited by M. Willander and H.L. Hartnagel. Published in 1997 by Chapman & Hall, London. ISBN 0 412 62510 5.

CMOS and perhaps bipolar circuits on SOI substrates present the opportunity to open the field for high temperature electronics up to 250 °C and above to silicon. SOI CMOS can fulfill the needs of high temperature electronics for mixed digital/analog circuits. For example, switched capacitor circuits and operational amplifier D/A converters have been operated successfully in this temperature range. Furthermore, for a wide range of ASIC (application-specific integrated circuit) applications high voltage devices are required. Even low density EEPROM (electronically erasable programmable read-only memory) cells are currently under investigation for use at high temperature.

These examples demonstrate that all the design skills and expertise well known for bulk silicon may be used for CMOS on SOI substrates. To support the design work, SOI device models have been developed and integrated into the CAD (computer-aided design) tools for circuit simulation and verification. Considerable further work is, however, necessary in order to attain the same level of confidence as for bulk designs.

In the past five years, the substrate preparation processes have made great progress, particularly concerning bond and etchback SOI (BESOI) and the separation by implanted oxygen (SIMOX) technologies. The crystal quality of these SOI substrates is approaching bulk standards. The investigations on reliability issues conducted thus far have revealed no insoluble problems.

In the temperature range above 200 °C the metallization process modules have to be modified to avoid pure aluminum. Sandwiches of aluminum and refractory metals like titanium or a substitution of aluminum by tungsten are under investigation.

Some questions related to packaging remain to be solved. The issues concern the choice of the material and processing steps for the bond metallization, the wire bonds, the die attach and the package itself. Once these problems are solved, the SOI devices will allow good high temperature performances up to 300 °C and above.

5.2 SOI SUBSTRATE PREPARATION TECHNOLOGIES

The first developments towards the realization of SOI structures were reported approximately 30 years ago [1]. The different approaches may be grouped roughly into six classes (Table 5.1). The first two groups are based on the epitaxial growth of silicon on either a crystalline insulator or a silicon wafer covered with an insulator. The formation and oxidation of porous silicon allows the isolation of a thin silicon film from the substrate. The recrystallization of silicon films on top of an insulator has been examined in various forms (class 4). Two silicon wafers, at least one of which is covered with an insulator, may be bonded to form an SOI structure. Finally, ion

Table 5.1 SOI substrate preparation technologies

Method	Acronym	Reference
1. Heteroepitaxial techniques		
Silicon-on-sapphire	SOS	43, 44
Solid phase epitaxy and regrowth	SPEAR	45
Silicon-on-diamond	SOD	46, 47
2. Homoepitaxial techniques		
Epitaxial lateral overgrowth	ELO	48, 49
Tunnel epitaxy		50
3. Buried oxidation		
Full isolation by porous oxidized silicon	FIPOS	51
Depleted lean channel	DELTA	52
4. Recrystallization		
Laser recrystallization		53, 54
Zone melt recrystallization	ZMR	55, 56
5. Silicon removal		
Epitaxial passivated integrated circuit	EPIC	57, 58
Wafer bonding	WB	8, 12, 59, 60
6. Ion beam synthesis		
Separation by implanted oxygen	SIMOX	61, 62
Separation by implanted nitrogen	SIMNI	63, 64
Separation by implanted oxygen and nitrogen	SIMON	65, 66

beam synthesis allows the formation of a buried insulator. A detailed description of the different approaches may be found in the book by Colinge [2].

The specific application determines which approach results in the best performances. For instance, SIMOX is regarded as the ideal candidate for VLSI, rad-hard and high temperature applications whereas wafer bonding is more suitable for bipolar and power applications.

The following sections briefly describe wafer bonding and SIMOX. These techniques present the best opportunities for meeting the requirements of industrial CMOS technology – producibility and reliability at an acceptable cost.

5.2.1   Wafer bonding

The starting point is two silicon wafers. One or both are covered with silicon dioxide. If these two wafers are brought into contact instantaneously, bonding occurs even at room temperature. The bonding forces depend strongly on the quality of both surfaces. Hydroxyl groups or van der Waals forces are responsible for the attracting forces [3]. In order to ensure a complete bonding, it is also imperative that the contact surfaces be free of particles. A particle of $1\,\mu m$ diameter may cause unbonded areas (voids) covering several millimeters [4].

In principle, bonding also occurs between insulators like quartz or sapphire and silicon. The differences in thermal expansion coefficients between silicon and these insulators, however, represent the reason why these combinations cannot withstand the temperature cycles of a CMOS process [5]. Therefore, for CMOS or bipolar applications, two silicon wafers covered with silicon dioxide are used. The oxide can be thermal oxide or deposited reflow glasses like phosphorous doped or boron and phosphorous doped oxides [6,7]. At room temperature, the bonding forces are weak. During an annealing at temperatures above $700\,^{\circ}C$, however, Si–O–Si bonds will be formed.

After bonding, one silicon wafer (called bond wafer) is subsequently polished or etched down to a thickness suitable for SOI applications [8]. The second wafer (handle wafer) serves as a mechanical substrate. For CMOS on SOI the typical silicon film thickness is in the range of several hundred nanometers. Standard polishing techniques allow a homogeneity of $\pm 0.5\,\mu m$ in film thickness. Variations in the silicon film thickness exercise an influence on the device parameters (section 5.4.1) and must therefore be minimized. Oxide islands brought into the bond wafer before bonding may be used as etch stops to improve the homogeneity to $\pm 0.1\,\mu m$ [9–11]. The difference in etch rate betweeen heavily boron doped and low doped silicon offers the opportunity to use a heavily boron doped silicon layer as an etch stop. Although the resulting lattice mismatch causes mechanical stress, the stress can be minimized by boron germanium doped silicon [12]. Local plasma etching with *in situ* film thickness measurement provides the best published values of film thicknesses of $100\,nm \pm 10\,nm$ [13]. A summary of the different methods may be found in Colinge [2] and Gassel [14].

5.2.2 SIMOX

Separation by implanted oxygen (SIMOX) is based on ion beam synthesis of a buried oxide. The process is performed in two steps. A high dose oxygen implantation is followed by a high temperature annealing step. Major process parameters are the implantation dose, implantation temperature, annealing temperature, time and atmosphere, all of which determine the quality of the material.

An implantation dose of $1.8 \times 10^{18}\,cm^{-2}$ oxygen atoms is commonly used for the SIMOX synthesis. This dose is a factor of approximately 200 higher than the implantation doses which are used for the source–drain area doping (about $10^{16}\,cm^{-2}$) in the standard bulk technology. In order to avoid amorphization of the silicon film during the oxygen implantation, the substrate temperature has to be greater than 600 °C. The annealing is performed at temperatures above 1300 °C in an atmosphere of argon and oxygen. These parameters produce an SOI structure of 200 nm silicon on top of 400 nm silicon dioxide above silicon. A detailed description of the SIMOX process and its history may be found in Colinge [2] and Gassel [14].

In contrast to wafer bonding, SIMOX delivers very thin films with excellent homogeneity of ±5 nm. Section 5.4 explains the advantages of such thin films for high temperature electronics. The drawbacks of SIMOX are the crystal damage and possible contamination during the high dose implantation. But considerable improvements over the past five years enable the crystal quality of the silicon film to approach that of bulk material.

The thickness of the buried oxide is limited by the maximum possible implantation dose at a given implantation energy. With today's oxygen implanters the maximum dose is approximately 2.0 to $2.5 \times 10^{18} \, \text{cm}^{-2}$, which produces a maximum oxide thickness of about 500 nm. In principle, all SOI techniques deliver substrates which allow the integration of high temperature circuits with better performances than bulk silicon circuits. Market forces will judge between them.

5.3 CMOS ON SOI SUBSTRATES

There are several reasons why the next few sections concentrate on CMOS technology:

- CMOS is the mainstream silicon technology which allows the highest integration density of all semiconductor technologies.
- CMOS devices can be built in thin silicon films which provide advantages regarding the high temperature characteristics of the devices.
- MOS devices exhibit a negative temperature coefficient for their current characteristics. Therefore, thermal runaway, which is a major problem for bipolar devices, does not affect MOS devices.

Nevertheless, good high temperature characteristics up to 300 °C are also reported for bipolar SOI processes [15].

Well-approved bulk silicon processes may be adopted for uses in the SOI CMOS process. Some process steps may even be omitted (e.g. well definitions) or become less critical (e.g. contact spiking). Other parameters must naturally be optimized, in order to reach the best high temperature performance. The following process flow represents the basic steps of a process which the authors have optimized for high temperature operation by using SIMOX substrates:

1. After the preparation of the SOI substrates, an oxidation of the silicon film and subsequent etchback is used to adjust the thickness of the silicon film (Fig. 5.1a). Within the discussion of the temperature characteristics of the threshold voltage and the leakage currents, the advantages of very thin films will be explained in section 5.4.

2. Now to define the active areas. Two alternatives can be used; the simplest way to achieve this is by etching away the undesired silicon film, thus creating mesa-type silicon islands to build in the active devices. The sharp corners of the mesas may cause problems during gate oxide growth and possible inversion channels at the sidewalls are difficult to control [16].

   Alternatively, local oxidation of silicon (LOCOS) (Fig. 5.1b and c) can be used as a lateral isolation method which preserves bulk silicon like process architecture. A stack of thin oxide and nitride is deposited and structured on top of the silicon film. A boron implantation masked by the nitride/oxide stack and a subsequent drive-in step may be used to control the sidewall channels mentioned above. This corresponds to a field implantation in a bulk process. During the field oxidation which follows, the nitride functions as a diffusion barrier for the oxygen, so that only the uncovered silicon is oxidized. Generally, the time needed for complete oxidation of the thin SOI film is less than the time required for field oxidation in a bulk technology.

3. After removing the nitride/oxide stack, two implantations are performed to adjust the threshold voltages of n- and p-channel devices (Fig. 5.1d).

4. Gate oxide growth, deposition, doping and structuring of the polysilicon gates are identical with the process steps in a bulk technology (Fig. 5.1e).

5. Major drawbacks of SOI devices are the reduced breakdown voltages and the intrinsic bipolar effects (section 5.4.7). These effects require a careful 'drain and source' engineering. Lightly doped drain (LDD) structures (Fig. 5.1f, g and h) have to be introduced for some channel lengths; they may not be necessary in a bulk technology. LDDs are helpful if the target supply voltages are close to 5 V.

   High temperature applications require electronics which often runs at higher supply voltages, e.g. readout electronics for sensors. Therefore, a high voltage option has been implemented in this process (section 5.4.6). A lateral low doped drift region is introduced between the polysilicon gate and highly doped drain contact.

6. The spacer oxide etchback process of the LDD regions is followed by the source drain implantations for p- and n-channel devices (Fig. 5.1i).

7. The deposition of a reflow glass, reflow and contact etching can be carried out identical to bulk technology (Fig. 5.1g).

8. Concerning reliability issues, a metallization based purely on aluminum is not suitable for operation at temperatures of 200 °C and above.

a) **thinning oxide**

silicon film

buried oxide

substrate

b) nitride

oxide

buried oxide

c) field oxide   field oxide

buried oxide

d) **implantation for threshold adjust**

field oxide   field oxide

buried oxide

e) polysilicon gate

gate oxide

field oxide

f) **LDD implantation**

polysilicon

buried oxide

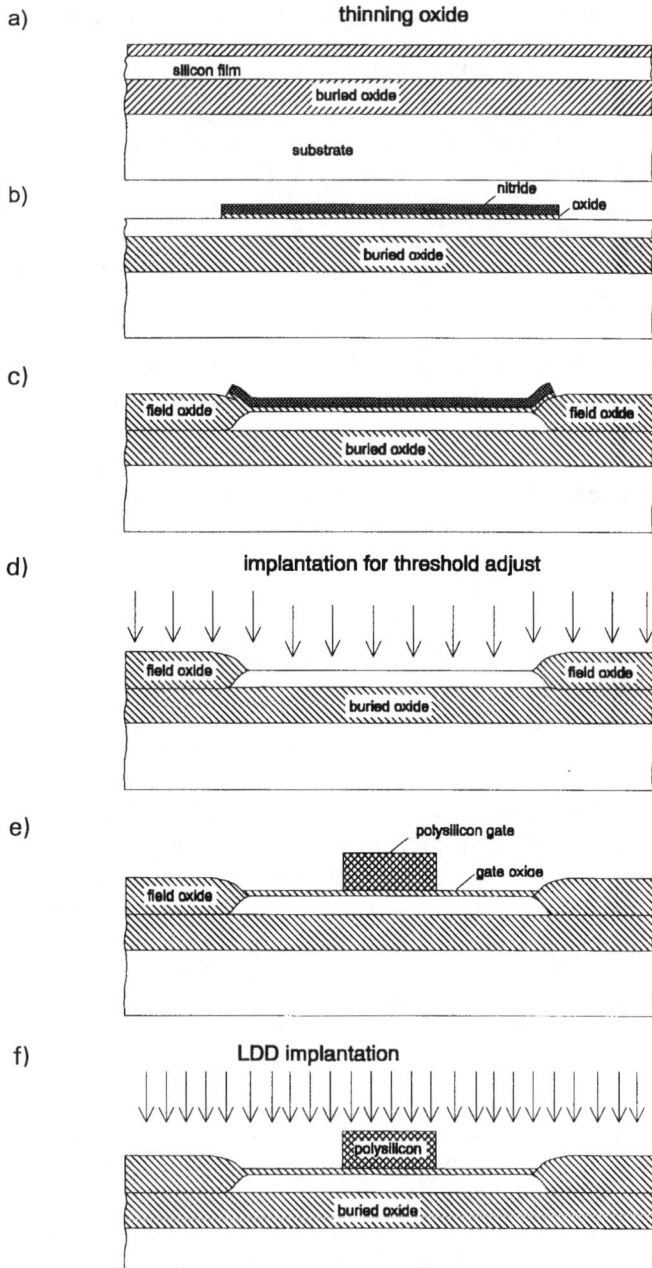

Figure 5.1 Process flow of CMOS on SOI substrates.

g)

CVD oxide

polysilicon

n-   n-

h)

etch back

polysilicon

buried oxide   oxide spacer

i)

source-drain implant

N+ N-   N- N+

buried oxide

j)

BPSG

n-   n-

buried oxide

k)

Ti/TiN layer

n-   n-

buried oxide

l)

source   drain

—tungsten—   Ti/TiN layer

BPSG   gate   BPSG

field oxide   n+   n-   n-   n+   field oxide

gate oxide   buried oxide

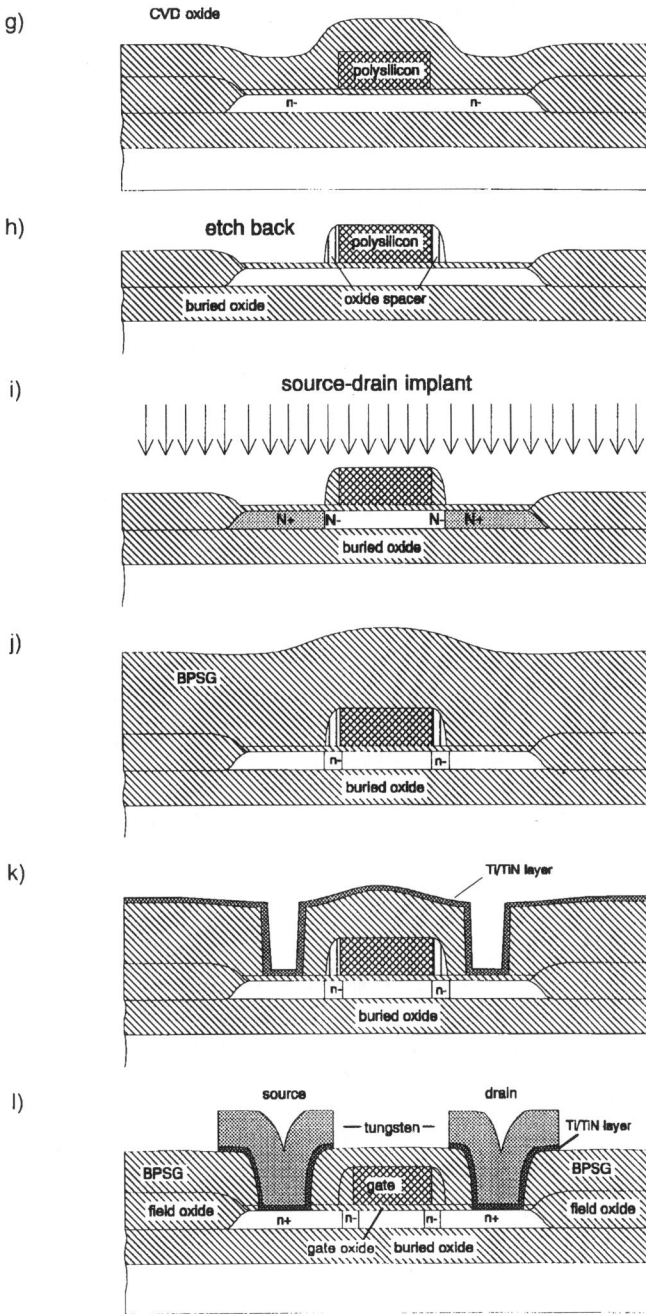

Figure 5.1 (*Continued*).

Sandwiches of aluminum and refractory metals like titanium may be used up to 250 °C. In order to explore the possibilities of SOI devices and circuits in the temperature range up to 300 °C and above, a metallization based on tungsten has been implemented (Fig. 5.1k and l) [17].

9. Depending on the circuit complexity, a second or third metal level may be integrated.
10. The deposition of a passivation layer and the pad opening finalize the process.

For a tungsten metallization on top of the pads, special layers have to be deposited so that standard wire bonding techniques can be used. For example, a sandwich of titanium, palladium and gold allows wire bonding with gold wires. Such a complex pad metallization allows the use of gold as the only material for the bond pad, the bond wire and the package leads, thereby minimizing possible failure due to interdiffusion. Initial examinations indicate some promising results with respect to long-term stability.

5.4 SOI DEVICE CHARACTERISTICS

This section presents a summary of the characteristics of SOI devices and the differences between SOI and bulk devices. First, the threshold voltage will be discussed. The discussion helps explain the temperature characteristics of the leakage current. The respective behaviors of the carrier mobility and the drive current will be shown. The section ends with the observations concerning the intrinsic bipolar effects and the breakdown characteristics.

5.4.1 Threshold voltage of SOI devices

The channel area of SOI devices forms a metal–oxide–silicon–oxide–silicon (MOSOS) structure. The top silicon film has two oxide interfaces. The upper interface to the gate oxide is called the front interface, the interface to the buried interface is called the back interface. At each interface, accumulation, depletion or inversion may occur, depending on the potential of the front and/or back gate. Furthermore, the thicknesses of the gate oxide, the buried oxide and the silicon film plus the number of interface states play an important role. The discussion of the SOI threshold voltage can be separated into the discussion of a thick film, partially depleted and a thin film, fully depleted device.

If the silicon film t_b is greater than twice the maximal depletion width x_{dmax}:

$$x_{dmax} = \left(\frac{2\varepsilon_{Si}2\Phi_F}{qN_A}\right)^{1/2} \tag{5.1}$$

where ε_{Si} = dielectric constant of silicon
Φ_F = Fermi potential
q = magnitude of electronic charge
N_A = dopant concentration

Under all external bias conditions, no overlap exists of the depletion regions near the front and back interface. Between the depletion layers, there remains a region where the potential $\Psi = 0\,V$ and the electrical field $E = 0\,V\,cm^{-1}$, i.e. the silicon is quasi-neutral. The front (back) threshold voltage, defined as the onset of strong inversion, can be calculated independently as in the bulk silicon theory:

$$V_{tf(b)} = V_{FB}^{f(b)} + 2\Phi_F + (2\varepsilon_{Si}qN_A 2\Phi_F)^{1/2}/C_{of(b)} \tag{5.2}$$

If the silicon thickness t_b is less than x_{dmax} the depletion region caused by the front gate expands across the whole silicon film from weak to strong inversion at the front interface. Such devices are called fully depleted devices. For film thicknesses in the range $x_{dmax} \leqslant t_b \leqslant 2x_{dmax}$ the situation depends on the backgate bias V_{Gb}. A critical V_{Gb}, V_{Gb}^{crit} can be defined where the behavior switches from a fully depleted to a partially depleted (bulk-like) mode.

Fossum and Lim [18] have developed a threshold voltage model for fully depleted devices. For the calculation of the front threshold voltage, the states of accumulation, depletion and inversion at the back interface must be considered separately.

If an accumulation or an inversion layer is formed at the back interface, a further decrease (increase) of the back gate bias will modulate the charge in the accumulation (inversion) layer, but will have no influence on the surface potential Ψ_{sf} at the front interface. In this case, the front threshold voltage becomes independent of back gate bias. If the depletion region fills the whole silicon film, the surface potential at the front and back interface are coupled. Abel [19] and Fossum and Lim [18] demonstrated that straightforward calculation results in the following equations:

Case 1

Accumulation at the back interface will be present for n-channel devices at back gate voltages of less than V_{Gb}^A. For p-channel devices the sign of the voltages in the following equation should be changed appropriately.

$$V_{Gb}^A \cong V_{Fb}^b - \frac{C_b}{C_{ob}} \cdot 2\Phi_F - \frac{Q_b}{2C_{ob}} \tag{5.3}$$

where V_{Fb}^b = flat band voltage at back interface
$(C_b = \varepsilon_{Si} t_{Si})$ = depletion capacitance
C_{ob} = back gate capacitance
Q_b = bulk (silicon film) charge

The threshold voltage can be calculated from

$$V_{tf} \cong V_{tf}^A = V_{Fb}^f + \left(1 + \frac{C_b}{C_{ob}}\right) \cdot 2\Phi_F - \frac{Q_b}{2C_{of}} \tag{5.4}$$

where V_{Fb}^f = flat band voltage at front interface
        C_{of} = front gate capacitance
        C_{sb} = capacitance related to interface states

Case 2

Inversion at the back interface will be achieved if for n-channel devices the back gate voltages are greater than V_{Gb}^I, which is given by

$$V_{Gb}^I \cong V_{Fb}^b + \frac{C_{sb}}{C_{ob}} \cdot 2\Phi_F - \frac{Q_b}{2C_{ob}} \tag{5.5}$$

The threshold voltage then becomes

$$V_{tf} \cong V_{tf}^I = V_{Fb}^f + 2\Phi_F - \frac{Q_b}{2C_{of}} \tag{5.6}$$

Case 3

In the back gate voltage range between $V_{Gb}^A \leqslant V_{Gb} \leqslant V_{Gb}^I$ the back interface is depleted. In this case a linear dependency correlates the front threshold voltage and the back gate voltage:

$$V_{tf} \cong V_{tf}^A - \frac{C_b C_{ob}}{C_{of}(C_b + C_{ob} + C_{bs})} \cdot (V_{Gb} - V_{Gb}^A) \tag{5.7}$$

As can be seen in Fig. 5.2, the agreement between the model and the measured data at room temperature is very good. The switch from accumulation to depletion at the back interface is smoother than the theoretical corner function. The reason for this lies in the abrupt depletion approximation in the model, which is only valid within the Debye length

$$L_D = \sqrt{\frac{\varepsilon_{Si}kT}{q^2 N_A}} \tag{5.8}$$

where k = Boltzmann constant
        T = absolute temperature.

The Debye length increases with increasing temperature; this causes the discrepancies between model and theory to increase. The case of inversion at the back interface is not shown in Fig. 5.2 because a current path exists between source and drain at the back interface, making it difficult to measure the threshold voltage. This state is to be avoided in all operating

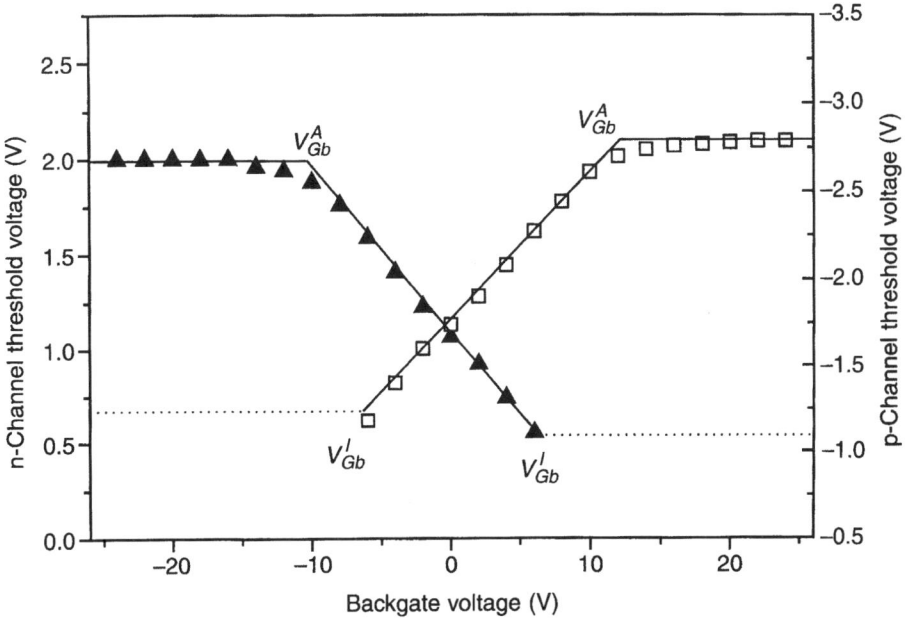

Figure 5.2 Back gate dependence of the threshold voltage for n- and p-channel devices, $W/L = 48/3.2$, silicon film 85 nm, gate oxide 40 nm: (▲) n-channel and (□) p-channel.

conditions and need never occur if the process parameters are carefully chosen. The last terms of equations (5.5) and (5.6) lead to the following approximate relationship between the front gate and the back gate threshold voltage:

$$\frac{V_{tf}}{V_{tb}} \approx \frac{t_{of}}{t_{ob}} \tag{5.9}$$

The thickness of the front gate is defined by the reliability issues for a given supply voltage of the application.

The front threshold voltage as a function of the back gate voltage in the state of a depletion region at the back interface is given by the slope b, which is approximately proportional to the ratio of gate and buried oxide:

$$b \equiv \frac{C_b C_{ob}}{C_{of}(C_b + C_{ob} + C_{bs})} \propto \frac{t_{of}}{t_{ob}} \tag{5.10}$$

Figure 5.3 shows the slope of n-channel devices for different gate oxide thicknesses and a fixed buried oxide thickness.

Because n-channel and p-channel devices have a common back gate (no wells like in bulk), the back gate for the n- or p-channel device is biased

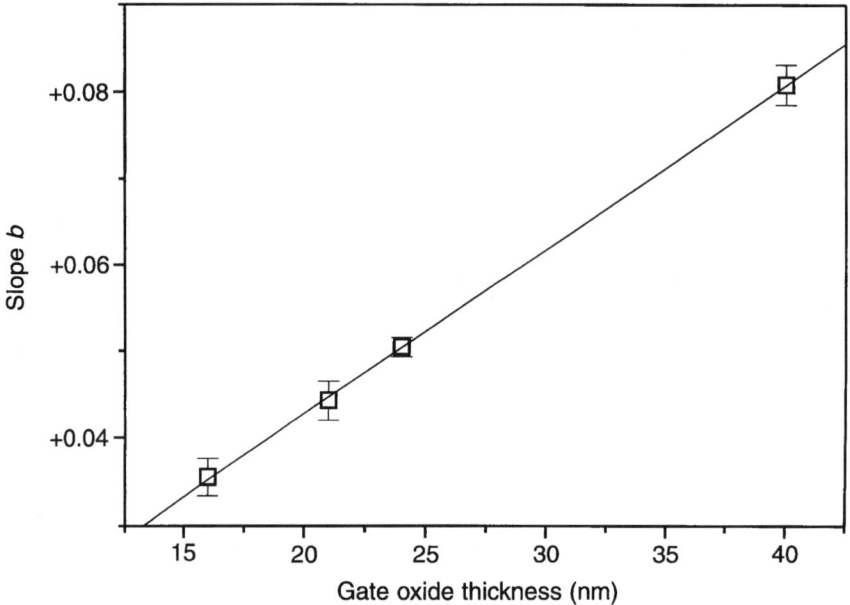

Figure 5.3 Slope of n-channel threshold voltage dependence on back gate voltage for different gate oxides with a buried oxide of 380 nm.

towards inversion if the back of the die is connected to one of the power supply voltages. The buried oxide should therefore be as thick as possible. However, in the case of SIMOX technology, for example, the maximum thickness of the buried oxide is defined by the maximum possible implantation dose. Omura et al. [20] show that increasing short channel effects set an upper limit to the buried oxide thickness. Trade-offs have to be made, bearing in mind the requirements of the application. The examples in this section show that the thicknesses of the gate oxide, the silicon film and the buried oxide have to be optimized according to these requirements.

5.4.2 Temperature dependence of the threshold voltage

For thick silicon films, the temperature dependence can be calculated according to bulk theory [21, 22]:

$$\frac{dV_{tf}}{dT} = \frac{d\left(\Phi_{ms} + 2\Phi_F - \frac{Q_{ox}}{C_{of}} - \frac{Q_b}{C_{of}} \right)}{dT} \tag{5.11}$$

which leads to the equation

$$\frac{dV_{tf}}{dT} = \frac{d\Phi_F}{dT} \cdot \left(2 + \frac{1}{C_{of}}\sqrt{\frac{\varepsilon_{Si}qN_A}{\Phi_F}}\right) \tag{5.12}$$

The first term describes the temperature dependence of the Fermi potential whereas the second shows the variation of the depletion capacitance with temperature. Equation (5.12) remains valid even in thin silicon films with an accumulation layer at the back interface.

In fully depleted devices, however, the depletion charge is limited by the film thickness $Q_{depl} = qN_At_{Si}$. As long as the silicon film is fully depleted, from equation (5.12), only the temperature dependence of the Fermi potential remains [22]

$$\Rightarrow \frac{dV_{tf}}{dT} = \frac{d\Phi_F}{dT} \tag{5.13}$$

Figure 5.4 shows the threshold voltage dependence on the back gate bias for various temperatures. The different temperature coefficients of the threshold voltage for a depleted or an accumulated back interface are clearly indicated.

Figure 5.5 shows the temperature dependence for different back gate biases. At $V_{bs} = -10\,\text{V}$ the back interface is forced into accumulation. For $V_{bs} = -5\,\text{V}$ the back interface is depleted up to $150\,^\circ\text{C}$. Above $150\,^\circ\text{C}$ an accumulation layer is formed at the back interface, producing a steeper temperature slope. For $V_{bs} = 0\,\text{V}$ the back interface remains depleted up to $230\,^\circ\text{C}$ before the influence of an increasing accumulation layer takes over the temperature behavior of the device.

Figure 5.6 shows the temperature coefficient of the threshold voltage for different gate oxide thicknesses. If the back interface is biased in accumulation, the coefficient increases linearly in accordance with equation (5.12). For fully depleted devices the temperature coefficient is much smaller. The slight increase of the coefficient with the oxide thickness is caused by the different dopant concentrations in the film, which cause different Fermi potentials in the examined samples.

For high temperature applications, the fully depleted mode of the devices promises the best performances. This means that thin silicon films with thicknesses of less than x_{dmax} should be chosen. For a typical doping concentration (e.g. $5 \times 10^{16}\,\text{cm}^{-3}$) the calculated maximal depletion width x_{dmax} is 144 nm at room temperature, decreasing with increasing temperature. On the other hand, the necessary implantation dose to adjust the threshold voltage to a certain value depends on the silicon film thickness, as can be seen if equation (5.6) is rewritten

$$V_{tf} = V_{Fb}^f + 2\Phi_F - \frac{qN_At_{Si}t_{of}}{\varepsilon_{ox}} \tag{5.14}$$

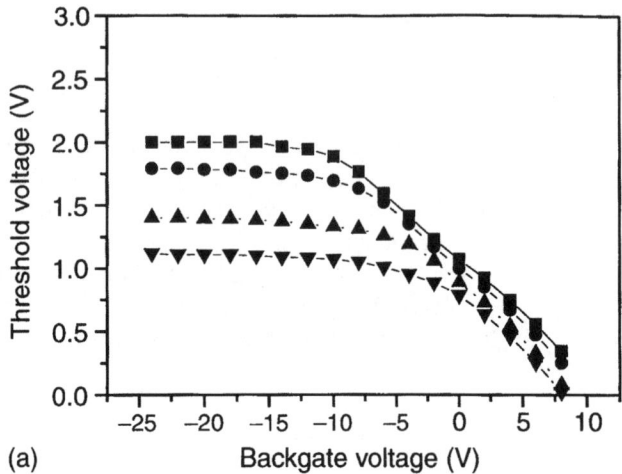

(a)

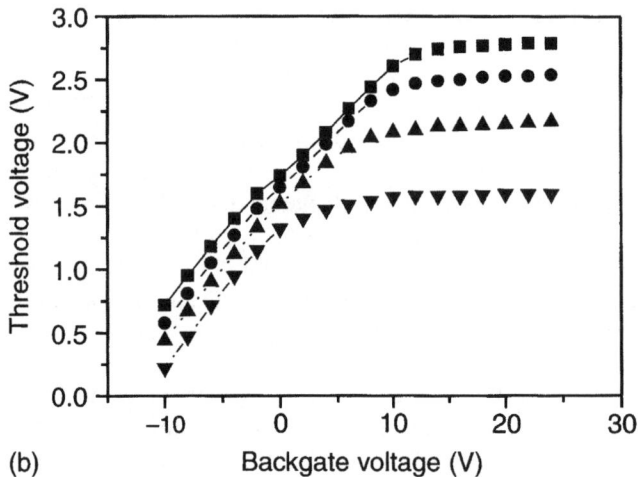

(b)

Figure 5.4 Threshold voltage dependence on the back gate voltage for a silicon film of thickness 85 nm at different temperatures (a) for n-channel devices and (b) for p-channel devices: (■) 26 °C, (●) 100 °C, (▲) 200 °C and (▼) 280 °C.

It can be shown that the temperature budget in typical CMOS processes is sufficient to provide a homogeneous doping profile in the film. However, the film doping concentration N_A in equation (5.14) is not simply given by the total implantation dose divided by the thickness of the film because not all implanted ions can be stopped in the film. For this reason, variation in the film thickness will cause severe variation in the threshold voltage if the doping profile is not well centered. Sherony *et al.* [23] show that a careful choice of implantation parameter together with the excellent homogeneity

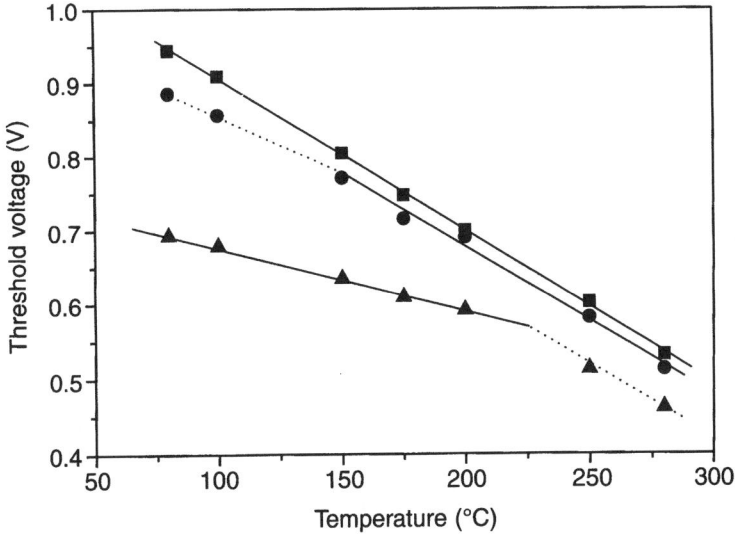

Figure 5.5 Temperature characteristics of the n-channel threshold voltage for 40 nm gate oxide at different conditions of the back interface: (■) $V_{bs} = -10$ V, (●) $V_{bs} = -5$ V, (▲) $V_{bs} = 0$ V.

of layer thicknesses in SIMOX substrates can eliminate this problem for practical purposes.

Figure 5.7 shows the doping concentration required for a threshold voltage $V_{tf} = 1$ V; it depends on the silicon film thickness and is plotted for gate oxide thicknesses of 20 nm and 40 nm. Figure 5.7 illustrates how the threshold voltage variation as a function of the film thickness dramatically increases if silicon films around 50 nm are used. The border (maximum depletion width x_{dmax}) between full depletion and partial depletion is also shown for 26 °C and 250 °C.

5.4.3 Temperature characteristics of leakage currents

In the early stage of SOI technology development the advantages of the drastically reduced source and drain p–n regions were demonstrated for a low leakage current behavior [24]. And the large well–substrate p–n regions along with the related leakage currents do not exist in SOI technology. These circumstances have been demonstrated as beneficial for high temperature operation, as the leakage currents in SOI devices at elevated temperatures are several orders of magnitude lower than in their bulk counterparts.

Figure 5.8 shows that the leakage current of the bulk device is a factor of approximately 1000 greater than for the SOI device. In the past, only the

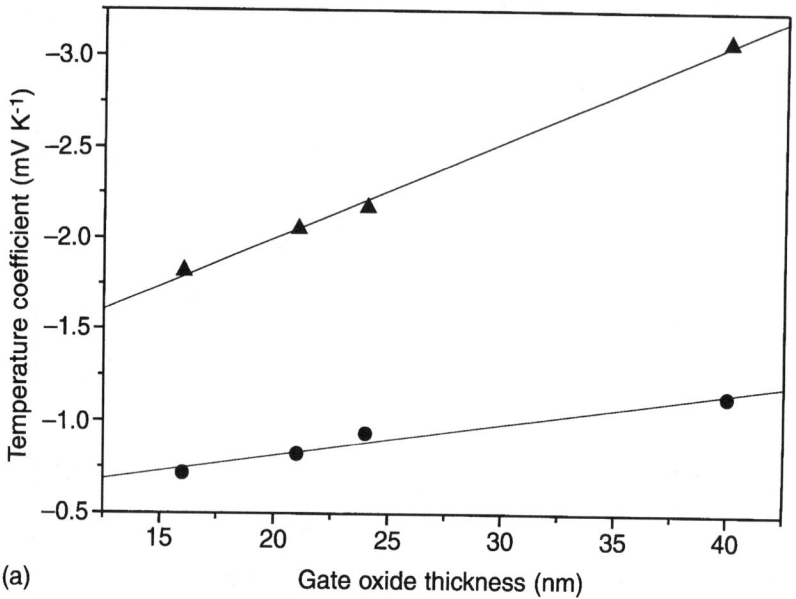

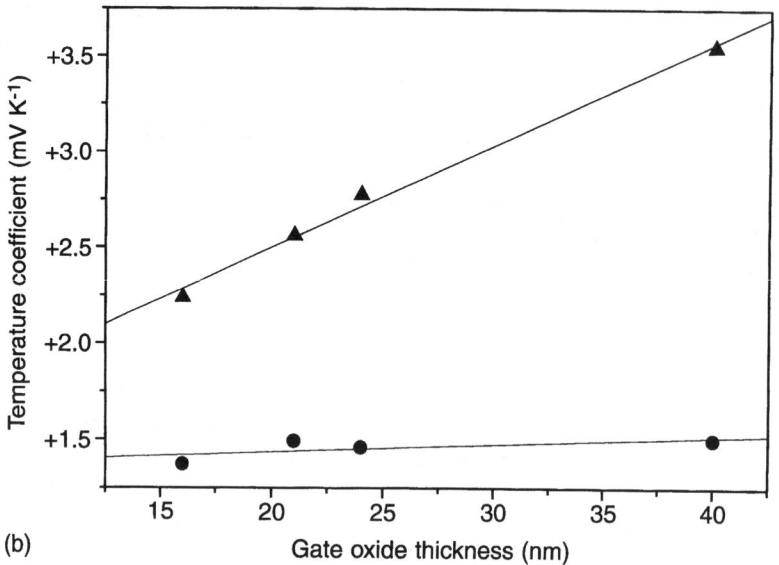

Figure 5.6 Temperature coefficient of the threshold voltage depending on the gate oxide thickness for n- and p-channel devices with $W/L = 50/3.2$: (a) n-channel, (●) $V_{sb} = 5\,V$ (fully depleted) and (▲) $V_{sb} = -5\,V$ (accumulated back interface); (b) p-channel, (●) $V_{sb} = -5\,V$ (fully depleted) and (▲) $V_{sb} = 5\,V$ (accumulated back interface).

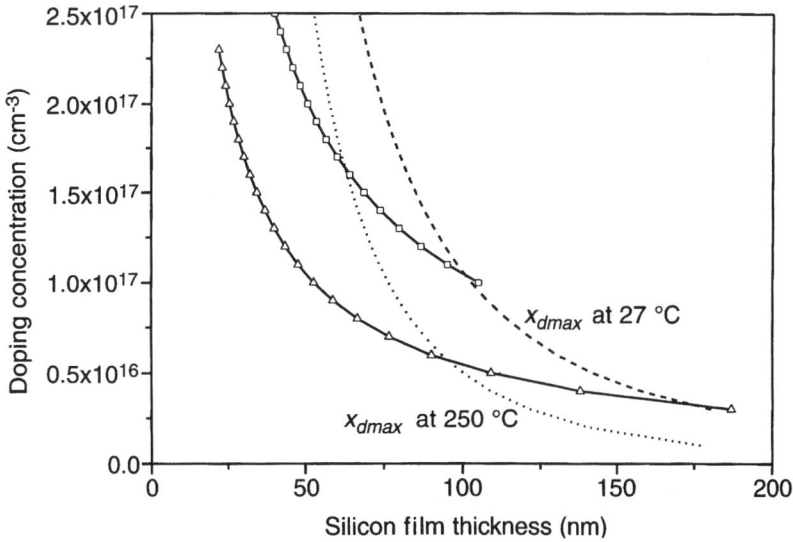

Figure 5.7 Dopant concentration required to adjust the n-channel threshold voltage to $V_{tf} = 1$ V as a function of the silicon film thickness and gate oxide thickness: ($\square$) gate oxide 20 nm and ($\triangle$) gate oxide 40 nm.

leakage currents caused by p–n junctions were compared with bulk devices. These p–n currents cannot be controlled by the gate voltage but can be increased significantly by the parasitic bipolar transistor effects, especially for decreasing channel length (section 5.4.7). For practical purposes, the leakage current should be defined as the drain current at a front gate bias of $V_{gs} = 0$ V, which has more relevance for integrated circuit design. With this definition the following components contribute to the leakage current:

- p–n junction leakage;
- subthreshold currents at the front and back interfaces;
- currents caused by the intrinsic bipolar effects.

The effects of the components are illustrated in Fig. 5.9 and will be explained in the following sections.

5.4.4 Leakage at a p–n junction

According to the discussion in Sze [21], the drain current for a bulk silicon MOSFET biased below the threshold is given by the sum of the following contributing elements (Fig. 5.10) [25]:

- the subthreshold current $I_{subth} \propto \exp(V_{gs})$ (see next section);
- the generation current I_{gen};
- the diffusion current I_{diff}.

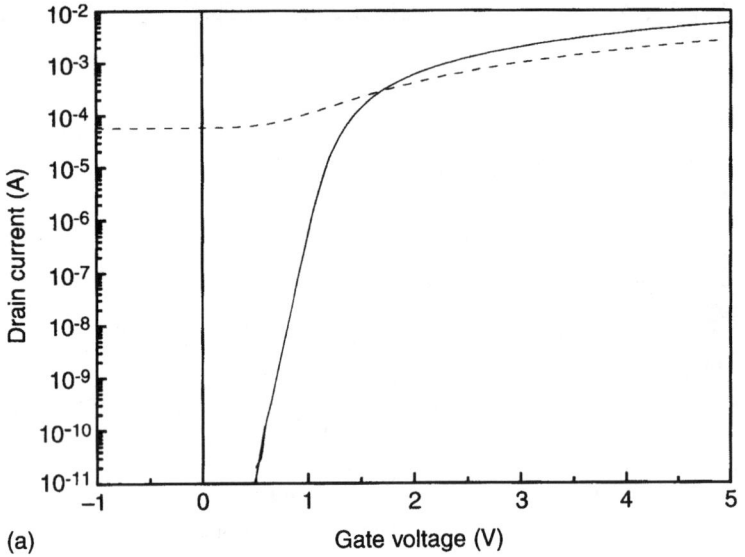

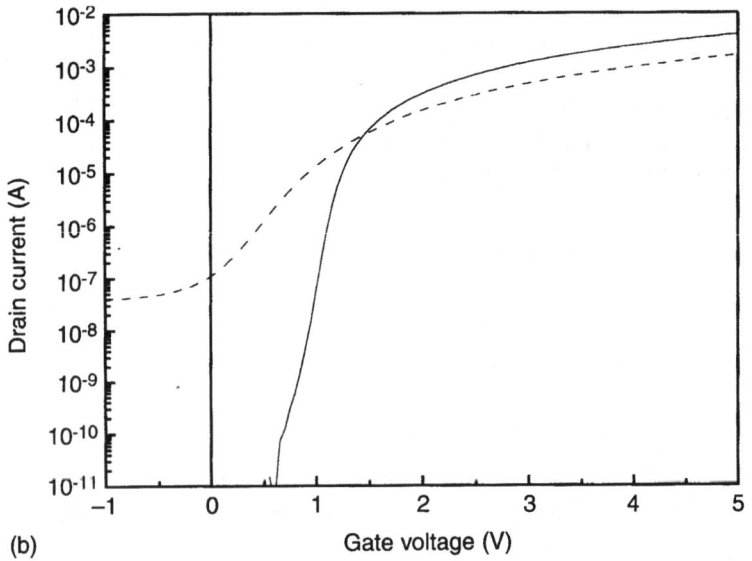

Figure 5.8 Transfer characteristics of (a) bulk ($W/L = 50/2$) and (b) SOI n-channel ($W/L = 60/2$) devices at (—) room temperature (26 °C) and (---) 280 °C with $V_{ds} = 5.1$ V.

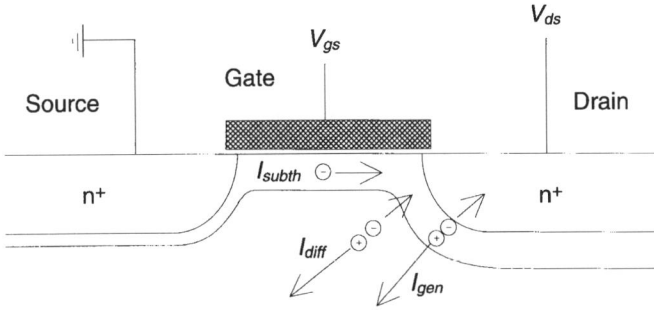

Figure 5.9 Sketch of the components of the leakage current of a bulk MOSFET.

The generation current consists of electron–hole pairs thermally generated inside the drain depletion region and subsequently separated in the electrical field of the depletion region. This component can be calculated by

$$I_{gen} = \frac{qn_i}{\tau_e} \times \text{volume (depletion region)} \tag{5.15}$$

where n_i = intrinsic carrier density
        τ_e = generation lifetime

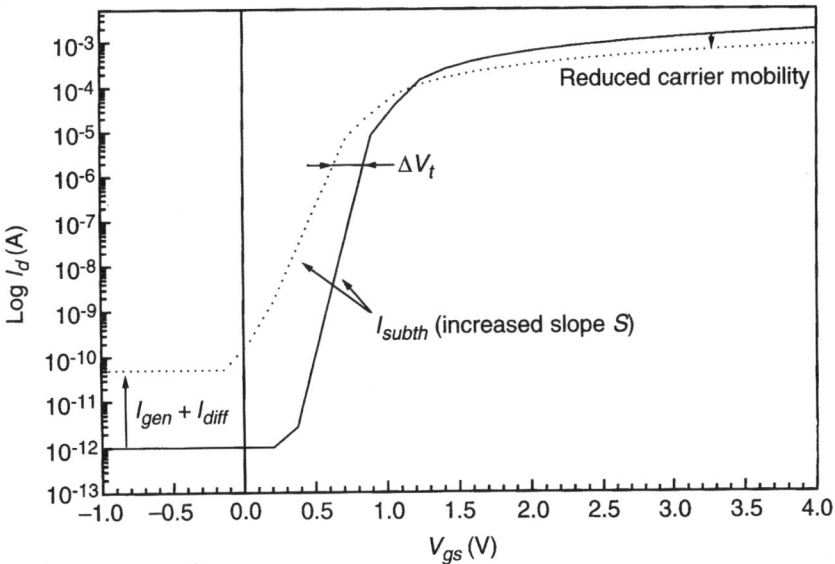

Figure 5.10 Sketch of transfer characteristics demonstrating the influence of the different physical mechanisms on the leakage current: (—) room temperature and (---) 250 °C.

The source area of the diffusion current I_{diff} is the quasi-neutral silicon under the depletion region. Minority carriers diffuse into the depletion zone where they are collected in the electrical field. This component may be calculated for an n-channel device with the aid of the following equation:

$$I_{diff} = \sqrt{\frac{D_e}{\tau_e} \frac{n_i^2}{N_A}} \times \text{area (device)} \qquad (5.16)$$

where D_e = diffusion constant of electrons

The exponential increase of the intrinsic carrier density n_i with temperature means that, for higher temperatures, the leakage current is dominated by the diffusion component; this leads to an increase of the leakage current by a factor 2 every 6 °C.

The temperature behavior of SOI devices depends on the silicon film thickness. No quasi-neutral silicon exists in fully depleted devices, so the diffusion current is totally suppressed. Even in non-fully depleted devices the area of the quasi-neutral silicon is much smaller than in bulk. The Arrhenius plots (Fig. 5.11) of the leakage currents of a bulk MOSFET and a fully depleted SOI MOSFET demonstrate the different behaviors. The leakage current of the bulk device is dominated by the diffusion current at a temperature of about 100 °C then increases proportionally to n_i^2 up to a temperature of 300 °C. The leakage current of the SOI device only increases proportionally to n_i as long as the device remains in the fully depleted mode (here up to 200 °C). This explains the strong difference in the leakage currents of the two different technologies. Researchers [26–28] have shown that for silicon film thicknesses of less than 50 nm the transistor function still exists at temperatures above 400 °C.

5.4.5   Subthreshold currents at the front and back interfaces

The threshold voltage decreases with increasing temperature. This is valid for the front and the back interface in SOI devices. Below the threshold voltage, the channel current consists of a diffusion current at the interface which decreases exponentially with the gate voltage. It is of primary importance to keep the subthreshold current below the p–n junction leakage.

For bulk silicon MOS devices, Sze [21] shows that the subthreshold current can be calculated by

$$I_{subth} = \left(\mu_n \frac{W}{L} q \left(\frac{kT}{q} \right)^2 \frac{n_i^2}{N_A} (1 - \exp(kTV_D/q)) \right) \cdot \frac{\exp(kT\Psi_S/q)}{-\left.\frac{\partial \Psi_S}{\partial x}\right|_S} \qquad (5.17)$$

where μ_n = carrier mobility
W/L = ratio of device width to channel length
Ψ_S = surface potential

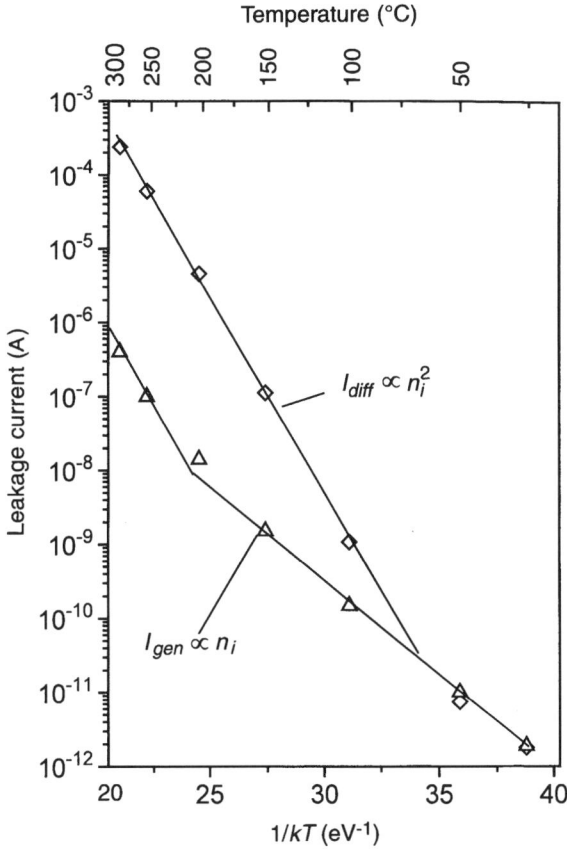

Figure 5.11 Arrhenius plot of the leakage current for a fully depleted octagonal n-channel device ($W/L = 600/10$) in ($\diamond$) bulk and ($\triangle$) SOI.

For $V_d \gg kT/q$ the drain current is nearly independent of the drain voltage. In order to compare different devices or different technologies, the decrease of the drain current as a function of the gate voltage below threshold can be evaluated. The slope S commonly used as a measure of the subthreshold behavior is equal to the change in V_{gs} necessary to decrease the subthreshold current by one decade. With this definition of S, equation (5.17) suggests the slope is given as

$$S \equiv \ln 10 \left(\frac{d\ln I_d}{dV_{gs}} \right)^{-1} = \frac{kT}{q} \ln 10 \left(1 + \frac{C_d + C_{it}}{C_{ox}} \right) \tag{5.18}$$

The last term $1 + (C_d + C_{it})/C_{ox}$ is equal to the ratio of the fraction of the gate voltage used to increase the surface potential to the fraction used to increase the depletion width. For fully depleted devices, the depletion width already occupies the whole silicon film. Consequently, a further increase in the gate voltage cannot cause an increase in the depletion width. In this case, the slope S is given by

$$S = \frac{kT}{q} \ln 10 \tag{5.19}$$

Figure 5.12 shows the temperature behavior of the front gate subthreshold slope for SOI devices with two different silicon film thicknesses. The SOI device in the 85 nm silicon film is fully depleted up to 200 °C. The subthreshold slope almost reflects the ideal value given by equation (5.19). Above 200 °C the fully depleted mode cannot be preserved, which produces a drastic steepening of the slope. The device in the 150 nm film is partially depleted.

It becomes obvious that the subthreshold slope of the partially depleted device is greater than for the fully depleted device and increases more rapidly with temperature. The same behavior may be observed at the back interface. Owing to the thickness of the buried oxide, the absolute value of the slope is strongly influenced by the second term of equation (5.18).

For a SIMOX substrate with a buried oxide thickness of about 400 nm and a silicon film thickness of 100 nm, the back gate slope is approximately

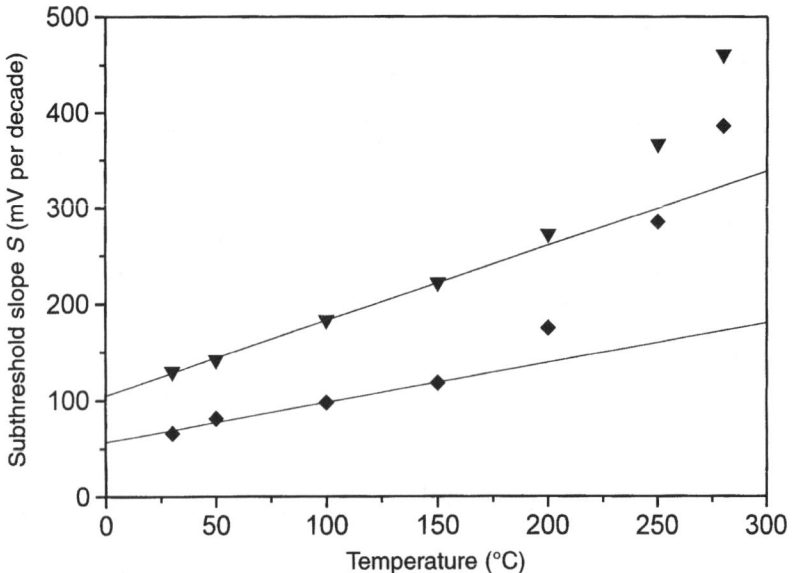

Figure 5.12 Temperature dependence of the subthreshold slope for n-channel devices (W/L = 50/3.2) with film thicknesses of (▼) 150 nm and (◆) 85 nm.

one order of magnitude higher than the corresponding front gate slope. Both the front gate and the back gate subthreshold slopes increase with rising temperature. The influence of this increase can be much more significant for the evaluation of the off-state leakage current than the temperature-related decrease of the threshold voltage.

This is demonstrated in Fig. 5.13, where the back gate threshold voltage of a p-channel device determined in two different ways is plotted versus the temperature. For these measurements, the front gate voltage is biased at 0 V. The lower curve corresponds to the threshold voltage, determined by the usual tangent method. This represents the onset of strong inversion at the back interface. The drain current value of 40 nA lies considerably above the p–n leakage current for the examined device, even in the high temperature range.

The back gate voltage which causes a drain current of 40 nA is shown in the upper curve in Fig. 5.13. At 280 °C, the threshold voltage range still lies below the value of -10 V, which corresponds to the negative supply voltage, whereas the upper curve, which corresponds to the back gate voltage at a fixed drain current of 40 nA, hits the line at -10 V at a temperature of about 220 °C. The back gate voltage required above 220 °C to keep the drain current below the threshold current of 40 nA is, in this case, less than the supply voltage $V_{dd} = 10$ V.

This aspect should be regarded carefully, for the reason illustrated in Fig. 5.14, representing a simple SOI inverter structure. The n- and p-channel

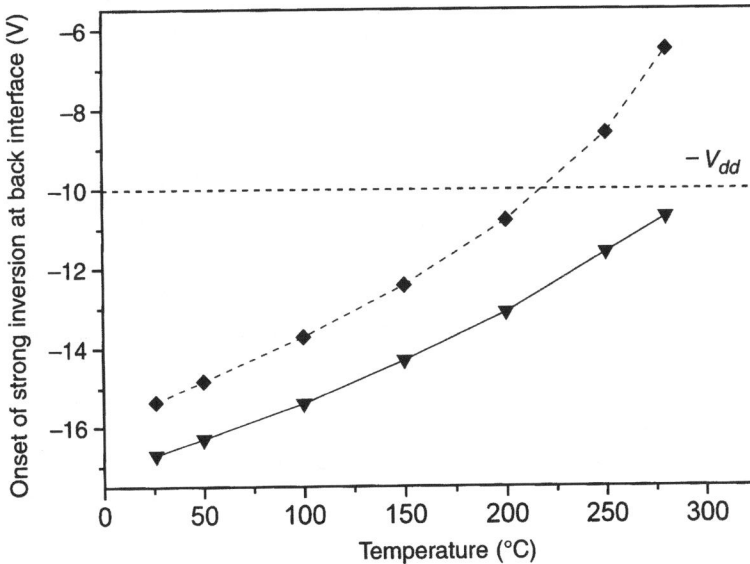

Figure 5.13 Onset of strong inversion at the back interface of a p-channel device measured (▼) with the tangent method and (◆) for a fixed threshold current I_{ds} of 40 nA. The front channel is switched off ($V_{gs} = 0$ V).

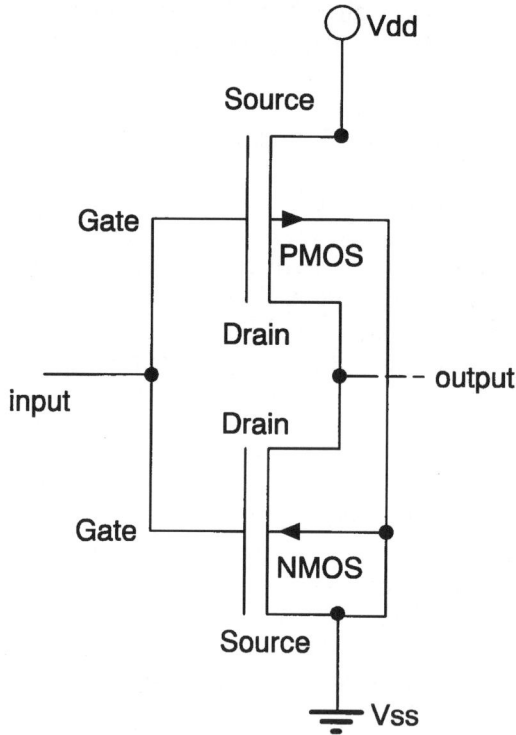

Figure 5.14 Sketch of an SOI inverter to demonstrate the common back gate.

devices have a common back gate, which in Fig. 5.14 is connected to V_{ss}. This means that for the p-channel devices the back gate voltage is $V_{bs} = -V_{dd}$. This in turn implies that for the shown p-channel MOSFET used in the inverter, biased at a supply voltage of 10 V and heated to a temperature above 200 °C, a permanent drain current flows at the back interface which cannot be controlled by the front gate. The functionality of the circuit is not lost but the static power consumption is increased.

The temperature behavior of the threshold voltage, the leakage currents and the subthreshold behavior have shown the advantages of fully depleted devices. In the discussion of the threshold voltage behavior, it has been shown that the fully depleted mode of the devices can only be preserved at high temperatures if ultrathin films are used.

The dependence of the threshold voltage on the film thickness, however, requires an excellent homogeneity of the substrates, which can hardly be guaranteed by the manufacturing processes. In ultrathin films, the sheet resistance of the source–drain areas increases, which produces a decrease of the drive current. A trade-off must be made between the excellent high temperature behavior of fully depleted SOI devices, the manufacturing considerations and the increase of sheet resistance.

Furthermore, the thicknesses of the gate oxide and the buried oxide must be carefully balanced with respect to their influence on the reliability, the ratio of the front and back gate threshold voltages and the temperature behavior of the subthreshold currents.

5.4.6 Temperature characteristics of the carrier mobility

As is known for bulk silicon devices, the carrier mobility is the parameter which may be strongly influenced by temperature. A decrease in the mobility is responsible for a decrease in the maximum drive current of the devices. In comparison with the well-understood behavior of bulk silicon devices, no surprising effects are known to occur in SOI devices. The normalized mobility $\mu(T)/\mu_0$ follows a similar relationship which may be represented by the equation

$$\frac{\mu(T)}{\mu_0} = \left(\frac{T}{T_0}\right)^n \tag{5.20}$$

The exponent n varies typically between -1.5 and -2. Figure 5.15 shows the normalized mobility for p-channel devices with a film thickness of 85 nm and 150 nm. As a result of the full depletion, the device with the 85 nm film exhibits a slightly different value of n ($n = -1.6$) as compared to the thicker film ($n = -1.5$). Fully depleted devices generally have a higher transconductance than partially depleted devices, owing to the reduced electrical field perpendicular to the current path in the channel [29]. This reduced electrical field can also be claimed as the reason for the slight variations in the temperature behavior.

5.4.7 Intrinsic bipolar effects

The intrinsic bipolar effects, also called the floating body effects of the SOI devices, represent the major drawbacks of the SOI technology. These effects are responsible for the well-known kink effect [30], the so-called self-latching effect [31, 32] and the reduced breakdown voltages of SOI devices [33, 34] compared to their bulk counterparts. In a bulk device the channel area is connected to the substrate or to the well region. For SOI devices (Fig. 5.16) the channel area floats, if no further measures are taken.

As may be seen in Fig. 5.16, an intrinsic bipolar transistor is realized in each SOI MOSFET. The base of the bipolar transistor is floating. If the drain–source voltage is considerably high, electron–hole pairs are generated in the depletion region of the drain area. The electrons are collected at the

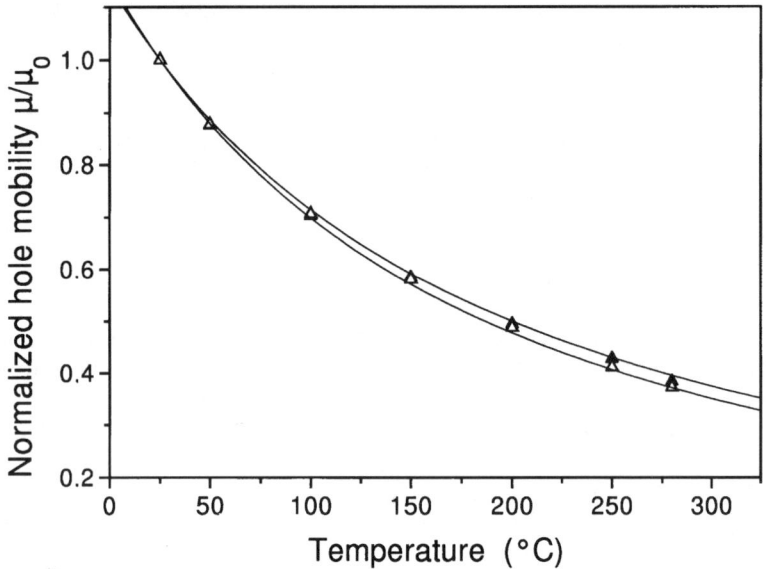

Figure 5.15 Temperature behavior of the normalized carrier mobility for a p-channel device with two different silicon film thicknesses: (▲) 150 nm, $n = -1.5$ and (△) 85 nm, $n = -1.6$.

drain, but the holes move to the source contact since they cannot flow to the substrate, as they do in a bulk device. This hole current produces an increase of the film potential accompanied by several unwanted effects. The first of these effects is the so-called kink effect in the output characteristics of a partially depleted device (Fig. 5.17). This effect is self-limiting because

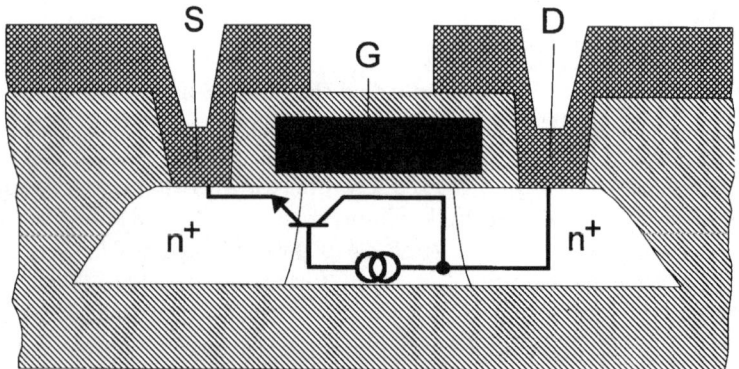

Figure 5.16 Sketch of an SOI MOSFET to demonstrate the intrinsic bipolar transistor.

the current gain of the bipolar transistor decreases with a further increase of the film potential. The kink effect is greatly reduced in fully depleted devices because the potential barrier from the channel to the source contact is reduced.

The second unwanted effect which may be observed in fully depleted devices is an abnormal subthreshold behavior and subsequent self-latching or snapback (Fig. 5.18). A similar explanation may be offered for this phenomenon.

The bipolar current gain, which is significantly higher at low gate voltages, leads to an abnormal increase of the drain current in the subthreshold regime. This is shown in the transfer characteristics at a drain voltage of $V_{ds} = 5$ V (Fig. 5.18). If the drain voltage is increased, a hysteresis may be observed between the switch-on and switch-off behavior. This is demonstrated for a drain voltage of $V_{ds} = 6.7$ V (Fig. 5.18). With a further increase of the drain voltage, a state will be reached when the injected carriers can no longer be controlled by the gate voltage. Then the device can no longer be turned off by decreasing the gate voltage. This effect is commonly called the self-latching or snapback effect.

The bipolar effect in SOI MOSFETs has given rise to a number of very detailed and theoretical analyses which may be found in the literature [31, 32, 35–38]. Note that the bipolar action also has a major effect on the off-state leakage current, even if the drain current is still controlled by the gate voltage.

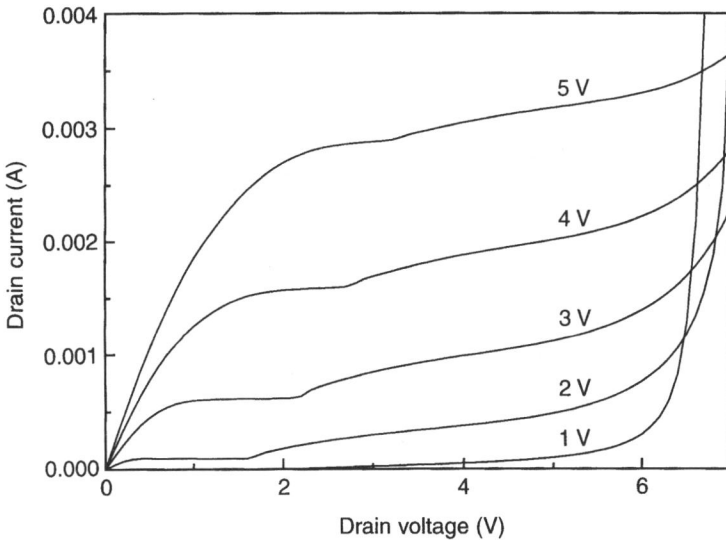

Figure 5.17 Output characteristics of a partially depleted n-channel device showing the kink effect.

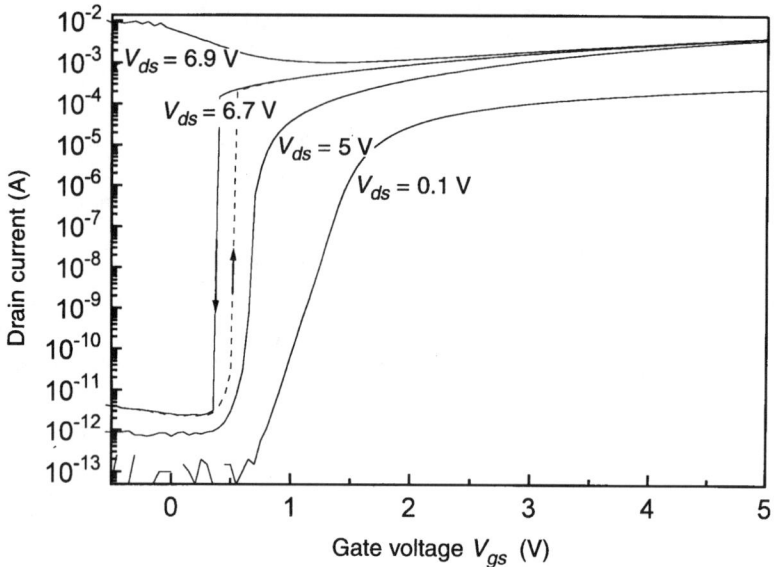

Figure 5.18 Weak inversion characteristics of an n-channel SOI device for different drain voltages.

A further disadvantage related to the bipolar action concerns the reduced breakdown voltage of the SOI MOSFETs. Figure 5.19 compares the breakdown voltage of bulk and SOI devices with different channel lengths. The advantages of the bulk devices are clear. Simulations [39], theoretical studies [40] and experiments reach the same conclusion, namely that the reduced breakdown voltage is caused by the intrinsic bipolar transistor. Only marginal improvements may be made by changing process parameters, such as the silicon film thickness, lightly doped source region, the gate oxide thickness or silicided source–drain regions.

In order to reach a reliable value for the breakdown voltage and to get rid of the kink effect for analog circuit design, the bipolar effect must be significantly reduced. This is achieved by changing the layout of the transistor in such a way that the floating body can be contacted. So-called body ties may be created without any additional process steps. As shown in Fig. 5.20, small stripes are introduced into the source area which are doped with the opposite type. The opposite doping for n-channel devices may be performed at the same time as the drain–source implant for the p-channel devices, and vice versa. The kink effect and the snapback effect can be suppressed with these split-source transistors. The improvements in the breakdown voltage are illustrated in Fig. 5.21. In the case of fully depleted devices, the high resistivity of the channel region limits the efficiency of the

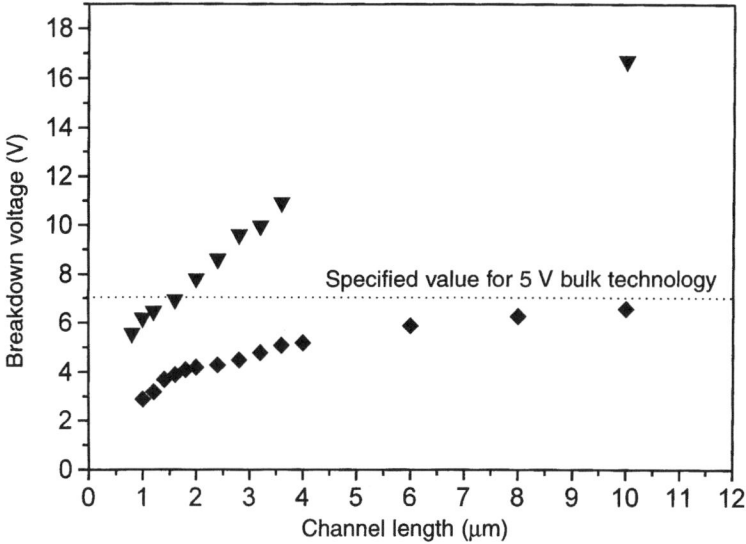

Figure 5.19 Comparison of the breakdown voltages for (▼) bulk and (◆) SOI n-channel devices as a function of the channel length.

splits. Taking this into account, design rules have to be developed which address the maximum distance between adjacent body ties, in order to assure a sufficiently high breakdown voltage.

Drain engineering is required for a further improvement in breakdown voltage. Concepts which are used in submicron bulk technologies should be introduced for devices with channel lengths greater than $1\,\mu$m, in order to achieve a comparable breakdown behavior. The low doped region near the source contact reduces the gain of the intrinsic bipolar transistor. However, a trade-off against the drive current must nevertheless be accepted.

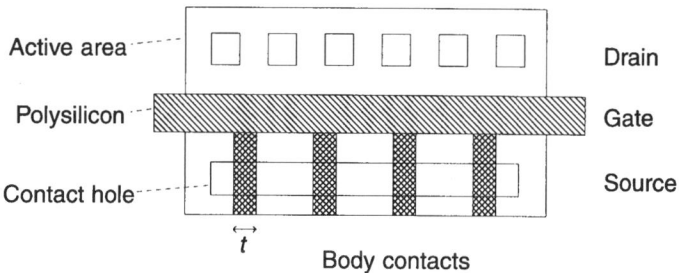

Figure 5.20 Sketch of a split-source transistor layout.

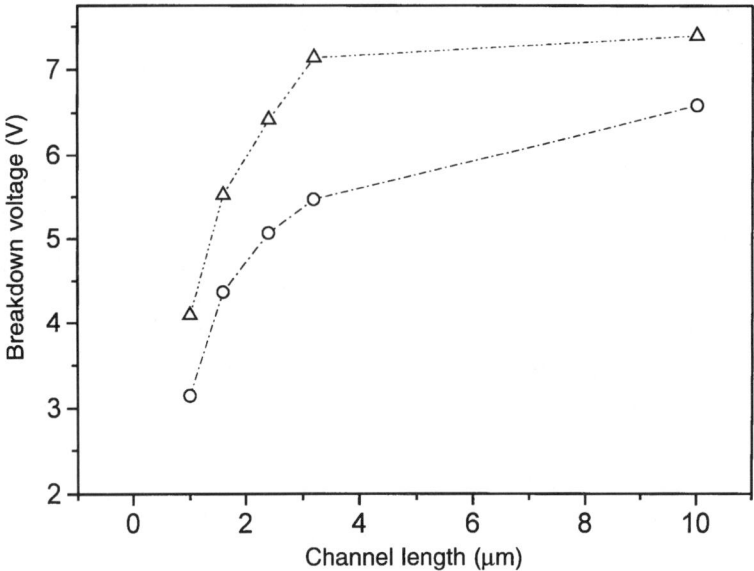

Figure 5.21 Breakdown voltages of n-channel devices (△) with and (○) without split sources for different channel lengths.

LDD devices can be used for supply voltages up to 5 V. But a new concept has to be developed for higher supply voltages, recommended for advanced analog sensor readout electronics. One additional lithography step and one implantation allow the introduction of a low doped drift region between the gate edge and the highly doped drain contact. The improvement in the breakdown voltage depends on the length of such a drain extension (Fig. 5.22). As illustrated, breakdown voltages up to 40 V can be achieved for devices with a channel length of 4 μm. A trade-off between the desired breakdown voltage and the on-resistance of the device must be accepted. The output characteristics of an n-channel device with a channel length of 10 μm is shown in Fig. 5.23. As may be seen, no kink effect occurs even with a drain–source voltage of up to 30 V.

Generally speaking, the described drain engineering has to be made only for n-channel devices. The ionization coefficient for holes is smaller than that for electrons by about a factor of 2, so that for p-channel devices no drain engineering is required in the power supply range up to 10 V. The ionization coefficients and the mobility of the charge carriers decrease with increasing temperature. This means that the base current for the intrinsic bipolar transistor also decreases, which reduces the intrinsic bipolar action at higher temperatures. Figure 5.24 shows the almost constant breakdown voltage in the temperature range from room temperature up to 300 °C.

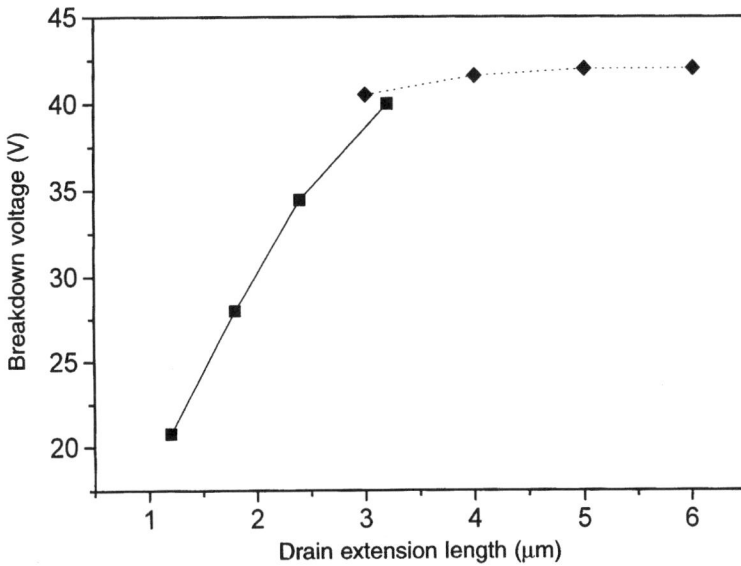

Figure 5.22 Breakdown voltages of an n-channel device with a split-source layout and an additional drain extension: (■) $L = 4\,\mu m$ and (◆) $L = 10\,\mu m$.

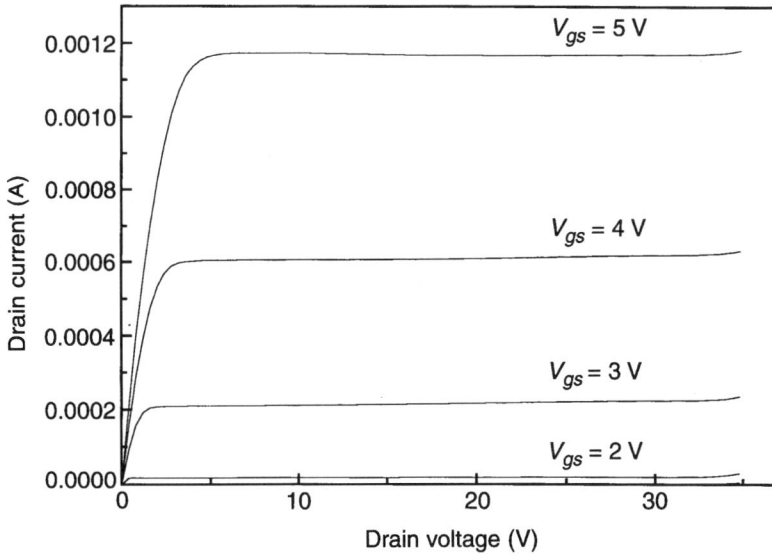

Figure 5.23 Output characteristics of an n-channel high voltage device with a channel length $L = 10\,\mu m$; extension length $= 3.6\,\mu m$.

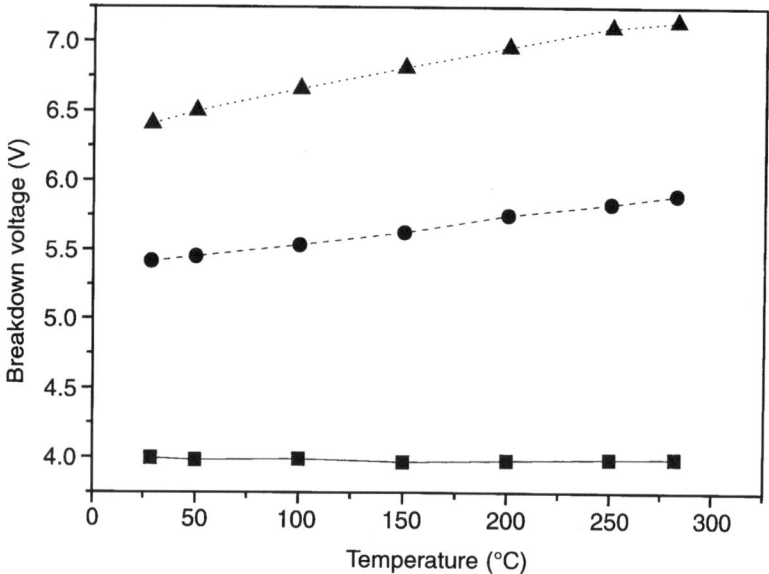

Figure 5.24 Temperature dependence of the n-channel breakdown voltage with and without split-source layout: (■) $L = 1.2\,\mu$m, without; (●) $L = 3.2\,\mu$m, without; and (▲) $L = 3.2\,\mu$m, with.

5.5 DESIGNING ANALOG CIRCUITS

The previous sections described the specific characteristics of SOI devices and compared them to bulk silicon devices. This section summarizes the devices which may be used in a CMOS SOI process and outlines their capabilities.

CMOS transistors with conventional bulk layout can only be used for significantly low supply voltages, since the intrinsic bipolar effects limit these capabilities. CMOS transistors with body contacts like split-source – or H-gate – transistors (Fig. 5.25) can be used to suppress the kink effect and improve the breakdown voltage. The split-source transistor, however, is no longer symmetric with respect to a change of source and drain. This is further accentuated if an additional drain extension is used.

A bidirectional function may be required but the number of applications requiring bidirectionality is usually small. For those applications which do require it, devices with an H-shaped gate structure may be used. Additional wiring is needed if the sidewall contacts should be fixed to a certain potential; extra wiring will reduce the circuit density. The efficiency of the sidewall contacts decreases with the increasing width of the devices. Therefore, transistors with a large channel width can only be realized as a mesh of many elementary transistors by increasing the wiring effort.

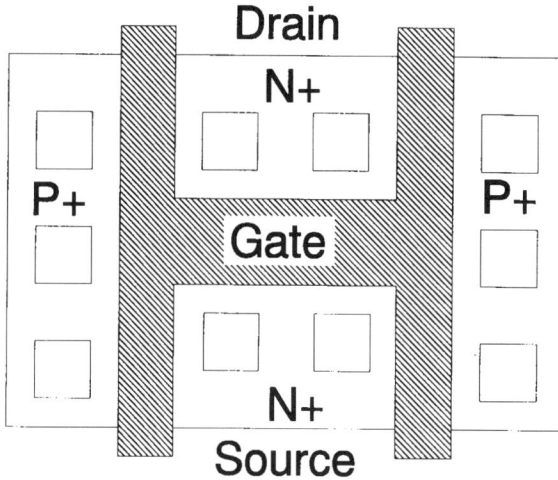

Figure 5.25 The layout of an H-gate transistor.

Devices with body ties and drift regions become necessary for the range above 5 V. They allow kink-free operation up to breakdown voltages of 40 V and can also be operated in the temperature range up to 300 °C. A lot of additional options then become available such as EEPROM cells, high voltage input and output stages.

The drain – source implants of the n- and p-channel devices allow the integration of PIN and Zener diodes. An n^+ and a p^+ area directly connected realize a Zener diode with $V_{Zener} \approx 5.1$ V and a considerably low temperature coefficient of 65 ppm K^{-1}. Lateral bipolar transistors have been realized in a thin silicon film and can be used at high temperature.

Passive components are needed for analog design. Capacitors for switched-capacitor circuits can be realized with the silicon film and the polysilicon gate layer serving as the electrodes and the gate oxide serving as the dielectric. For a voltage-independent capacitance, an additional lithography step and an implant have to be implemented in the process flow of section 5.3.

Resistors in SOI technology have the advantage that they are free of any leakage current and can be provided from $14\,\Omega/\square$ up to several k$\Omega/\square$. Attention has to be paid to the back gate voltage dependence of low doped resistors. A summary of some particularly exotic devices which may be realized in thin silicon films is provided in Colinge [2].

5.6 SOI CIRCUITS AT HIGH TEMPERATURE

The excellent performance of thin film devices makes SOI CMOS technology suitable for high temperature applications. It is reported that the digital cells like inverters, ANDs, NANDs and NORs are fully functional up to 320 °C [24]. The digital cells based on the described 12 V technology could be operated up to 350 °C before severe performance degradation could be observed. The functionality of larger circuits such as 16K and 256K SRAMs has been demonstrated up to a temperature of 300 °C.

Building blocks for analog circuits like operational transconductance amplifiers (OTAs) have been examined. It has been demonstrated that the DC gain is almost entirely temperature independent from 25 °C up to 300 °C. The offset voltage of the amplifier remains in the range 0–2 mV for temperatures between 25 °C and 300 °C.

CMOS technology on SIMOX has already reached a considerably mature state. The first products – a quad operational amplifier and a quad analog switch circuit on SIMOX – have been announced by Honeywell and will be sold with a guaranteed lifetime of 5 years for 225 °C operation. These products are claimed to be operatable at 300 °C with reduced lifetime. Switched-capacitor circuits are under development using the described process. A seventh-order low-pass ladder filter has been published recently. The simplified schematic is shown in Fig. 5.26. The filter has a Bessel characteristic with its 3 dB frequency at 20 Hz at a clock rate of 100 kHz [41]. Figure 5.27 shows the Bode diagrams of the filter (solid line) and, as a reference point, the ideal LCR prototype ladder filter (dashed lines) at room temperature and at 300 °C. The switched-capacitor filter characteristic matches the characteristic of the ideal prototype. This is still valid at a temperature of 300 °C.

5.7 SUMMARY

A bulk CMOS process can easily be implemented on SOI substrates. Thin film SOI devices have several advantages for high temperature applications: the reduced shift of the threshold voltage, the almost ideal subthreshold

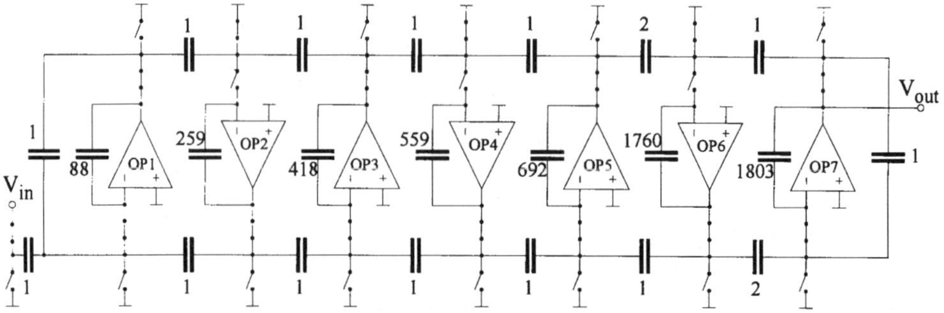

Figure 5.26 Schematic of a switched capacitor seventh-order low-pass filter.

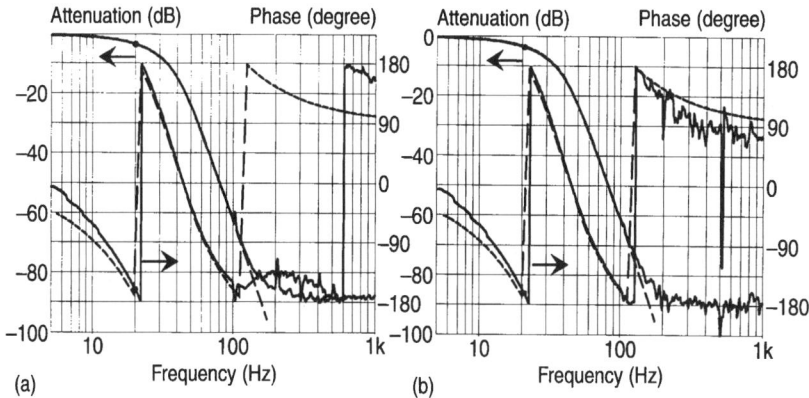

Figure 5.27 Bode diagram of a switched-capacitor ladder filter (a) at room temperature (25 °C) and (b) at 300 °C.

behavior and, most important, the lower leakage currents of the devices. The drawbacks of the SOI devices – lower breakdown voltages, the snapback and kink effects – can be overcome by the appropriate design of body ties. With the high voltage option, SOI circuits get interesting for automotive applications.

Optimization of the process requires a trade-off between device performance reliability and high temperature functionality. The reliability of the process is greatly enhanced by the use of refractory metal (tungsten) for interconnecting layers. ESD protection for SOI circuits has been successfully demonstrated [42]. Issues related to packaging still require a lot of work.

SOI CMOS technology has demonstrated the ability for circuit operation at temperatures well above the limits for conventional bulk silicon circuits. The higher costs for the SOI substrates have to be considered and balanced against the increased high temperature performance and the gain in process simplification.

REFERENCES

1. Maxwell, D.A., Beeson, R.H. and Allison, D.F. (1965) The minimization of parasitics in integrated circuits by dielectric insulation. *IEEE Trans. on Electron Devices*, **12**(1), 20–25.
2. Colinge, J.-P. (1991) *Silicon-on-Insulator Technology: Materials to VLSI*, Kluwer Academic, Boston.
3. Mitani, K. and Gösele, U.M. (1992) Wafer bonding technology for silicon-on-insulator technology for silicon-on-insulator applications: a review. *J. Electron. Materials*, **21**(7), 669–676.
4. Stengl, R., Ahn, K.Y. and Gösele, U. (1988) Bubble-free silicon wafer bonding in a non-cleanroom environment. *Jpn. J. Appl. Phys.*, **27**(12), 2364–2366.

5. Abe, T., Ohki, K., Mitani, K., Yoshizawa, K. and Nakazota, Y. (1993) Bonded SOI wafers with various substrates for ULSI use. *ECS Spring Meeting Extended Abstracts*, **93**(1), 1179.
6. Maszara, W.P. (1990) SOI by wafer bonding: a review. *Silicon-on-Insulator Technology and Devices*, **90**(6), 199–212.
7. Fleming, J.G., Roherty-Osmun, E. and Godshall, N.A. (1992) Low temperature, high strength, wafer-to-wafer bonding. *J. Electrochem. Soc.*, **139**(11), 3300–3302.
8. Benggtsson, S. (1992) Semiconductor wafer bonding: a review of interfacial properties and applications. *J. Electron. Mater.*, **21**(8), 841–862.
9. Matsushita, T., Satoh, H., Shimanoe, M., Nieda, A., Hashimoto, M., Ogasawara, A., Yamagishi, M., Yagi, A., Saitoh, Y. and Sakai, S. (1989) High-quality thin-film SOI technology using wafer bonding and selective polishing for VLSIs. *IEEE Trans. on Electron Devices*, **36**(11), 2621.
10. Wada, S., Takahashi, S. and Hayashi, Y. (1992) Silicon device thinning using preferential polishing: progress in flatness and electrical properties. *Semicond. Sci. Technol.*, 7, A243–A248.
11. Arimoto, Y., Horie, H., Higaki, N., Kojima, M., Sugimoto, F. and Ito, T. (1993) Advanced metal oxide semiconductor and bipolar devices on bonded silicon-on-insulators. *J. Electrochem. Soc.*, **140**(4), 1138–1143.
12. Maszara, W.P. (1992) Semiconductor wafer bonding: an overview. *Semiconductor Wafer Bonding: Science, Technology and Applications*, **92**(7), 3–17.
13. Mumola, P.B., Gardopee, G.J., Clapsis, P.J., Zarowin, C.B., Bollinger, L.D. and Ledger, A.M. (1992) Plasma thinned SOI bonded wafers, in *Proc. IEEE SOS/SOI Technology Conf.*, pp. 152–153.
14. Gassel, H. (1995) *Untersuchung der SIMOX-Technologie zur Erzeugung anwendungsspezifischer SOI-Substrate*, Shaker, Aachen.
15. Post, I.R., Jerome, R.C., Williams, D.R., Pawlikiewicz, A.H. and Steidinger, G.L. (1994) High temperature (21–300 °C) electrical characterization of NPN and PNP polysilicon emitter bipolar transistors for a high voltage (35 V) analog technology with SOI/trench isolation, in *Trans. 2nd Int. High Temperature Electronics Conf.*, Vol. I, pp. II9–II15.
16. Elewa, T. *et al.* (1992) Detailed analysis of edge effects in SIMOX–MOS transistors. *IEEE Trans. on Electron Devices*, **39**(4), 874–881.
17. Werner, R., Burbach, G., Leiberg, W., Lukat, K. and Dreizner, A. (1995) Wolframmetallisierung zum Einsatz bei hohen Temperaturen, in *Proc. GME-Fachtagung Baden-Baden*, VDE-Verlag, Offenbach.
18. Lim, H. and Fossum, J.G. (1983) Threshold voltage of thin-film silicon-on-insulator (SOI) MOSFETs. *IEEE Trans. on Electron Devices*, **30**(10), 1244–1251.
19. Abel, H.B. (1989) A general model of the thin-film SOI–MOSFET, in *Proc. IEEE SOI/SOS Conf.*, pp. 105–106.
20. Omura, Y., Nakashima, S., Izumi, K. and Ishii, T. (1991) 0.1 μm gate, ultrathin-film CMOS devices using SIMOX substrate with 80 nm thick buried oxide layer. *Tech. Dig. IEEE Int. Electron Devices Meeting*, p. 675.
21. Sze, S.M. (1981) *Physics of Semiconductor Devices*, 2nd edn, Wiley, New York.
22. Groeseneken, G., Colinge, J.P., Maes, H.E., Alderman, J.C. and Holt, S. (1990) Temperature dependence of threshold voltage in thin-film SOI MOSFETs. *IEEE Electron Device Letters*, **11**(8), 329–331.
23. Sherony, M.J., Su, L.T., Chung, J.E. and Antoniadis, D.A. (1994) Minimization of threshold voltage variation in SOI MOSFETs, in *Proc. IEEE Int. SOI Conf.*, pp. 131–132.
24. Francis, P., Terao, A., Gentinne, B., Flandre, P. and Colinge, J.P. (1992) SOI technology for high temperature applications. *Tech. Dig. IEDM 1992*, pp. 310–312.
25. Seegebrecht, P. (1992) High temperature operation of SOI devices and circuits, in *Proc. Euroform Seminar: High Temperature Semiconductor Electronics, Darmstadt*.
26. Grzybowski, R.R. (1993) High temperature testing of SOI devices to 400 °C, in *Proc. IEEE SOI Conf.*, pp. 176–177.
27. Tyson, S.M. and Grzybowski, R.R. (1994) High temperature characteristics of silicon-on-insulator and bulk silicon devices to 500 °C, in *Proc. 2nd High Temperature Electronics Conf.*, Vol. II, p. 9.
28. Karulkar, P.C. (1993) Ultra-thin SOI–MOSFETs at high temperature, in *Proc. IEEE SOI Conf.*, pp. 136–137.
29. Tack, M. and Claeys, C. (1990) The mobility and transconductance behaviour in thin film SOI transistors. *Silicon-on-Insulator Technology and Devices*, **90**(6), 532–544.

30. Tihany, J. and Schlötterer, H. (1975) Influence of the floating substrate potential on the characteristics of ESFI MOS transistors. *Solid State Electronics*, **18**, 309–314.
31. Haond, M. and Colinge, J.P. (1989) Analysis of drain breakdown-voltage in SOI n-channel MOSFETs. *Electronics Letters*, **25**(24), 1640–1641.
32. Choi, J. and Fossum, J.G. (1991) Analysis and control of floating-body bipolar effects in fully depleted submicrometer SOI MOSFETs. *IEEE Trans. on Electron Devices*, **38**(6), 1384–1391.
33. Fossum, J.G. (1990) SOI design for competitive CMOS VLSI. *IEEE Trans. on Electron Devices*, **37**(3), 724–729.
34. Young, K.K. (1988) Avalanche-induced drain–source breakdown in silicon-on-insulator n-MOSFETs. *IEEE Trans. on Electron Devices*, **35**(4), 426–431.
35. Davis, J.R. *et al.* (1986) *IEEE Electron Device Letters*, **7**, 570.
36. Chen, C.E.D. *et al.* (1988) A single transistor latch in SOI MOSFETs. *IEEE Electron Device Letters*, **9**, 636.
37. Huang, J.S.T. (1992) Modeling of output snap-back characteristics in n-channel SOI MOSFETs. *IEEE Trans. on Electron Devices*, **39**(5), 1170–1178.
38. McKitterick, J.B. (1994), The floating body in SOI, in *Proc. 6th Int. Symp. on Silicon-on-Insulator Technology and Devices*, pp. 278–289.
39. Smeys, P. and Colinge, J.P. (1992) Analysis of drain breakdown voltage in enhancement-mode SOI MOSFETs. *Solid State Electronics*, **36**(4), 569–573.
40. Kistler, N. (1992) Dependence of fully depleted SOI MOSFET breakdown voltage on film thickness and channel length, in *Proc. IEEE SOI/SOS Conf.*, pp. 128–129.
41. Verbeck, M., Bahr, J., Burbach, G., Fiedler, H.L., Werner, R. and Zimmermann, C. (1995) Hochtemperaturtaugliche Mikrosysteme in SOI-Technologie, in *Proc. GME-Fachtagung, Baden-Baden*, VDE-Verlag, Offenbach.
42. Verhaege, K., Groeseneken, G. and Maes, H.E. (1992) Analysis of snapback in SOI mMOSFETs and its use for an SOI ESD protection circuit, in *Proc. IEEE Int. SOI Conf. 1992*, pp. 140–141.
43. Sato, T., Iwamura, J., Tango, H. and Doi, K. (1984) CMOS/SOS VLSI technologies. *Mat. Res. Soc. Symp. Proc.*, **33**, 25–34.
44. Lawrence, R.K. and Hughes, H.L. (1989) Technology comparison of interface states as measured by charge pumping: bulk vs. SIMOX vs. SOS, in *Proc. IEEE SOS/SOI Technology Conf.*, pp. 89–90.
45. Vasudev, P.D. and Mayer, D.C. (1984) Characterization of CMOS devices in $0.5\,\mu m$ silicon-on-sapphire films modified by solid phase epitaxy and regrowth (SPEAR). *Mat. Res. Soc. Symp. Proc.*, **33**, 35–39.
46. Landstrass, M.I. (1990) Diamond based silicon-on-insulator structure, in *Proc. IEEE SOS/SOI Technology Conf.*, p. 128; Landstrass, M.I. and Fleetwood, D.M. (1990) Total dose radiation hardness of diamond-based silicon-on-insulator structures. *Appl. Phys. Lett.*, **56**(23), 2316–2318.
47. Karulkar, P.C. and Annamalai, N.K. (1992) Silicon on diamond (SOD) MOSFETs, in *Proc. IEEE SOS/SOI Technology Conf.*, pp. 108–109.
48. Jastrzebski, L., Corboy, J.F. and Soydan, R. (1989) Issues and problems involved in selective epitaxial growth of silicon for SOI fabrication. *J. Electrochem. Soc.*, **136**(11), 3506–3513.
49. Ishiwara, H. (1990) Current status of solid phase epitaxy, in *Silicon-on-Insulator Technology and Devices*, **90**(6), 72–82.
50. Ogura, A., Furuya, A. and Koh, R. (1993) 50 nm thick silicon-on-insulator fabrication by advanced epitaxial lateral overgrowth: tunnel epitaxy. *J. Electrochem. Soc.*, **140**(4), 1125–1130.
51. Imai, K. and Unno, H. (1984) FIPOS (full isolation by porous oxidized silicon) technology and its application to LSIs. *IEEE Trans. on Electron Devices*, **31**(3), 297–302.
52. Hisamoto, D., Kaga, T. and Takeda, E. (1991) Impact of the vertical SOI 'DELTA' structure of planar device technology. *IEEE Trans. on Electron Devices*, **38**(6), 1419–1424.
53. van der Wel, W., Buchner, R., Haberger, K., Seitz, S., Weber, J. and Seegebrecht, P. (1991) Avoidance of substrate damage upon laser recrystallization of an SOI layer. *J. Electrochem. Soc.*, **138**(4), 1117–1122.
54. Takao, Y., Shimada, H., Suzuki, N., Matsukawa, Y. and Sasaki, N. (1992) Low-power and high-stability SRAM technology using a laser-recrystallized p-channel SOI MOSFET. *IEEE Trans. on Electron Devices*, **39**(9), 2147–2152.

55. Zavracky, P.M. (1990) ISE technology: a flexible SOI solution. *Silicon-on-Insulator Technology and Devices*, **90**(6), 49–60.
56. Liu, L., Tsien, P.-H. and Li, Z. (1990) Defect-free silicon-film on SiO$_2$ formed by zone melting recrystallization with high scanning speed. *IEEE Trans. on Electron Devices*, **37**(4), 952–957.
57. Beasom, J.D. (1973) A process for simultaneous fabrication of vertical NPN and PNPs, Nch and Pch MOS devices. *Int. Electron Devices Meet. Tech. Dig.*, pp. 41–43.
58. Itoh, S., Usui, T., Akahane, K., Ishikawa, N., Yokoyama, T. and Maeda, Y. (1990) High speed polysilicon deposition process for dielectric isolation technology, in *Proc. 2nd Int. IEEE Symp. on Power Semiconductor Devices and ICs*, pp. 174–179.
59. Haisma, J., Spierings, G.A.C.M., Biermann, U.K.P. and Pals, J.A. (1989) Silicon-on-insulator wafer bonding–wafer thinning technological evaluations. *Jpn. J. Appl. Phys.*, **28**(8), 1426–1443.
60. Abe, T., Nakano, M. and Itoh, T. (1990) Silicon wafer-bonding process technology for SOI structures. *Silicon-on-Insulator Technology and Devices*, **90**(6), 61–71.
61. Izumi, K., Doken, M. and Ariyoshi, H. (1978) CMOS devices fabricated on buried SiO$_2$ layers formed by oxygen implantation into silicon. *Electron. Lett.*, **14**(18), 593–594.
62. Guerra, M.A. (1990) The status of SIMOX technology. *Silicon-on-Insulator Technology and Devices*, **90**(6), 21–47.
63. Vogt, H. (1986) Hochdosisimplantation von Stickstoff in Silizium: Eine Technik zur Erzeugung von dielektrisch isolierten CMOS-Bauelementen mit hoher Integrationsdichte, PhD Thesis, Universität Dortmund.
64. Belz, J. (1986) Elektronenmikroskopische Charakterisierung von durch Ionenimplantation synthetisiertem Siliziumnitrid, PhD Thesis, Universität Dortmund.
65. Danilin, A.B., Drakin, K.A., Kukin, V.V., Malinin, A.A., Mordkovich, V.N., Petrov, A.F., Saraykin, V.V. and Vyletalina, O.I. (1991) Peculiarities of buried silicon oxynitride layer synthesis by sequential oxygen and nitrogen ion implantation in silicon. *Nuclear Instr. & Meth. B*, **58**, 191–193.
66. De Veirman, A., Van Landuyt, J. and Skorupa, W. (1991) TEM study of combined oxygen and nitrogen implanted silicon. *Philosophical Magazine A*, **64**(3), 513–531.

PART THREE

Gallium Arsenide

High temperature electronics based on compound semiconductors 6

H. L. Hartnagel

6.1 INTRODUCTION

High temperature electronic circuitry can be based on semiconducting materials with wide energy gap. It is therefore also of interest to consider the wide field of compound semiconductors. The particular advantage here is that single-crystal heterostructure transitions can enhance the electronic capability at high temperatures. Lattice-matched or nearly matched sandwiches with large energy gaps are relevant here, especially if the energy offsets for conduction or valence band are also large.

6.2 ENERGY GAP DATA AND OFFSET VALUES

Let us begin by reviewing the energy gap data and then the offset values when considering the transition from one semiconductor to the next. A selection of III−V compound semiconductors is given in Table 6.1 and Fig. 6.1. Here the gap values are given for the conduction band minima at the points Γ, L and X in the momentum space for the crystalline structure (Fig. 6.2) and in greater detail as energy contours for GaAs and AlAs (Fig. 6.3).

If we mix these two semiconductors to produce $Al_xGa_{1-x}As$, the three valleys at Γ, X and L depend on the mixing ratio x (Fig. 6.4). This is, however, only a small part of the many details required for the full energy

High Temperature Electronics. Edited by M. Willander and H.L. Hartnagel. Published in 1997 by Chapman & Hall, London. ISBN 0 412 62510 5.

Table 6.1 Bandgap and lattice data ($T = 300\,\mathrm{K}$)

Material	Lattice constant (nm)	$E_g(\Gamma)$ (eV)	$E_g(L)$ (eV)	$E_g(X)$ (eV)
AlP	0.5463 5	3.6	3.5	2.45
AlAs	0.5660 5	3.00	2.36	2.15
AlSb	0.6135 8	2.3	2.21	1.612
GaP	0.5451 0	2.78	2.6	2.272
GaAs	0.5653 25	1.424	1.708	1.900
GaSb	0.6096 02	0.725	0.810	1.032
InP	0.5868 9	1.350	1.95	2.15
InAs	0.6058 3	0.355	1.43	2.0
InSb	0.6479 43	0.170	1.0	1.7
Si	0.5431 1	4.1	2.0	1.124
Ge	0.5657 9	0.805	0.666	0.89

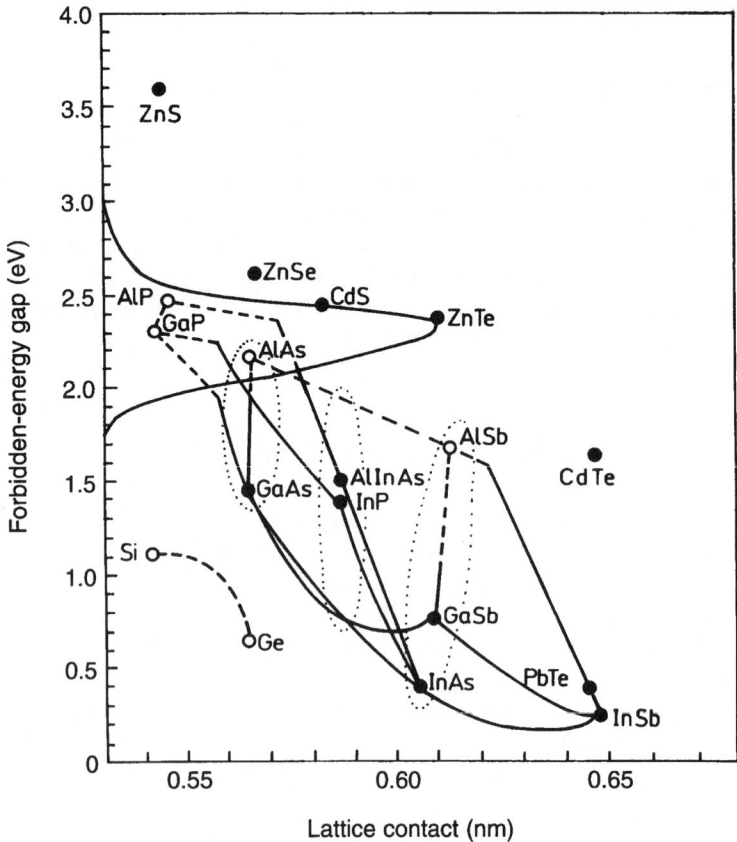

Figure 6.1 Bandgap data as a function of lattice constant: GaN (3.4 eV, 0.451 nm) SiC (hex) (-3 eV, 0.308 nm); (●) direct and (○) indirect.

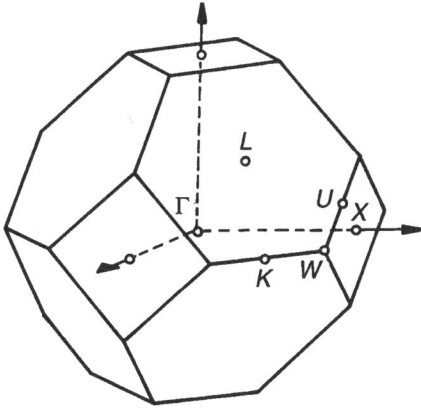

Figure 6.2 First Brillouin zone of the lattice of GaAs and AlAs.

contour information when going through the mixing ratios. A useful first approach is to extrapolate between the two end figures like Fig. 6.3a and b for $x = 0$ and $x = 1$ in $Al_xGa_{1-x}As$. Similar extrapolation approaches are often useful for ternary or even quaternary mixtures like InGaAsP.

The information of bandgap versus lattice constant is also available for II–VI compound semiconductors (Fig. 6.5). Some typical examples of

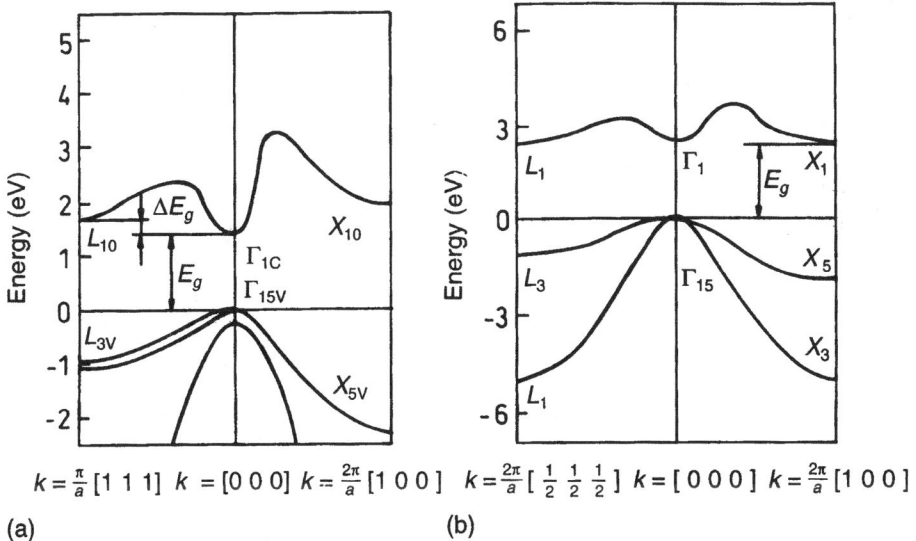

$$k = \frac{\pi}{a}[1\ 1\ 1] \quad k = [0\ 0\ 0] \quad k = \frac{2\pi}{a}[1\ 0\ 0] \qquad k = \frac{2\pi}{a}[\tfrac{1}{2}\ \tfrac{1}{2}\ \tfrac{1}{2}] \quad k = [0\ 0\ 0] \quad k = \frac{2\pi}{a}[1\ 0\ 0]$$

(a)   (b)

Figure 6.3 Band structure of (a) GaAs and (b) AlAs.

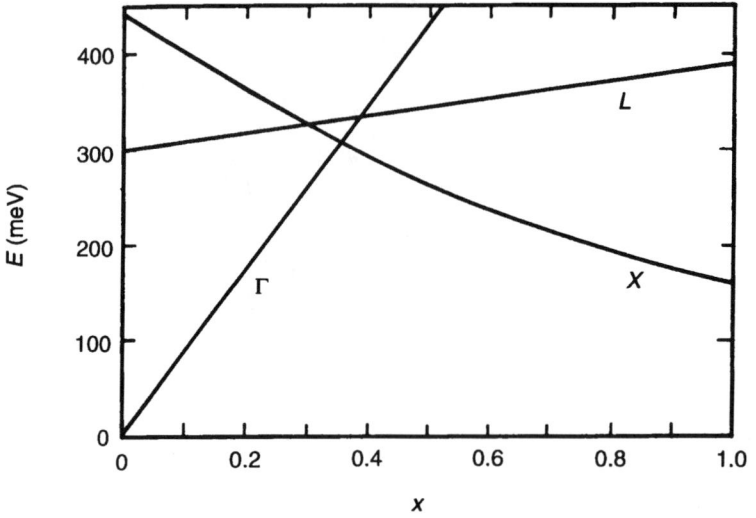

Figure 6.4 Energy minima for $Al_xGa_{1-x}As$ compositions X, L and Γ as a function of x.

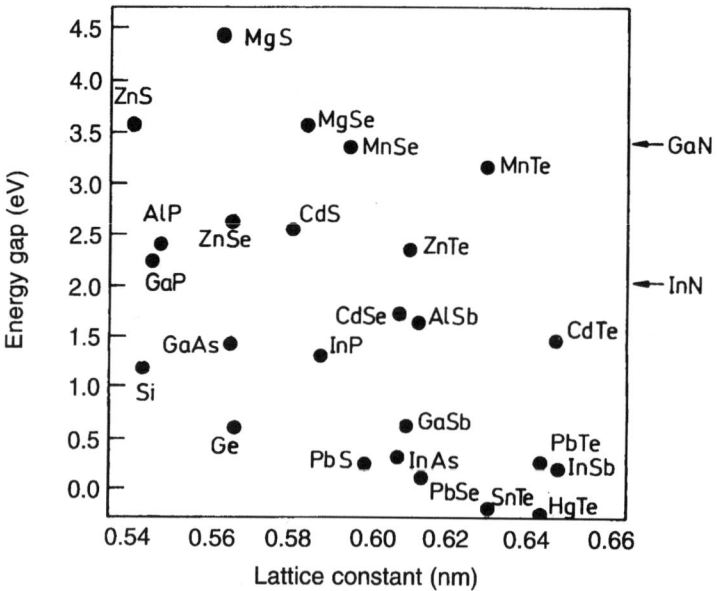

Figure 6.5 Bandgap versus lattice constant for II–VI wide gap semiconductors.

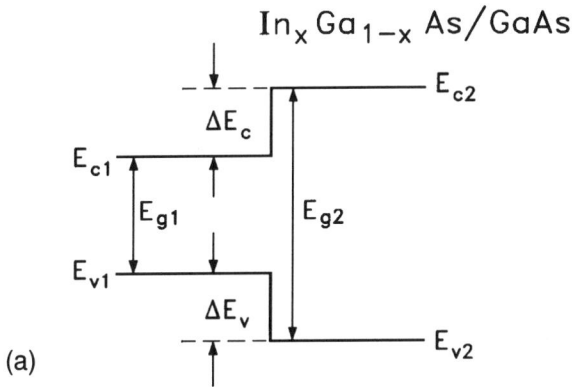

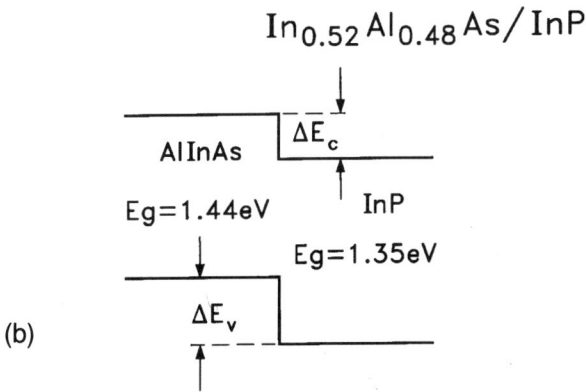

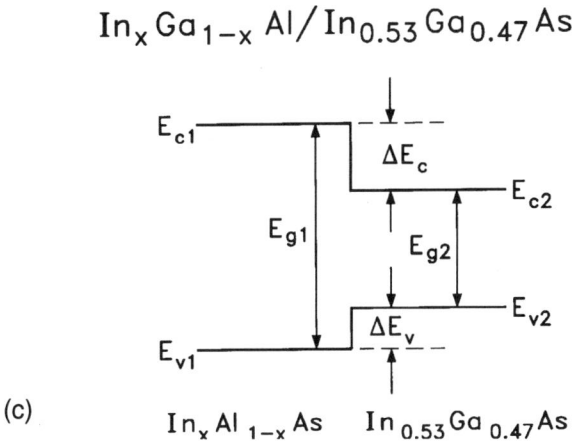

Figure 6.6 Typical examples of bandgap alignment for heterostructure transitions: (a) In_xGa_{1-x} on GaAs; (b) $In_{0.52}Al_{0.48}As$ on InP (note the offset ΔE_v is now in the other direction and therefore of negative type); (c) $In_xGa_{1-x}Al$ on $In_{0.53}Ga_{0.47}As$.

bandgap alignment for heterostructure transitions are given in Fig. 6.6. A more systematic description of the important offset data is given in Fig. 6.7 and Table 6.2. This type of information is only partly presented by a number of relevant handbooks; recent offset data is given in Tiwari and Frank [1]. One can clearly see from Fig. 6.1 that there are the following families of heterostructures based on the three types of commercially available substrate materials: GaAs, InP and GaSb. This is also the basis of the listing in Table 6.2.

It is of course possible to accept a certain mismatch of lattice constants, provided that either a superlattice structure is employed, consisting of a sequence of several thin films (typically 10 nm thick, about 20 atomic layers) of alternating material, or that the active film is only very thin and remains severely stressed. In the latter case, it can be expected that the electronic properties are slightly modified due to the compression or enlargement of the resulting lattice of this stressed layer. This means that epitaxial layers can be grown on large wafers of cheap material such as Si, GaAs or InP. Lattice distortion produces for electron diffraction the type of pattern normally seen with amorphous material. Therefore this slightly lattice-mismatched structure is also called pseudomorphic layers. After all, the epitaxial layer is crystalline but only distorted.

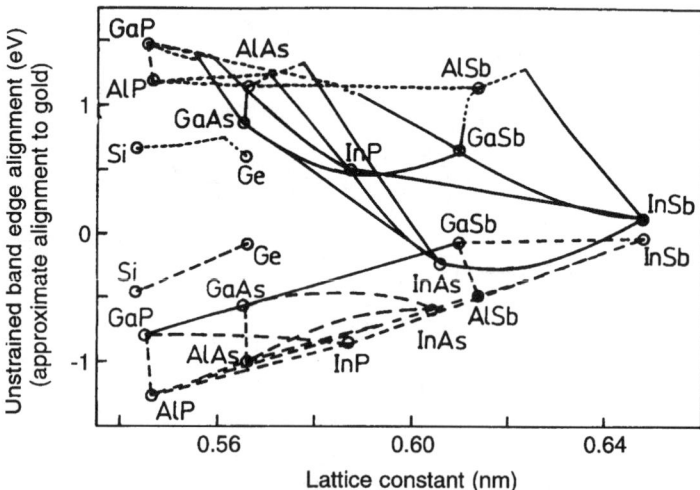

Figure 6.7 The conduction band edge and valence band edge energies plotted as a function of the lattice constant of semiconductors at $T = 300$ K. The circles indicate the band edges of the binary semiconductors and the lines show the band edges of the ternary alloys. The two endpoints of each ternary line are the binary constituents of that ternary. Discontinuities between two lattice-matched or nearly matched semiconductor alloys may be found from the difference in energy between their band edge energies. The point of zero energy represents the approximate gold Schottky barrier position in the bandgap of any given alloy [1]. Key to symbols: (—) $E_c(\Gamma)$, (----) $E_c(L)$, (———) $E_c(X$ or $\Delta)$ and (———) E_v.

Table 6.2 Discontinuity data[a]

Lattice constant of group (nm)	Material system	ΔE_g (eV)	ΔE_c (eV)	ΔE_v (eV)
0.544	GaP/Si	1.148	0.798	0.35
0.566	GaAs/Ge	0.758	0.268	0.49
	AlAs/Ge	1.48	0.55	0.93
	$Ga_{0.7}Al_{0.3}As/GaAs$	0.395	0.263	0.132
	AlAs/GaAs	0.73	0.29	0.44
	$Ga_{0.51}In_{0.49}P/GaAs$	0.463	0.223	0.24
0.587	$Al_{0.48}In_{0.52}As/InP$	0.106	0.30	−0.194
	$InP/Ga_{0.47}In_{0.53}As$	0.600	0.20	0.40
	$Al_{0.48}In_{0.52}As/Ga_{0.47}In_{0.53}As$	0.706	0.50	0.206
0.610	AlSb/GaSb	0.887	0.487	0.40
	GaSb/InAs	0.370	0.88	−0.51
	AlSb/InAs	1.257	1.367	−0.11

[a]ΔE_g = bandgap difference; ΔE_c = conduction band offset; ΔE_v = valence band offset; temperature = 300 K.

Some of the traps leading to strong $1/f$ noise and other similar problems are associated with particular energetic positions. An example is the DX centre in AlGaAs. This trap is near the energy level of the X valley. The mixture of $Al_xGa_{1-x}As$ shows this trap to be effective down to $x = 0, 3$ (Fig. 6.4). This trap therefore limits the use of $Al_xGa_{1-x}As$ above $x = 0, 3$. Many of these types of detail are not yet known. However, sufficient information is available to design new device structures for the many applications of interest, particularly those suitable for operation at high ambient temperatures.

6.3 METALLIZATIONS AND INSULATOR DEPOSITS

Other considerations for high temperature electronic circuitry concern the relevant metallizations and insulator depositions such as used for ohmic, Schottky, heat-sinking and packaging structures for capping and capacitors. But high temperature stable resistors are also important, such as those fabricated by using NiCr thin films.

The metallizations must be chosen in such a manner that they remain stable and continue to operate at the elevated temperatures without any degradation for periods of at least 1000 h. It is also required that, after fabrication, no physical or chemical interaction is possible, such as the reaction of solid Au thin films on GaAs, producing deep spiking metal into the semiconductor [2]. This requires diffusion barriers such as WSi or LaB_6

[3–5]. It is necessary to have reasonable thermal expansion matching and good adhesion of the metal. All these features can only be realized by a complex sandwich of various metals. A wealth of experience is available in the literature [6–10].

The hetero- and homojunction transitions must not degrade during long-term application of elevated temperatures. Dopants and other compounds must be highly stable with respect to diffusion and its effects.

The total device structure, but particularly the free semiconductor surfaces with such volatile components as As, must be covered by a suitable capping layer such as Si_3N_4 deposited by plasma enhanced chemical vapor deposition (PECVD). This must be achieved without any damage to the underlying semiconductor and with excellent adhesion [11]. The normal gas-phase plasma reactor can cause difficulties here. It is usually required to keep the free GaAs surfaces as far away as possible from the plasma region with its high ion energies.

To avoid substrate leakage currents, relatively thick high bandgap material (such as $Al_xGa_{1-x}As$) buffer layers are useful between the substrate with its usual heat-sinking metallization requirements (this is easier on GaAs than on $Al_xGa_{1-x}As$) and the active layer. This is then a similar approach as silicon-on-insulator (SOI) structures obtained by deep oxygen implantation into the Si semiconductor, for which operating temperatures of around 200 °C are possible. Heterostructuring permits us to fabricate the various sensor configurations on the same chip in a monolithic approach due to the extreme etch selectivity possible with different heteroepitaxial layers.

Structuring can be made possible by suitable deep ion implantation. This may not only be due to the damage introduced by ion bombardment but also by material modification such as the production of GaN deep in the GaAs.

6.4   THERMAL PROPERTIES

It is clear, then, that an essential property is the absence of leakage currents due to thermal electron–hole pair generation for temperatures up to 500 °C. This means that the energy gap needs to be larger than 1.4 eV. However, the material also needs to remain stable. This requires that the vapor pressure of the volatile component remains low up to the operating temperatures. Phosphorus is therefore often not satisfactory; at the InP surface it may produce such a high vapor pressure that no simple capping layer (such as Si_3N_4) can be found that will resist being blown off at 400 °C. The situation is different with suitable monocrystalline heterostructures. Evaporation of one of the components leads to defect generation near the surface and therefore severe lifetime limitations of the devices.

Table 6.3 Compound materials and their parameters as required for high temperature considerations[a]

Material	E_G (eV)	Lattice constant (nm)	Melting point (°C)	Thermal expansion coefficient ($10^{-6} K^{-1}$)	Thermal conductivity ($W cm^{-1} K^{-1}$)
GaAs	1.424	0.5653	1238	6.0	0.54
AlAs	3.00	0.5660	1740	5.2	0.08
AlSb	2.3	0.6135	1080	4.88	0.56
GaSb	0.725	0.6096	712	6.7	0.33
InP	1.350	0.5868	1060	4.5	0.7
AlP	3.6	0.5463	2000	–	0.9
BN (cubic)	8.0	0.36	2700	3.5	–
BN (hex)	3.8	0.25	3000	40.5/−2.9	0.8
AlN	5.9	0.31	2400	4.0	0.3
GaN	3.4	0.32 (IIc axis)	600 (dissoc.)	5.6	–

[a]The GaN structure may be hexagonal or cubic.

A further consideration is a reasonable matching of the thermal expansion coefficients, particularly important for the large temperature range used in this field. Table 6.3 shows a selection of interesting materials with their parameters.

GaAs and $Al_x Ga_{1-x} As$ are relatively mature compound semiconductors. There are many properties which are only available with certain compound materials, such as strong piezoelectricity, light generation, HF performance, heterostructuring for strong etch selectivity and for numerous other possibilities.

Interesting examples are given in Fig. 6.8 where one can select heterostructure sandwiches in view of such sensor capabilities as the maximization of the thermal resistivity for high thermal gradients in order to achieve a high local potential due to the Seebeck effect. This is a typical case for a thin membrane sensing the absorbed radiation or indicating the cooling due to a gas passing along a heated resistor.

6.5 HIGH TEMPERATURE REQUIREMENTS

The device requirements concerning high temperature of operation are as follows: a conventional FET such as a MESFET requires (1) a temperature-stable transition between the active n layer and a nonleaking semi-insulating substrate and (2) an equally stable depletion layer modulation of this n region under the gate. An FET based on a two-dimensional electron

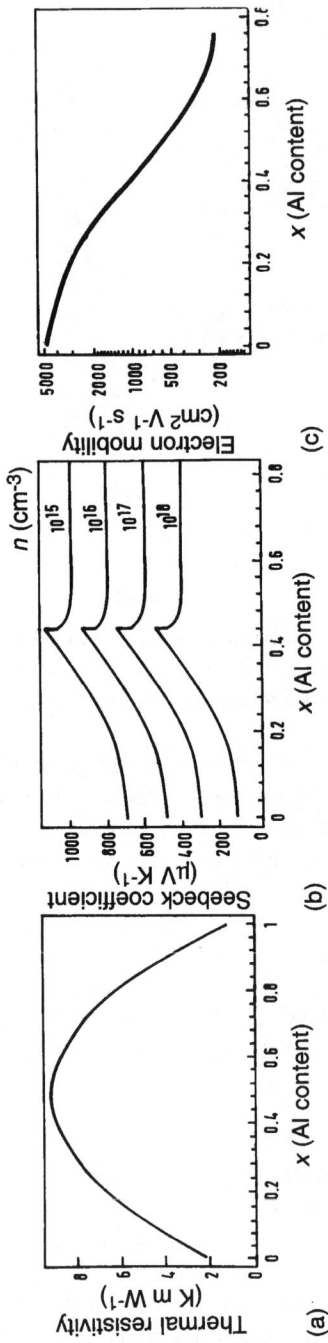

Figure 6.8 Material properties suitable for sensor optimization of $Al_xGa_{1-x}As$: (a) thermal resistivity, (b) Seebeck coefficient and (c) electron mobility.

(hole) gas requires a large conduction (valence) band offset so that the charge carriers do not discharge when high lattice temperatures exist. Finally, a hetero bipolar transistor (HBT) requires a large offset, so the majority carriers of the base do not discharge into the emitter. The frequently realized npn HBT thus needs a large valence band offset.

At high operating temperatures, HBTs experience easy generation of electron–hole pairs in the depletion layer of the reverse biased base–collector junction, leading to a nonsaturating collector current versus emitter–collector voltage. This can be avoided by employing a wide gap layer of the collector region underneath the gate. This is then the double-heterojunction HBT of the high temperature electronic circuitry.

There exist similar requirements for small leakage currents, even for high lattice temperatures, in transmission lines and passive components of high temperature integrated circuits. Layers having wide bandgaps are needed, or perhaps insulating films. Some of the subsequent chapters consider the relevant device concepts and the important field of metallization more systematically.

REFERENCES

1. Tiwari, S. and Frank, D.J. (1992) Empirical fit to band discontinuities and barrier-heights in III–V alloy systems. *Appl. Phys. Lett.*, **60**(5), 630–632.
2. Zanoni, E., Callegari, A., Canali, C., Fantini, F., Hartnagel, H.L., Magistrali, F., Paccagnella, A. and Vanzi, M. (1990) Metal GaAs interaction and contact degradation in microwave MESFETs. *Quality Reliab. Eng. Int.*, **6**, 29–46.
3. Böttner, T., Fricke, K., Goldhorn, A., Hartnagel, H.L., Rappl, A., Ritter, S. and Würfl, J. (1991) Technology and performance of a high temperature stable (up to 300 °C) operational amplifier on GaAs, in *Proc. 1st Int. High Temp. Electronics Conf.*, Albuquerque NM, June 16–20, 1991, pp. 77–82.
4. Würfl, J., Fricke, K. and Hartnagel, H.L. (1990) Nonalloyed, high temperature stable ohmic contacts to GaAs based on LaB$_6$ diffusion barriers, in *Proc. Int. Symp. GaAs and Related Compounds*, Jersey, UK, September 24–27, 1990, pp. 239–244.
5. Fricke, K., Hartnagel, H.L., Schütz, R., Schweeger, G. and Würfl, J. (1989) A new GaAs technology for stable FETs at 300 °C. *IEEE Electron Device Lett.*, **10**(12), 577–579.
6. Murakami, M., Price, W.H., Shih, Y.C., Braslau, N., Childs, K.D. and Parks, C.C. (1987) Thermally stable ohmic contacts to n-type GaAs, II: MoGeInW contact metal. *J. Appl. Phys.*, **62**(8), 3295–3303.
7. Murakami, M., Shih, Y.C., Price, W.H., Wilkie, E.L., Childs, K.D. and Parks, C.C. (1988) Thermally stable ohmic contacts to n-type GaAs, III: GeInW and NiInW contact metals. *J. Appl. Phys.*, **64**(4), 1974–1982.
8. Dubon-Chevallier, C., Glas, F., Henoc, P., Hugon, M.C., Agius, B. and Blanconnier, P. (1990) Ge(As)MoW and NiInW n-type refractory ohmic contacts on GaAs: a comparison, in *Proc. Int. Symp. GaAs and Related Compounds*, Jersey, UK, September 24–27, 1990, pp. 245–250.
9. Würfl, J. and Hartnagel, H.L. (1989) Thermal stability and degradation mechanisms of Ti–Pt–Au, Ti–W–Au and WSi$_2$–Ti–Pt–Au Schottky contacts on GaAs, in *Proc. 4th Int. Conf. on Quality in Electronic Components, Prevention, Detection and Failure Analyses*, March 1989, Vol. 1, DVS, pp. 10–12.

10. Würfl, J., Merkl, B. and Nold, U. (1989) Suitability of GaAs Schottky metallizations for continuous device operation at elevated temperatures up to 300 °C: a comparative study. *Int. J. Electron.*, **66**(3), 437–444.

11. Krawczyk, S.K., Gendry, M., Tardy, J., Krafft, K., Viktorovitch, P., Abraham, P., Bekkaoui, A., Monteil, Y., Schütz, R., Riemenschneider, R., Richter, R., Hartnagel, H.L. (1991) Improved surface treatments and passivation procedures of GaAs crystals controlled by photoluminescence measurements. Paper presented at INFOS91, Liverpool, UK, April 2–6, 1991.

Contacts for GaAs devices | 7

K. Fricke and W.-Y. Lee

7.1 INTRODUCTION

GaAs has become a very popular material for the fabrication of high frequency, low noise [1] and microwave power devices [2]. As shown in Chapter 8, GaAs devices are also well suited for high temperature operation because of the large band gap of this material [3]. However, the standard GaAs technology has to be modified for stable operation at increased temperatures. Special high temperature technologies and designs have been published for GaAs FET [4–7], for GaAs/AlGaAs heterostructure bipolar transistors (HBTs) [8, 9] and for GaAs sensors [10].

What are the requirements for a GaAs device to provide stable operation at high ambient temperatures? Two problems have to be solved:

- The circuit has to maintain its electrical specifications in the whole temperature range. The active devices have to be optimized for constant gain, constant input and output impedances and for minimum noise in the operational temperature range. This electrical stability means that, in the case of an ohmic contact, the contact resistance does not change with temperature. The experimental investigations show that the contact resistance has nearly no temperature dependence for p-type GaAs or even decreases with temperature if the doping is lower than $N_A = 10^{20}\,\text{cm}^{-3}$ [11, 12]. Similar results are obtained for n-type GaAs material [13, 14].
- Another problem is the degradation of the device. According to the Arrhenius law, most of the degradation mechanisms are accelerated exponentially with the temperature. The diffusion coefficient D is given by

$$D = D_0 e^{(E_a/kT)} \tag{7.1}$$

High Temperature Electronics. Edited by M. Willander and H.L. Hartnagel. Published in 1997 by Chapman & Hall, London. ISBN 0 412 62510 5.

This means that degradation mechanisms may cause reliability problems in a high temperature environment, depending on the activation energy E_a. A special technology has therefore to be developed to maintain the reliability of GaAs devices even at increased temperatures.

Since the first of the two aspects does not cause any problem, the reliability of the ohmic contacts remains as the problem to be solved. The reliability problems of GaAs devices at increased temperatures are as follows:

- Degradation of the ohmic contacts. Stable ohmic contacts are a severe problem for the device reliability. Since ohmic contacts are often alloyed they have a complicated metallurgical structure, which is difficult to stabilize against further reaction.
- The Schottky contacts may degrade at high temperatures.
- A problem in III–V devices is the outgassing of the group V element. A suitable passivation (SiN) has to be specially optimized for this purpose.
- Electromigration effects are also increased at higher temperatures. Therefore the metallizations have to be operated at lower current densities.
- The doping profiles may degrade at high current densities and increased temperatures. An appropriate choice of donors and acceptors is required to solve this problem.

7.2   HIGH TEMPERATURE STABLE OHMIC CONTACTS ON GaAs

Semiconductor devices require ohmic contacts for the input of an electrical current. The ohmic contact is considered to be good if its series resistance is small compared to the resistance of the device. The ohmic contact should not degrade the device performance.

The reliability of high temperature GaAs devices depends to a large extent on the technology of the ohmic contacts [15]. Although GaAs devices have become very mature in the past few years [16], formation of the ohmic contacts is still not fully understood. This is because of the complicated chemical and physical nature during the formation of the ohmic contact, which is normally performed by a heat treatment. With silicon there is only one element to consider, Si; with gallium arsenide there are two, Ga and As.

An excellent review of state-of-the-art ohmic contact fabrication has been published by Shen *et al.* [17]. A review especially on new concepts of ultrastable contacts was published by Hartnagel [18].

The degradation processes leading to problems are different interdiffusion processes, electromigration, formation of intermetallic phases and changes in the structure of metallic films [19].

Ohmic contacts for high temperature devices have to meet the following requirements:

- low contact resistance;
- good homogeneity of the contact resistance;
- good morphology for the realization of submicron structures;
- long lifetime at ambient temperatures of about 600 K;
- formation temperatures lower than approximately 1000 K, to be compatible with the processing of GaAs devices;
- it is advantageous to avoid noble metals like Au and Pt in the contact in order to avoid problems with the catalytically induced, accelerated etch rates in the vicinity of these metals at self-aligned wet etching processes;
- simple technology by liftoff;
- compatibility to the device processing technology;
- low sheet resistance of the contact, usually achieved by an Au top layer; at increased temperatures electromigration will limit the current capability of an Au layer [20, 21].

In fact, it is difficult to meet all these goals with one contact system. Now for some examples.

The emitter contacts of HBTs experience a high current density (at pulsed operation up to $1 \times 10^6 \, \text{A cm}^{-2}$). They require a very homogeneous contact system with low contact resistance which need not be structured by liftoff, since the emitter is located on the top layer of the device.

For the collector contact, a metallization which can be structured by liftoff is advantageous. In most cases, two different metallizations are needed for n- and p-type GaAs.

It is sufficient to restrict the investigations to high temperature stable ohmic contacts on GaAs and $In_x Ga_{1-x} As$ because it is possible to choose the layer structure of the device (bipolar or FET) in such a way that either a GaAs or an $In_x Ga_{1-x} As$ layer is under the contact metallization. It is difficult to fabricate contacts to $Al_x Ga_{1-x} As$ because of its high band gap and its affinity to oxygen.

Several different physical principles are known for the fabrication of ohmic contacts. One may distinguish two aspects:

1. The principle of the processing

- The alloyed contacts melt during the processing. They therefore exhibit a relatively inhomogeneous morphology and an inhomogeneous distribution of the current. Deep spikes into the semiconductor are observed in this kind of contact. The well-known AuGeNi contact belongs to this group.
- Contacts reacted out of a solid phase are called nonalloyed ohmic contacts or sintered contacts. They have better homogeneity than alloyed contacts.

- Implanted contacts can be fabricated by implantation and subsequent annealing of the wafer. This type of contact has an inherently homogeneous morphology and electrical performance.
- Ohmic contacts can also be fabricated by epitaxial growth of a layer with low band gap, like $In_xGa_{1-x}As$, or with metals. The quality of this layer is excellent.

2. The physical principle of the operation

- Tunneling contacts work due to a high doping direct under the contact metal. The resulting thin depletion layer leads to a high tunneling current through the contact.
- Contact metallizations with a very low barrier height make it possible for a current to flow by thermal field emission.
- Ohmic contacts are based on the lowering of the semiconductor band gap under the metallization.

For high temperature stable ohmic contacts the sintered ohmic contacts, epitaxial contacts and implanted contacts are well suited. These systems have their individual applications: the sintered contact is for general purpose, the epitaxial contact is applicable for the emitter metallization of self-aligned and nonself-aligned HBTs. Although the implanted ohmic contacts are relatively new, in future they may become an alternative for sintered contacts.

7.3   INVESTIGATION TECHNIQUES FOR OHMIC CONTACTS

Measurement of the ohmic contact resistance can be performed by the transmission line method (TLM) [22, 23]. A mesa structure of the doping type to be investigated is needed on a semiinsulating substrate. Contacts on top of this mesa structure at different distances allow extrapolation of the resistance measured to zero distance [17, 23].

Another method is the Kelvin cross [17, 24, 25]. A horizontal current flow through a metal–semiconductor junction is achieved by feeding the current in the semiconductor through an implanted conductor and extracting it by a metal conductor via a window in the passivation on top of the ohmic contact. The voltage drop is measured using two extra connections to the implanted conductor and to the metal conductor. This method is very precise because it measures only the voltage drop across the ohmic contact.

The following methods have been used for chemical and physical investigation:

- secondary ion mass spectroscopy (SIMS)
- transmission electron microscopy (TEM)
- Auger electron spectroscopy (AES)

- X-ray photoelectron spectroscopy (XPS)
- Rutherford backscattering spectroscopy (RBS)
- Bragg–Brentano diffractometry
- X-ray reflectometry
- energy dispersive X-ray analysis (EDX)

7.4 n-TYPE OHMIC CONTACT SYSTEMS

The standard contact system of GaAs technology is the AuGe/Ni alloyed system. It uses Au and Ge in eutectic composition. The layer sequence commonly used is GaAs–Ni/AuGe/Ni/Au. The top Au layer improves the horizontal conductivity of the contact.

The main stability problem concerned with the eutectic AuGe/Ni contact is caused by an Au–Ga interaction. For a low contact resistance the formation of a β-AuGa phase is important, since the Ga vacancy thus formed in the GaAs is needed to incorporate the n-type donor Ge in the proper site [26, 27]. β-AuGa melts at 630 K. At temperatures of about 500 K the contact degrades quickly [19]. This is due to the high solubility of Ga in Au which leads to a continuous interdiffusion process of Ga and Au [28]. The vacancies in the GaAs, caused by the outdiffusion of Ga, are initially filled only with the dopant Ge. Later in the degradation process, the Ga vacancies are filled by Au atoms, which act as acceptors and compensate the n-doping. Another aspect of the Au–Ge interaction is the spiking of the reaction deep into the semiconductor. The contact is therefore not very homogeneous. The eutectic AuGeNi contact is not suitable for high temperature systems since the AuGa reaction does not stop because of the large amount of Au in the layers.

The morphology of the contact is not very good, but it can be improved using rapid thermal annealing (RTA).

The disadvantages of the eutectic AuGe/Ni contact gave rise to many new developments of ohmic contact systems. Table 7.1 summarizes a number of publications on high temperature stable ohmic contacts for n-type material. The following attempts to solve the stability problem can be distinguished:

- Variations of the commonly used AuGeNi contact. A compromise between optimum stability and low contact resistance is achieved by the reduction of the Au content.
- Pd is used in some contacts in spite of Au in order to get the Ga out of the GaAs. n-Doping is only achieved if a Ga vacancy is created in the lattice and if this site is occupied with a group IV element (often Ge). And Pd is able to penetrate the native oxide of GaAs [29].

Table 7.1 High temperature stable ohmic contact systems on n-GaAs from literature[a]

Contact system	$T_F(t_F)$	ρ_C or R_C (N_D)	Thermal stability	Remark
Ge/Au/Ni/ W$_{0.5}$Si$_{0.2}$/Au	920 K (5 s)	$1 \times 10^{-5}\,\Omega\,\text{cm}^2$ $(1 \times 10^{17}\,\text{cm}^{-3})$	670 K (550 h)	Stable; evaporated WSi diffusion barrier; used very small amount of Au [19]
Ni/Ge/Au/ Ni/W	920 K (30 s)	$0.23\,\Omega\,\text{mm}$ $(6 \times 10^{13}\,\text{cm}^{-2})^d$	670 K (20 h)	Stable; used very small amount of Au; 1000 times reduction in the β-AuGa phase [57]
Ni/In–Ni/Ni/ W/Al[b]	1070 K (7 s)	$0.3\,\Omega\,\text{mm}$ $(3.5 \times 10^{13}\,\text{cm}^{-2})^d$	670 K (100 h)	Stable; simultaneous evaporation of In and Ni; regrown InGaAs epitaxial layer [47]
Ni$_{0.9}$In$_{0.1}$/W	1120 K (10 s)	$2 \times 10^{-6}\,\Omega\,\text{cm}^2$ $(2 \times 10^{18}\,\text{cm}^{-3})$	670 K (100 h)	Stable; sputtered from In$_{0.1}$Ni$_{0.9}$ target; regrown InGaAs epitaxial layer [48]
Pd/Ge	500–645 K (30 min)	$4 \times 10^{-7}\,\Omega\,\text{cm}^2$ $(4.5 \times 10^{18}\,\text{cm}^{-3})$	570 K (1000 h)	Stable; homogeneous Ge doping and Ti/Pt/Au cap layer [51]; unstable at 400°C due to diffusion [13]
Pd/(Si, Ge)/ Pd/Si	720 K (30 s)	$4 \times 10^{-7}\,\Omega\,\text{cm}^2$ $(1 \times 10^{18}\,\text{cm}^{-3})$	670 K (50 h)	Unstable due to diffusion; small amount of Ge or Si for doping, but more stable than Pd/Ge contact [13]
Pd/Ge/Pd/ [20 × In/Pd]	920–970 K (5 s)	$1 \times 10^{-6}\,\Omega\,\text{cm}^2$ $(1 \times 10^{18}\,\text{cm}^{-3})$	670 K (30 h)	Stable; regrown InGaAs epitaxial layer [45]
Pd/(Ge)/Pd/In/ Pd/Ti/Pt/Au[c]	900–920 K (5–10 s)	$4.5 \times 10^{-7}\,\Omega\,\text{cm}^2$ $(4.5 \times 10^{18}\,\text{cm}^{-3})$	670 K (25 h)	Stable; regrown InGaAs epitaxial layer; small amount of Ge for doping [14]

[a] $T_F(t_F)$ = formation temperature (time); $\rho_C(N_D)$ = specific contact resistance, (n-doping); R_C = contact resistance.
[b] Al cap layer was evaporated after formation of the ohmic contact.
[c] TiPtAu cap layer was evaporated after formation of the ohmic contact.
[d] Dose of Si$^+$ ions [17] or SiF$^+$ ions [18] during the implantation for the surface doping.

- Recrystallization of an epitaxial InGaAs layer out of a solid phase reaction is used to form ohmic contacts without the use of Au.
- Diffusion barriers based on the refractory metals tungsten and platinum, or on LaB_6, W_5Si_2 and TiN, are used to avoid any interaction between the Au top layer and the metal–semiconductor contact.

7.4.1   AuGe–based contacts

Sintered ohmic contact systems for high temperature application and based on the AuGe/Ni system have been developed by Würfl [19, 30], Wittmer [31] and Shih et al. [32]. To achieve high temperature stability the sequence of layers was changed. In particular, the thickness of the Au layer was reduced. After the formation of the contact, the Au is consumed totally in stable intermetallic phases, so no further reaction under high temperature stress will take place.

In the approach of Würfl, an Au layer for high current density operation is evaporated on top of the contact system. The evaporated layers are shown in Fig. 7.1. In order to maintain the reliability of the contact, the Au layer is separated by a W_5Si_2 diffusion barrier. The W_5Si_2 diffusion barrier is

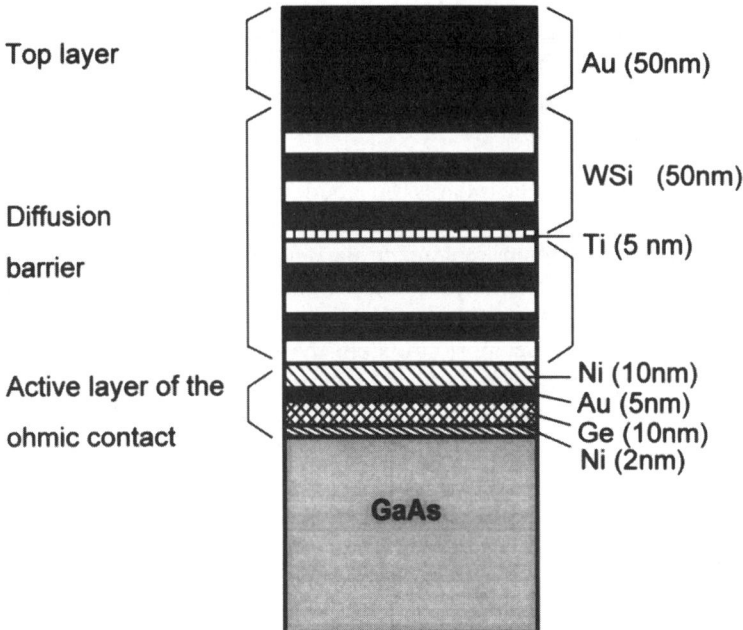

Figure 7.1 AuGe/Ni contact system developed by Fricke et al. [30] with W_5Si_2 diffusion barrier, realized by a multilayer structure.

Table 7.2 Contact systems obtained by a variation of the AuGeNi contact system

No.	Contact system	Thickness (nm)	Formation temperature (K)	Contact resistance ($\Omega\,cm^2$)
1	Ge–Ni–LaB$_6$–Au	10–10–100–100	1200	$\geqslant 10^{-3}$
2	Ge–Mo–LaB$_6$–Au	10–10–100–100	1200	$\geqslant 10^{-2}$
3	Ge–Au–Mo–LaB$_6$–Au	10–3–10–100–100	800	1×10^{-5}

formed in the same temperature step with the formation of the ohmic contact. The W_5Si_2 diffusion barrier is evaporated as a multilayer structure, so the fabrication is relatively complicated. An alternative approach would be sputter deposition. The resistance of this contact is $\approx 10^{-5}\,\Omega\,cm^2$ on GaAs with $N_D = 10^{17}\,cm^{-3}$. Lifetime tests at 670 K for 600 h demonstrated its reliability [33].

In another experiment [34] the AuGe contact was improved by reducing the Au content. The experiments are listed in Table 7.2. The contact was fabricated in two steps. First the layers for the ohmic contact and half the thickness of an LaB$_6$ diffusion barrier were evaporated then annealed. Afterwards the second half of the LaB$_6$ diffusion barrier and an Au top layer were evaporated. This ensures that the Au has no chance to penetrate the GaAs during formation. A good contact resistance with the penalty of low reliability would be obtained if any Au were to reach the GaAs.

The aim of the experiment is to achieve a high n-doping by Ge without the use of excessive Au, thus avoiding the Au–Ga interaction. This would mean that all the Au is bonded after the formation of the contact in stable (saturated) intermetallic phases. A good morphology of the contact is achieved by using compositions with melting points above the formation temperature. The contact remains in the solid phase during formation.

In the first experiment, according to Table 7.2, the Au layer was totally omitted. The result shows that a GeNi contact does not result in a good ohmic contact, because most probably the Ge is not incorporated in Ga sites as required for high n-doping. As shown by the second experiment, the contact resistance does not change significantly if the Ni is replaced by Mo. Mo has nearly the same catalytic performance as Ni.

Better results were achieved with the system Ge–Au–Mo–LaB$_6$–Au. The Au content was reduced below the eutectic composition. The melting point was therefore higher than the formation temperature. At a formation temperature of 800 K a contact resistance of $\approx 10^{-5}\,\Omega\,cm^2$ and a mirror-like surface was obtained. The XPS sputter profile in Fig. 7.2 shows that, to a large extent, Ge is bonded in a Ge–Mo phase. Only a small amount of Ge is diffused into the GaAs. An Au–Ga phase may be present at the metal–GaAs interface. It seems to be saturated since a lifetime test at 673 K

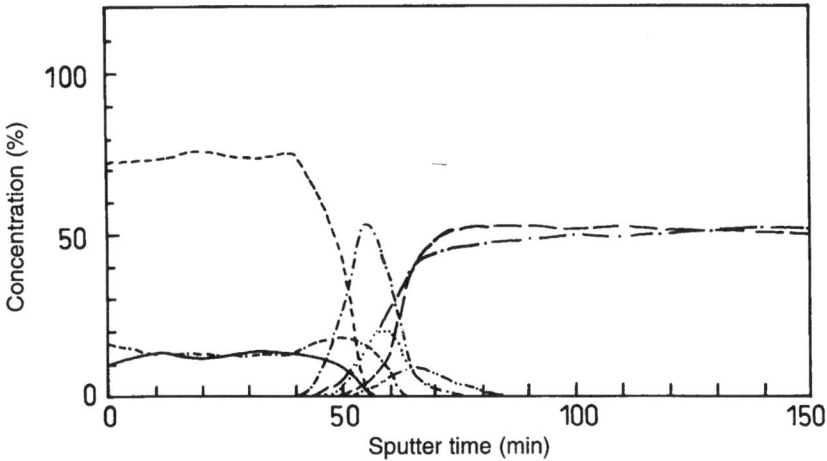

Figure 7.2 XPS sputter profile of a Ge–Au–Mo–LaB$_6$ sample after annealing at 800 K: (—) La 3d$_{5/2}$, (- - -) B 1s, (- · · · -) Mo 3d, (- ··· -) Au 4f, (· · · ·) Ge 2p$_{3/2}$, (- - - - -) O 1s, (- -) Ga 2p$_{3/2}$ and (- · -) As 2p$_{3/2}$.

for more than 600 h did not produce a degradation of the contact. It is assumed that the Ge, which is diffused into the GaAs, is occupying Ga sites because of the Ga depletion; this produces n-doping.

7.4.2   Pd/In/Pd/W contacts on n-type GaAs

Ohmic contacts may be produced by the lowering of the band gap of the semiconductor under the metal. The low band gap semiconductor is realized by a regrowth out of a solid phase (explained below). The band diagrams of the regrown InGaAs/metal and the GaAs/metal interface are shown in Fig. 7.3. They show that the metal/n^{++}-InGaAs/n-GaAs interface can be made ohmic if the In content is sufficient. This may be accomplished by an In$_x$Ga$_{1-x}$As regrown heterostructure. A flexible process can be achieved by evaporating a suitable In-containing layer structure.

During a thermal formation process, an InGaAs epitaxial layer has to be grown on the GaAs. The simplest form of this kind of ohmic contact [35], is to evaporate a thick In layer on top of the GaAs then sinter it at 800 K. During the heating process, In melts and a certain amount of Ga and As are dissolved in it. During the cooling, a solid layer containing InGaAs crystals and an In-rich layer are formed [29]. However, the contact is relatively inhomogeneous since the native oxide on the GaAs surface prevents the reaction. Only at pinholes in the oxide layer can the InGaAs be formed. Ding *et al.* [36] measured an InGaAs layer thickness of approximately 3 nm.

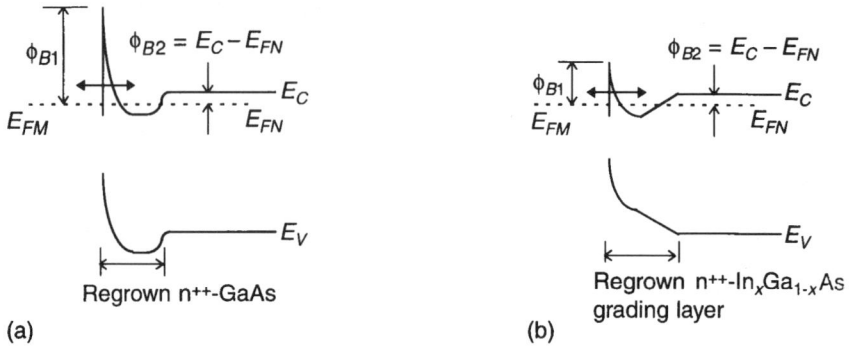

Figure 7.3 Schematic band diagram of (a) metal/regrown n^{++}-GaAs/n-GaAs substrate and (b) metal/regrown n^{++}-InGaAs/n-GaAs substrate.

A miscibility gap has been found in the InAs–GaAs pseudobinary system within the critical temperature range 875–950 K, over which the interesting In concentrations of about 50% will not be formed. In-based contacts have the following problems [29]:

- Indium is accumulated during the adhesion and forms little islands, if only thin layers are evaporated [37].
- The thin native oxide layer on GaAs prevents the reaction between In and GaAs.
- The surplus In melts at the very low temperature of 429 K causing a very inhomogeneous morphology.

To avoid these problems, reaction out of a solid phase is proposed. By the introduction of an additional Pd layer [38] or alternatively Pt [39] between In and GaAs, intermediate phases are formed during the heating, especially PdIn$_3$ and Pt$_3$In$_7$. Both of them have melting points well above the formation temperature of the contact. This ensures that, at the GaAs–contact interface, a layer remains in the solid phase. Pt as well as Pd are able to penetrate the native oxide of GaAs during the formation. A third metal may be chosen for this purpose, the transition metal Ni, which has comparable catalytic properties. Since relatively thick In layers are used (400 nm), a fraction of the In melts. This produces a minor surface morphology and a reduced reliability.

Although the authors cited above use thick In layers (400 nm), Murakami [40] reduced the In layer to thicknesses of a few nanometers (1–5 nm). High temperatures of 900–1300 K are required for the formation of the contact. A W top layer prevents balling of the contact during the temperature treatment. Murakami uses Ni at the GaAs–contact interface. Ni and In form a phase with high melting point. These contacts have an excellent

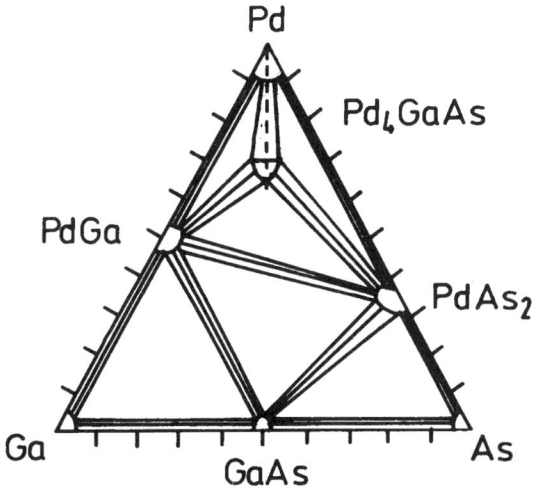

Figure 7.4 Ternary phase diagram of the system Pd–Ga–In; regions with a single phase are bordered by lines, regions with two phases are connected by lines. The dotted line indicates the course of the reaction. (After Sands [29])

morphology and a very good reliability. The high formation temperature is a disadvantage.

According to the experiments of Sands [29], Pd is more suitable for penetration of the GaAs native oxide than Ni. The function of Pd becomes clear in the ternary Pd–Ga–As phase diagram (Fig. 7.4). These diagrams are valid for a closed system and show the stable phases which are possible. Figure 7.4 shows that a Pd layer on GaAs will react to Pd_4GaAs. This layer can be detected already after the evaporation. A further reaction (at higher temperatures) will yield PdGa and $PdAs_2$. These phases are stable. Parallel to the reaction between GaAs and Pd, Pd and In are reacted to $PdIn_3$. During the cooling an epitaxial layer of InGaAs [41] grows on the GaAs.

The morphology of Pd/In contacts is relatively inhomogeneous [29, 42], but it can be improved if an additional layer, here called the intermediate layer, is inserted between the In and Pd. Suitable materials for the intermediate layer are Ag, Mg, Mo, Ni, Au, Ni, S and Te. Some of the materials act as dopants, others are purely to improve the morphology. Using the intermediate layer, the morphology of the contact remains mirror-like even after annealing. And this approach achieves a formation temperature below 1000 K, compatible with device processing. The contact resistance is about $1 \times 10^{-6} \, \Omega \, cm^2$ on GaAs with $N_D = 1 \times 10^{18} \, cm^{-3}$.

The improved layer structure is GaAs/Pd/In/Te/Pd/W/Au, using Te as intermediate layer and W as diffusion barrier. The formation process can be followed in Fig. 7.5. According to Fig. 7.5b, at 400 K, a temperature reached during the evaporation process, a homogeneous Pd_4GaAs layer is formed at the GaAs–metal interface. This can be seen in the XPS sputter profile (Fig. 7.6) and is also proven by TEM profiles [43, 44].

In the same temperature range, Pd and In react to $PdIn_3$ at the upper Pd–In interface (Fig. 7.5b). The temperature required for this reaction is about 400 K [45]. The second Pd layer prevents the In layer from melting by forming the $PdIn_3$ phase, which has a higher stability and does not melt during the formation process. At the lower Pd–In interface the thin Te layer, which reacts with Pd (most probably to PdTe), prevents a Pd–In reaction. The additional Te layer assures that the Pd–In reaction does not counteract the GaAs–In reaction, needed to form the Pd_4GaAs layer. The Pd_4GaAs layer is essential for regrowth of the epitaxial InGaAs layer.

According to the phase diagram, the Pd_4GaAs layer decomposes during further reaction to PdGa and $PdAs_2$ (Fig. 7.5c).

PdIn is formed at 850 K (Fig. 7.5d) and the surplus In in this reaction is necessary for the formation of the $In_xGa_{1-x}As$ layer. During this reaction the intermediate layer has to be penetrated, so there exists an optimum layer thickness; it depends on the element chosen but is usually a few nanometers.

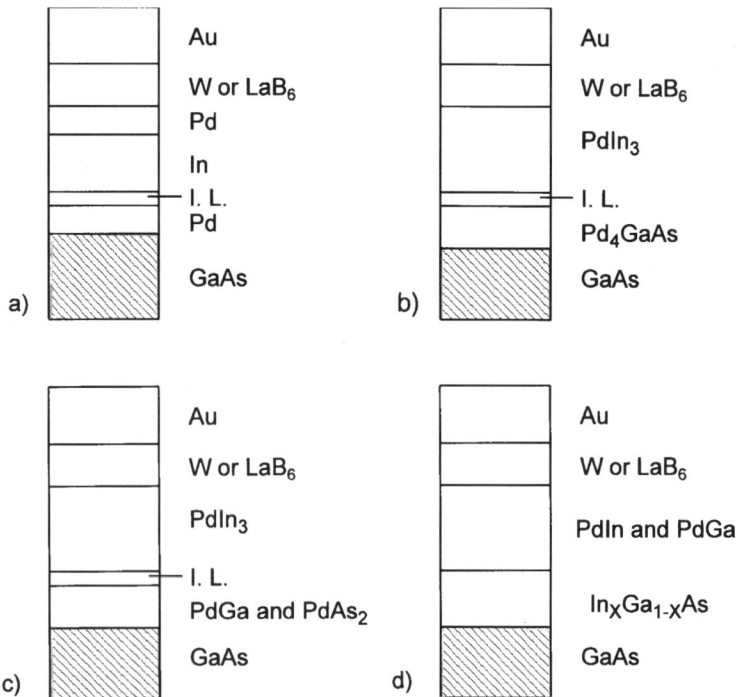

Figure 7.5 Layers of a Pd–In contact system: (a) evaporation sequence, (b) after evaporation, (c, d) during formation.

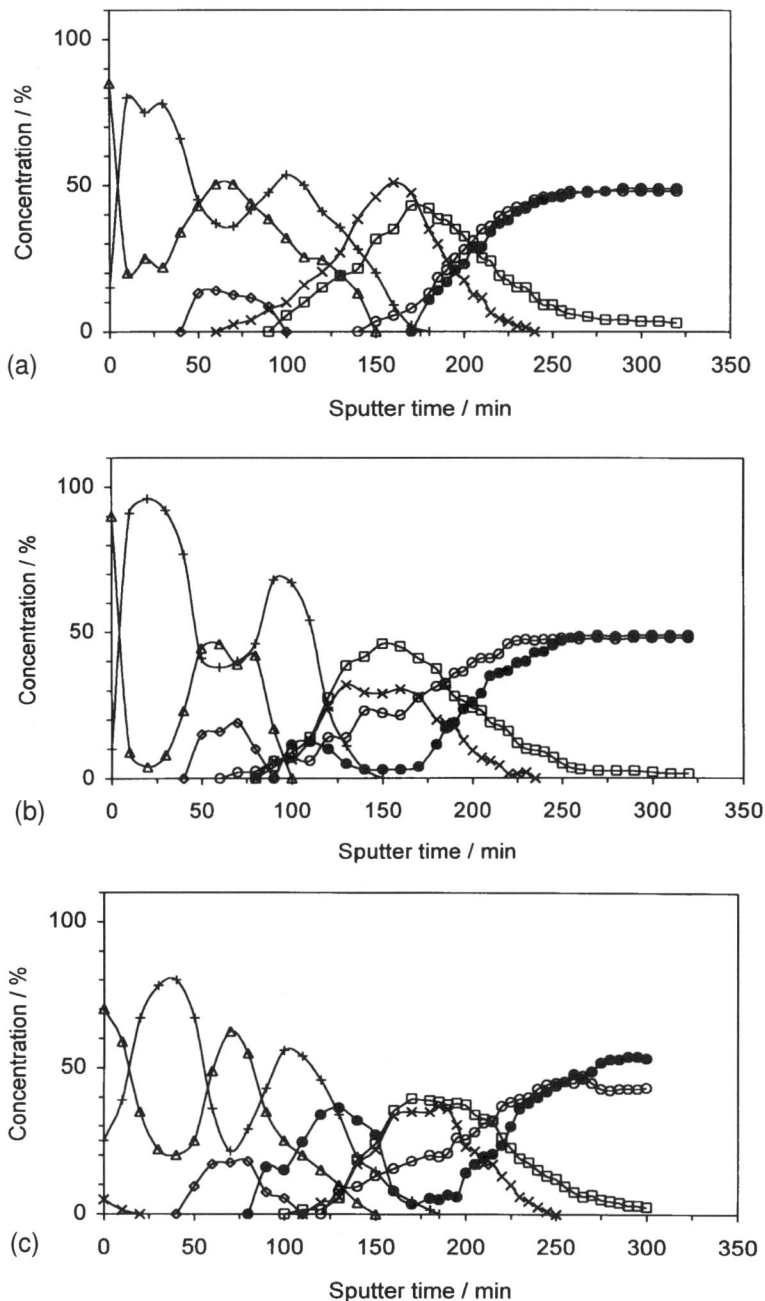

Figure 7.6 XPS sputter profile of a GaAs/Pd/Te/In /Pd/Mo/TiO/Mo sample: (a) after evaporation, (b) after annealing at 900 K and (c) after annealing at 900 K and 500 h stress test at 670 K: ($+$) Mo $3p_{3/2}$, ($\triangle$) O 1s, ($\diamond$) Ti $2p_{3/2}$, ($\square$) Pd 3d, ($\times$) In $3d_{5/2}$, ($\bullet$) As $2p_{3/2}$ and ($\bigcirc$) Ga $2p_{3/2}$.

The PdGa and $PdAs_2$ phases deliver Pd. The reaction can be schematically written as

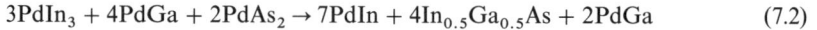

$$3PdIn_3 + 4PdGa + 2PdAs_2 \rightarrow 7PdIn + 4In_{0.5}Ga_{0.5}As + 2PdGa \qquad (7.2)$$

This equation assumes that $In_xGa_{1-x}As$ with an In content of $x = 0.5$ is regrown. The equation can be used to optimize the In/Pd ratio in the evaporated contact. The equation results in a Pd/In mole fraction of 1, so the Pd layer should be 0.6 times as thick as the In layer.

The diffusion barrier is important for the morphology of the contact. For instance, W is a diffusion barrier for As and prevents evaporation of As during the formation process. It is known that W can consume surplus In. This is important to avoid melting of pure In phases.

An XPS sputter profile of the GaAs/Pd/Te/In/Pd/Mo/TiO/Mo is presented in Fig. 7.6. The contact has an Mo/TiO/Mo diffusion barrier as described below. On top of this diffusion barrier an Au layer may be evaporated for the improvement of the sheet resistance. In Fig. 7.6a the contact is analyzed after the evaporation. The Pd_4GaAs layer can be identified by the flattening of the Pd, Ga and As concentration. The concentration of Te is lower than the resolution of XPS.

Figure 7.6b indicates that the diffusion barrier is stable during the formation process, since no outdiffusion of As can be observed. An outdiffusion of Ga into the Pd/In layer can be observed. This is an indication for the PdGa phase shown in Fig. 7.5d. Also the Pd/In phase can be identified.

Figure 7.6c shows the contact after a 500 h accelerated lifetime test at 670 K. The further outdiffusion of As is apparent, but the outdiffusion is effectively stopped by the TiOMo layer. This contact shows stable contact resistance up to 500 h at 670 K with a slight increase of 20% at the end of the stress test.

The Bragg–Brentano diffractogram of a GaAs/Pd/Te/In/Pd/W sample is presented in Fig. 7.7a. It is interesting to identify the peaks of GaAs (2 0 0) and InGaAs (2 0 0) of the CuK_α radiation in Fig. 7.7b. It can be seen clearly that the intensity of the InGaAs peak of the optimized formation temperature (1000 K/5 s) is higher and exhibits a higher lattice constant. A higher lattice constant is a result of a higher In concentration. Using Wigard's rule one calculates an In content of approximately 5%. This value is too low because the InGaAs layer is very thin, as demonstrated by TEM [43] and therefore the InGaAs layer is strained. Wigard's rule applies only for unstrained samples.

The reliability of a $Pd/Te/In/Pd/LaB_6/Au$ contact has been tested at 670 K. Figure 7.8 shows the contact resistance versus the stress time. After an initial stabilization the resistance increases slowly. In general Pd/Te/In/Pd/LaB_6/Au contacts have sufficient reliability for operation at 570 K [50]. A widely applied method to anneal these contacts is the RTA technique. Very stable results have been obtained using an e-beam annealing process [46].

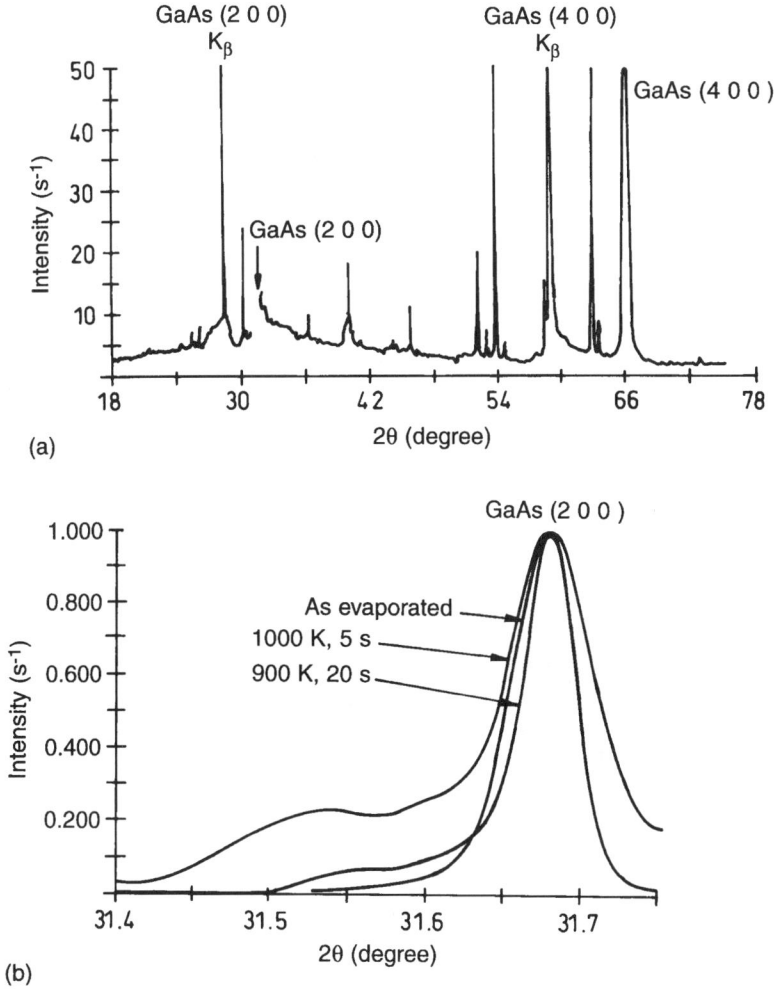

Figure 7.7 (a) Bragg–Brentano diffractogram of a GaAs/Pd/Te/In/Pd/W sample annealed at 1000 K for 5 s. (b) Detail of the Bragg–Brentano diffractogram of a GaAs/Pd/Te/In/Pd/W sample.

It was considered interesting to see how a donor would improve the contact resistance of a Pd/In contact. The In was mixed with a certain amount of Ge before evaporation; S was used as the intermediate layer. The contact resistance of the Pd/S/In(Ge)/Pd was measured after formation at 850 K. Figure 7.9 plots the minimum of the measured contact resistance for several of these contacts versus the Ge concentration. A concentration of 5% Ge in the In layer is optimum.

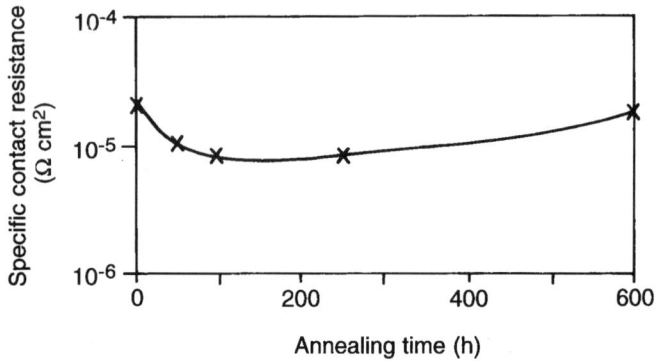

Figure 7.8 Accelerated lifetime test of the $Pd/Te/In/Pd/LaB_6/Au$ contact on n-GaAs as a function of the stress time at 670 K.

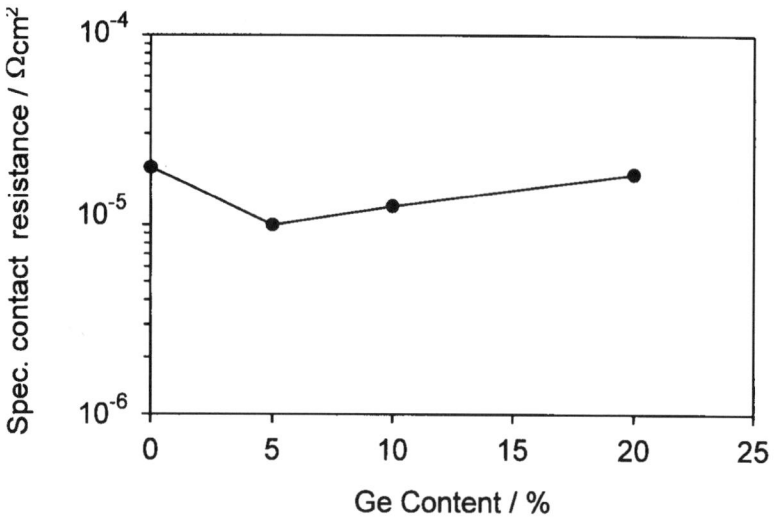

Figure 7.9 Contact resistance of a $Pd/S/In(Ge)/Pd$ contact versus the Ge concentration in the In layer; formation temperature 850 K, $N_D = 10 \times 10^{17} \, cm^{-3}$.

7.4.3   In/Ni contacts for n-type GaAs

Instead of using Pd in the Pd/In contact [55], Ni has been proposed for reduction of the native oxide film on the GaAs surface [34, 47–49].

A contact system developed especially for high temperature application is the layer structure $Ge–In–Ni–LaB_6–Au$ [34]. An LaB_6 diffusion barrier under the Au top layer was used for stabilization of the contact. With layer thicknesses of Ge(10 nm)/In(3 nm)/Ni(10 nm)/LaB$_6$(100 nm)/Au(100 nm) and an annealing temperature of 1100 K a contact resistance of $8 \times 10^{-6} \, \Omega \, cm^2$ was obtained. A lifetime test at 670 K for more than 600 h did not identify any degradation of the contact.

According to the experiments of Sands [29], the ability of Ni to penetrate the native oxide is less pronounced than for Pd. This suggests that the contact resistance and the homogeneity of Pd/In contacts is better than for Ni/In contacts.

7.4.4 Pd/Ge–based contacts

Another class of contacts is based on Pd/Ge and utilizes the principle of band gap lowering at the interface [13, 51, 52]. Ge is lattice matched on GaAs, the expansion coefficients of Ge and GaAs are nearly equal and Ge has a low band gap of 0.67 eV.

The formation process is similar to that for Pd/In contacts. Initially a Pd$_4$GaAs layer is formed during evaporation. At higher temperatures (~ 570 K) the Pd$_4$GaAs layer decomposes by reaction with Ge to produce PdGe and GaAs with the incorporated dopant GaAs(Ge):

$$x\text{Ge} + \text{Pd}_4\text{GaAs} \rightarrow 4\text{PdGe} + \text{Ge} + \text{GaAs(Ge)} \qquad (7.3)$$

The Ge phase is localized after the formation on top of the GaAs(Ge). Contact resistances of about $4 \times 10^{-7} \, \Omega \, cm^2$ on $2 \times 10^{18} \, cm^{-3}$ n-doped material have been achieved with this method [13, 51]. The required formation temperature was in the range 600–650 K. A stable contact system with excellent morphology was obtained. However, an instability due to an interdiffusion process at the semiconductor–metal interface was detected at 670 K stress temperature. The application of this contact is therefore limited to temperatures less than 600 K.

7.4.5 Epitaxial ohmic contacts on GaAs

Epitaxial ohmic contacts on GaAs can be fabricated by the epitaxial growth of In$_x$Ga$_{1-x}$As on top of GaAs. The method is reproducible, the contacts are very reliable and homogeneous. The different approaches [17] are as follows:

- A single n-type InAs layer is grown on top of the GaAs. With refractory and Ni/AuGe metallizations on GaAs contact resistances between $1 \times 10^{-6} \, \Omega \, cm^2$ and $1.7 \times 10^{-8} \, \Omega \, cm^2$, respectively, have been obtained on GaAs and InGaAs.
- This technique can be improved by using a graded layer of In$_x$Ga$_{1-x}$As instead of the InAs layer. This approach avoids the discontinuity of the

conduction band in the simple metal–InAs/GaAs contact. Such a system is discussed in detail in the next paragraph.

• Alternatively an InAs/GaAs superlattice can be used instead of the graded $In_xGa_{1-x}As$ layer.

The disadvantages of the epitaxial contact are found in its lateral restrictions. The epitaxial structure is homogeneous over the wafer (except if selective epitaxy is used), so the contact can be fabricated only on selected layers.

The method is useful for the fabrication of the emitter of HBTs and for the ohmic contacts of field effect devices. It is difficult to fabricate the collector of emitter-up HBTs by epitaxial InGaAs layers because the growth of InAs degrades the quality of the following epitaxial layers.

Multilayer structures of $In_xGa_{1-x}As$ can be used to obtain ohmic contacts. In this example, nine layers with increasing In content from the GaAs to the metal–semiconductor interface have been grown. The total thickness was 30 nm. On this layer system a high temperature stable LaB_6–Au metallization was evaporated, which has some advantages for this application:

• LaB_6 is able to consume the oxygen on the GaAs–LaB_6 interface. Since the oxygen is removed during an annealing step, the quality of the interface improves.
• Very good adhesion is obtained because of the oxygen removal.
• LaB_6 is an effective diffusion barrier and produces good reliability.
• The contact resistance measured with the 'transmission line method' is $2 \times 10^{-6} \, \Omega \, cm^2$.

A disadvantage of this contact system is the large barrier height of the LaB_6–semiconductor interface. However, because of the high doping of the InGaAs, tunneling is the dominant mechanism. A large barrier height is not a great disadvantage if tunneling is dominant. The 50 nm thick LaB_6 layer was covered by a 10 nm thick In layer in order to avoid oxidation of the LaB_6 during annealing.

The result of the annealing was studied by X-ray diffractometry (Fig. 7.10). The amplitudes and the reflectivity of the reflectogram demonstrate that the surfaces are very good. The surfaces are better than before, especially after annealing. A fitting of the layer thicknesses [44] produces 55.5 nm (LaB_6) and 11.9 nm (Ni) for the untreated sample and 50.1 nm (LaB_6) and 10.9 nm (Ni) for the annealed sample. The reduction of the thicknesses is an indication of the better crystallinity. Pure LaB_6 and Ni phases have been found by X-ray diffractometry.

After annealing, the contact resistance of the samples was slightly improved. This is because oxygen is removed from the interface into the

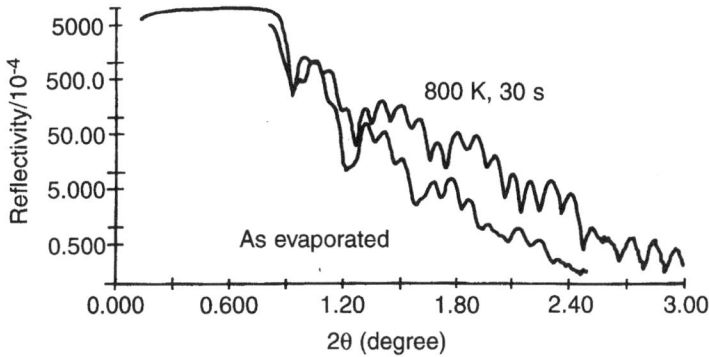

Figure 7.10 X-ray diffractometry of a GaAs/LaB$_6$(50 nm)/Ni(12 nm) sample before and after annealing at 800 K for 30 s in H$_2$.

volume of the LaB$_6$. The annealing step is not needed to achieve a good ohmic contact but it improves the quality of the diffusion barrier.

7.4.6 Implanted ohmic contacts on GaAs

A high doping for ohmic contacts may also be obtained by implantation. However, the damage caused by the high dose required makes it impossible to obtain an ohmic contact by implantation alone. Therefore implantation is always used with another method. The advantages of implantation for ohmic contact fabrication are as follows:

• high homogeneity;
• higher lateral flexibility compared with epitaxial methods;
• the process is not sensitive to surface oxides;
• high stability, if the implanted material has low diffusion constants in GaAs.

7.5 p-TYPE OHMIC CONTACTS ON GaAs

p-Type ohmic contacts are needed for heterostructure bipolar transistors, for p-channel FETs and for p–n diodes. Publications on p-type ohmic contacts are few compared to those on n-type material. The popular Ti/Pt/Au contact is suitable for high temperatures [53]. Zinc-based contacts cannot be used because of the high diffusion coefficient of Zn in GaAs. Table 7.3 gives a survey of contacts for high temperature operation. The following sections give more details about some of them and consider new approaches. Many other metallization schemes exist, e.g. Si/Ni(Mg) [60] and Mn/In/Mo [61].

Table 7.3 High temperature stable ohmic contact systems on p-GaAs from literature[a]

Contact system	T_F (t_F)	ρ_C or R_C (N_A)	Thermal stability	Remark
Ti/Pt/Au	920 K (30 s)	$1 \times 10^{-5}\,\Omega\,cm^2$ ($3 \times 10^{19}\,cm^{-3}$)	670 K (1200 h)	Stable; Ti–As reaction; thick (77 nm) Pt as diffusion barrier against Au and Ga; observed diffusion of small amount of Au and Ga [53]
Ni/Mg/Ni/Si	670 K (30 min)	$2 \times 10^{-6}\,\Omega\,cm^2$ ($1.3 \times 10^{18}\,cm^{-3}$)	620 K (20 h)	Stable; homogeneous and thin Mg doping in regrown GaAs layer, but unstable at 400 °C [54]
Pd/Mn/Sb/Pd	770 K (10 s)	$2 \times 10^{-6}\,\Omega\,cm^2$ ($2.5 \times 10^{18}\,cm^{-3}$)	670 K (20 h)	Stable; regrown GaAsSb epitaxial layer with Mn doping; unstable at 500 °C [49]
Pd/Ge/Ti/Pt	720 K (15 s)	$6.4 \times 10^{-7}\,\Omega\,cm^2$ ($5 \times 10^{19}\,cm^{-3}$)	570 K (20 h)	Stable; higher p-background doping than activated Ge concentration in GaAs; Ge diffusion in As place during long-term annealing [62]
Ni/In–Mn/W	1070 K (2 s)	$0.6\,\Omega\,mm$ ($1 \times 10^{15}\,cm^{-2}$)[b]	670 K (30 h)	Stable; regrown InGaAs epitaxial layer; evaporated with 10% Mn-doped In [49]

[a]T_F (t_F) = formation temperature (time); ρ_C (N_A) specific contact resistance (p-doping); R_C = contact resistance.
[b]Dose of Be$^+$ + F$^+$ [28] ions during implantation for surface doping.

7.5.1 Ti/Pt/Au contact on p-GaAs

The Ti/Pt/Au contact is widely used for the base contacts of HBTs. The Ti/Pt/Au contact shows stable contact resistance when operationating at 550 K. However, a degradation mechanism can occur, even if the contact resistance remains stable. The Ti/Pt/Au contact is observed to sink into the GaAs. This is a serious problem, especially for HBT devices. If the base is penetrated by the contact, the base–collector junction will degrade. An important prerequisite for the stability of the Ti/Pt/Au contact is the proper function of the Pt layer acting as a diffusion barrier for Au. The thickness of the Pt layer has to be chosen thick enough to be an effective barrier for Au. How is it possible to control the reaction of the contact? The Ti is reacting with the As at the GaAs surface, producing the ohmic behavior. The Ti layer should be very thin. This will stop the AsTi reaction after saturation of the phase. The Ga from the GaAs surface reacts with the Pt layer. A PtGa phase is formed [11, 12, 53] which consumes the surplus Ga. This is another reason for choosing a thick Pt layer. With this approach it is possible to control the stability of the contact up to about 650 K.

A homogeneous reaction at the surface is important for the formation of the contact. This can be achieved by an optimized surface treatment prior to the evaporation.

7.5.2 Ti/Mo/TiO/Mo/Au contacts on p-GaAs

A variation of the Ti/Pt/Au contact can be obtained by changing the diffusion barrier Pt by Mo/TiO/Mo. The contact resistances are in the range of $2 \times 10^{-5} \Omega \, cm^2$ to $3 \times 10^{-6} \Omega \, cm^2$ depending on the doping.

Figure 7.11 shows the XPS sputter profile before and after the formation at 800 K. The oxygen concentration in the Mo layer is about 20–25% and much more in the TiO layer. There is no remarkable reaction at the Ti–GaAs interface. It is difficult to avoid the oxygen in the first Ti layer. After annealing, the Ti and GaAs react at the Ti–GaAs interface and most probably form a Ti/As phase [56]. Ga has diffused to a small extent into the MoO layer and stops at the TiO layer. The formation of a Ti/As phase was also found by Katz [11] in Ti/Pt/Au contacts. A Ga-rich layer at the metal–GaAs interface was believed to be the cause of the p-doping in the GaAs. This contact does not degrade further as in the case of the Ti/Pt/Au contact, where a PtGa phase consumes the Ga [11, 53]. The TiO/MoO diffusion barrier effectively stops the Ga outdiffusion, as demonstrated by the XPS sputter profile, thus maintaining the Ga-rich interface. It can be seen clearly that the diffusion barrier is also effective in preventing Au indiffusion. The contact resistance of this contact remained almost constant in a stress test at 670 K for 600 h.

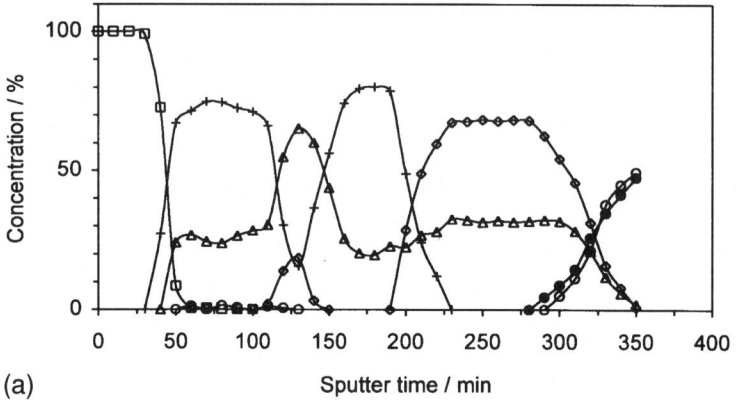

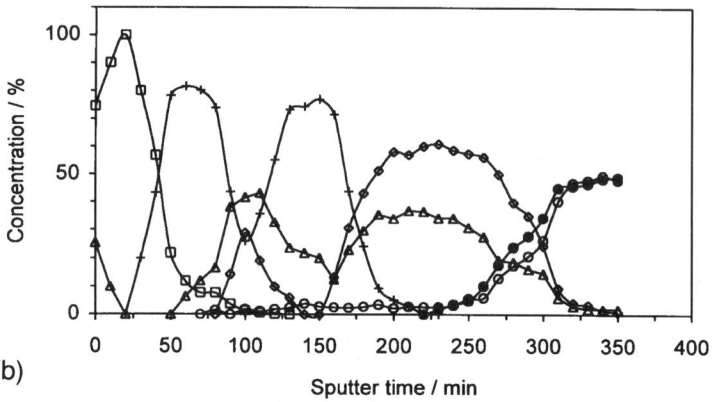

Figure 7.11 XPS sputter profile of a GaAs/Ti/Mo/TiO/Au ohmic contact (a) after the evaporation and (b) after annealing at 800 K: ($\square$) Au 4f, ($+$) Mo $3p_{3/2}$, ($\triangle$) O 1s, ($\diamond$) Ti $2p_{3/2}$, ($\bullet$) As $2p_{3/2}$ and ($\circ$) Ga $2p_{3/2}$.

7.5.3 Pd/Mg/In/LaB$_6$/Au contacts on p-GaAs

The Pd/Mg/In/LaB$_6$/Au contact is a development of the Pd/In system for application in heterostructure bipolar transistors (HBTs). With this contact metallization it is intended to achieve regrowth of an epitaxial InGaAs layer, which is p-doped in this case. The schematic band diagrams of InGaAs and GaAs regrown phases on GaAs are shown in Fig. 7.12. Mg acts as a p-dopant in this contact. The thicknesses of the layers are Pd(6 nm)/ Mg(2 nm)/In(12 nm)/LaB$_6$(100 nm)/Au(70 nm).

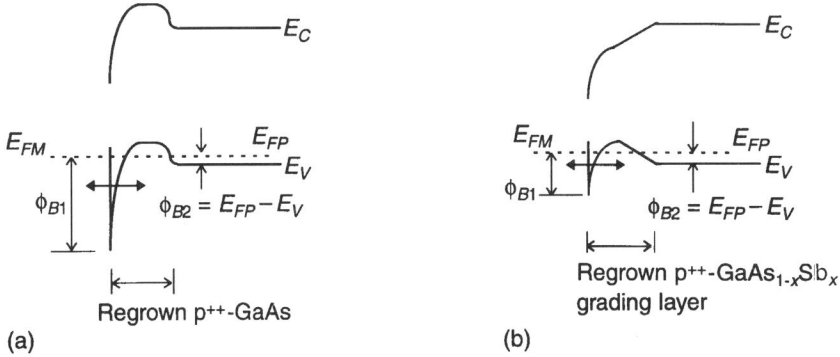

Figure 7.12 Schematic band diagram of (a) regrown metal/regrown p^{++}-GaAs/p-GaAs substrate and (b) regrown metal/regrown p^{++}-InGaAs/p-GaAs substrate.

The contact is formed at 870 K for 20 s (RTA) together with the n-type contacts of the HBT. The n-type contacts also utilize the principle of band gap lowering (Pd/In contacts). The contacts have a mirror-like surface before and after formation. The results of a lifetime test at 670 K are shown in Fig. 7.13. The test shows stable contact resistance after initial stabilization.

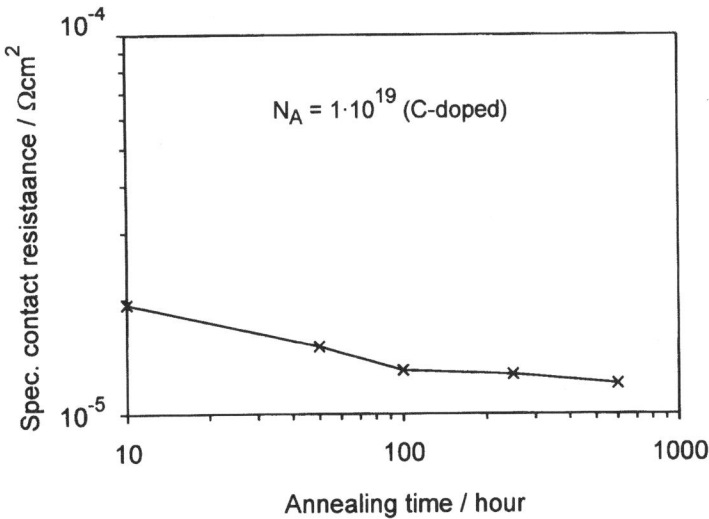

Figure 7.13 Lifetime test of the contact resistance of a Pd/Mg/In/LaB$_6$/Au contact on p-GaAs at 670 K.

7.6 OHMIC CONTACTS FOR n- AND p-TYPE GaAs

An interesting application of Pd/In contact systems is the simultaneous contact of n-GaAs, p-GaAs, n-InGaAs and p-InGaAs [63]. This is made possible by the lowering of the band gap, which allows n- and p-doped semiconductors to exhibit ohmic behavior. Simultaneous contact reduces the number of processing steps for the fabrication of heterostructure bipolar transistors.

For an optimum ohmic contact resistance the n- and p-type doping of the corresponding epitaxial layers of the semiconductor should be very high. A suitable metallization for this purpose is Pd/Si/In/Pd/W/Au, since Si does not act as a dopant in this contact system. Appropriate thicknesses of the layers are Pd(12 nm)/Si(3 nm)/In(36 nm)/Pd(6 nm)/Mo(30 nm)/Au(100 nm).

Figure 7.14 shows the contact resistance as a function of the formation temperature (measured using the TLM method) for a Pd/Si/In/Pd/W/Au contact. The optimum formation temperature is the same for p- and n-type

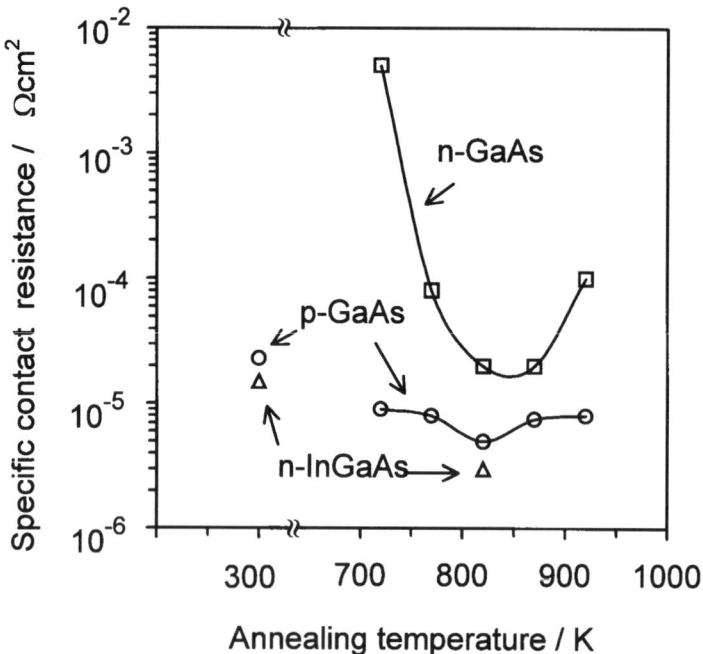

Figure 7.14 Contact resistance of a Pd/Si/In/Pd/W/Au contact on n-GaAs, p-GaAs, n-InGaAs as a function of the annealing temperature: n-GaAs, $N_D = 5 \times 10^{17} \, cm^{-3}$; p-GaAs, $N_A = 5 \times 10^{18} \, cm^{-3}$; and n-InGaAs, $N_D = 5 \times 10^{18} \, cm^{-3}$.

GaAs. This indicates that the ohmic behavior is caused by the band gap lowering and that doping effects can be neglected.

The contact resistance for n-type InGaAs is also given. For the optimum formation temperature of 720 K it is little improved compared to the contact resistance without annealing. The stability data are comparable with the other Pd/In contacts. The Pd/Si/In/Pd metallization is a good choice for simple processing of high quality n-type and p-type ohmic contacts. It has the advantages of high stability, excellent morphology and a reasonable contact resistance.

Another approach was made by Han *et al.* [62]. They tested the Pd/Ge/Ti/Pt system, originally designed for n-type GaAs, for application on p-GaAs. It exhibits good results on p-type material only if the doping is very high; this is because it has to compensate the n-type doping due to the Ge. The Al–Sn–Ni metallization shows similar ohmic behavior to p- and n-type GaAs, as demonstrated by Roedel *et al.* [63].

7.7 HIGH TEMPERATURE STABLE SCHOTTKY CONTACTS FOR GaAs

High temperature Schottky contacts have been developed for fabricating self-aligned GaAs MESFETs [16]. The gate was used as an implantation mask. The Schottky contact has to experience the following annealing step without degradation. The stability of the gate was achieved by refractory materials which suppress the interaction of the GaAs, the contact material and the Au top layer. Stability of Schottky contacts may be monitored by the measurement of barrier height, ideality factor, reverse saturation current, breakdown voltage and series resistance.

The classic metallization is Ti/Pt/Au. In a comparative study, Würfl *et al.* [59] investigated three different diffusion barriers for the Ti contact to p-type GaAs: Ti/Pt/Au, Ti/W/Au and WSi_2/Ti/Pt/Au. The result of a stress test at 570 K can be summarized as follows:

- Ti/W/Au contacts showed a slight improvement of the ideality factor during the first hours of the stress test. After about 100 h the ideality factor increased dramatically, whereas the barrier height reduced to values of about 0.5 eV. This indicates that the contact is not stable enough for operation at increased temperatures. The e-beam evaporated W diffusion barrier was made responsible for the degradation. Au–Ga interdiffusion and Ti outdiffusion were assumed to be the failure mechanism.

- The Ti/Pt/Au contact remained stable in the stress test with regard to barrier height (0.8 eV) and ideality factor (1.15) after 100 h. However, an increase of the reverse saturation current was observed after this time.

After longer stress times, the reverse saturation current increased further and produced nearly ohmic behavior.

- The WSi_2/Ti/Pt/Au Schottky contact also had stable barrier height (0.75 eV) and ideality factor (1.4) after an initial stabilization phase of 1 h stress. In contrast to the Ti/Pt/Au contact, the reverse saturation current remained stable. This was confirmed by a stress test at 670 K.

Canali *et al.* [73] demonstrated an instability of the Ti/W/Au gate metallization in MESFET devices, causing a gate-sinking effect.

Also LaB_6–Au Schottky contacts to GaAs have been proposed by Takatani [58, 64, 72]. LaB_6 is a refractory material which has a very dense hexagonal crystal structure. LaB_6 is a strong getter for oxygen and is therefore able to consume the oxygen at the GaAs–LaB_6 interface. LaB_6 is a barrier against the outdiffusion of As and interdiffusion of Au and Ga. An additional advantage, especially for high temperature MESFETs, is the barrier height of LaB_6, about 0.95 eV. LaB_6 exhibits excellent DC characteristics [58].

The reliability aspects of LaB_6–Au Schottky contacts to GaAs were studied by Würfl *et al.* [65]. A practical application of LaB_6–Au Schottky contacts in a high temperature GaAs IC were very reliable [33].

7.8 DIFFUSION BARRIERS

Diffusion barriers are an important prerequisite for increasing the reliability of GaAs high temperature devices. The most important function of the diffusion barrier is to prevent the outdiffusion of As, which has a high vapor pressure [74]. And an Au–Ga interaction has to be avoided, since an Au top layer is needed for low sheet resistance.

Materials for diffusion barriers have high melting points and a high density. Often very dense hexagonal crystal structures are used. Very important is the structure of the film. Diffusion barriers should be either amorphous or polycrystalline. Amorphous barriers are obtained by sputter deposition. The weak points of polycrystalline films are the grain boundaries. Decoration by oxygen is a common method to avoid diffusion along the grain boundaries. These diffusion barriers are called stuffed barriers.

The following materials have been proposed as diffusion barriers:

- W [59, 66, 73] is a suitable refractory material which may be deposited either by sputtering or by e-beam evaporation. The different results obtained in the literature [59, 73] indicate that the stability of the barrier is questionable.
- WN has been proposed by Nicolet *et al.* [66]. This material is an excellent diffusion barrier against Au–Ga interdiffusion.

- $W_{0.5}Si_{0.2}$ has been used by many authors. The diffusion barrier has proven effective against Au and Ga diffusion up to 670 K [18, 19, 67]. A multilayer structure and subsequent annealing as well as sputtering are suitable deposition mechanisms.
- The composition WSi_2 [68] has also been proposed for application in diffusion barriers.
- TiB_2 is a suitable material for diffusion barriers on GaAs [70, 71].
- A new solution is the Mo/TiO/Mo metallization. Oxygen is a suitable material for decoration of the grain boundaries in a polycrystalline film. The grain boundaries are known to be fast diffusion paths [29]. TiO and MoO are well-conducting metal oxides [31]. A TiO layer was deposited to control the oxygen. The function of such a barrier against Ga and Au diffusion was demonstrated above.
- TiN is known to be a very effective diffusion barrier [69].
- LaB_6 can be used as a diffusion barrier [58, 64, 72]. This barrier was proved to be stable up to 673 K for 600 h [65]. An annealing step is important for stability. After evaporation the diffusion barrier is stabilized for 24 h at 373 K. During this stabilization a segregation of oxygen in the grain boundaries takes place and the crystal structure changes. The grain boundaries are later decorated by oxygen.

7.9 CONCLUSION

This chapter has reviewed the technology of GaAs contacts. The most challenging work is to optimize the ohmic contacts. Although the eutectic AuGe/Ni contact to n-type GaAs exhibits the best contact resistance, its morphology makes it difficult to realize submicron technology. The stability is not sufficient for high temperature operation.

Some attempts have been made to optimize the AuGe/Ni contact by reducing the Au content and by the introduction of a diffusion barrier. These contacts exhibit reasonable contact resistance and high temperature stability with an acceptable morphology.

The ohmic contact using a Pd/In/Pd metallization is relatively new. The ohmic behavior is achieved by lowering the semiconductor band gap. An epitaxial InGaAs layer is regrown on the GaAs out of a solid phase for this purpose. The contact has optimum stability and morphology. An intermediate layer to control the formation reaction is essential to obtaining a mirror-like surface. The contact is useful for highly doped semiconductors because of its relatively high contact resistance. Since this contact is ohmic on p-type and n-type GaAs and InGaAs, a great flexibility in the application has been obtained.

The In/Ni contacts similarly reduce the band gap of the semiconductor. Because of the reduced ability of Ni to penetrate the native oxide of GaAs, the quality of these contacts is believed to be lower than for In/Pd contacts. Pd/Ge contacts use the principle of band gap lowering at the interface. Because of an instability, due to an interdiffusion process at the semiconductor–metal interface, the application of this contact is limited to temperatures less than 600 K.

Epitaxial ohmic contacts are a useful approach for the contacts to the top layer of a device: source and drain of a high electron mobility transistor (HEMT) and emitter of an emitter-up HBT. Epitaxial contacts exhibit excellent reliability and low contact resistance. Various InGaAs layer structures are possible.

Ti/Pt/Au and Ti/Mo/TiO/Mo/Au contacts have been discussed for p-type GaAs. Ti/Mo/TiO/Mo/Au has the better reliability because of the improved diffusion barrier. $Pd/Mg/In/LaB_6/Au$ contacts are the suitable variation of the Pd/In contact on p-type GaAs; they have the same reliability.

It is relatively easy to fabricate a high temperature stable Schottky metallization. Improved fabrication techniques mean that Ti/Pt/Au, $WSi_2/$ Ti/Pt/Au and LaB_6/Au contacts can be operated reliably at 560 K.

Several diffusion barriers have been summarized. Diffusion barriers require proper control of the deposition technique, since their function relies on the morphology of the evaporated film.

REFERENCES

1. Crozat, P. *et al.* (1992) Cryogenic behaviour of ultrashort gate AlGaAs/GaAs and pseudomorphic AlGaAs/InGaAs/GaAs HEMTs, in *Microelectronic Engineering 19*. Proceedings of ESSDERC 92, Leuven, Belgium, September 14–17, Elsevier, pp. 861–864.
2. Kraemer, B., Basset, R., Chye, P., Day, D. and Wei, J. (1994) Power PHEMT module delivers 12 watts, 40% PAE over the 8.5 to 10.5 GHz band, in *Proc. IEEE MTT-S Symposium, San Diego CA, 23–27 May*, pp. 683–686.
3. Hartnagel, H.L. (1995) III–V semiconductor properties for high temperature electronics. *Materials Science and Engineering B*, **29**(1–3), 47–53.
4. Papanicolaou, N.A., Anderson, W.T., Katzer, D.S., Jones, S.H. and Jones, J.R. (1994) All-refractory GaAs FET for high temperature applications, in *Proc. 2nd Int. High Temperature Electronics Conference, Charlotte NC, 5–10 June*, pp. V9–V14.
5. Baier, S., Nohava, J., Jeter, R., Carlson, R. and Hanka, S. (1994) High temperature electronics using complementary heterostructure FET (CHFET) technology, in *Proc. 2nd Int. High Temperature Electronics Conference, Charlotte NC, 5–10 June*, pp. V21–V26.
6. Sadwick, L.P., McDonald, R.M., Crofts, R.J., Koniak, J. and Hwu, R.J. (1994) 350 °C GaAs MESFET-based electronic technology, in *Proc. 2nd Int. High Temperature Electronics Conference, Charlotte NC, 5–10 June*, pp. V27–V32.
7. Fricke, K., Hartnagel, H.L., Schütz, R., Schweeger, G. and Würfl, J. (1989) A new GaAs technology for stable FETs at 300 °C. *IEEE Electron Device Letters*, **10**(12), 577–579.
8. Fricke, K., Krozer, V., Lee, W.-Y., Schüßler, M., Schweeger, G., Sigurdardóttir, A. and Hartnagel, H.L. (1994) Design and technology of GaAs/AlGaAs HBT for high temperature circuits, in *Proc. Gallium Arsenide Applications Symposium 94, Torino, Italy, 28–29 April*, pp. 123–126.

9. Lee, W.-Y., Schüßler, M., Fricke, K. and Hartnagel, H.L. (1994) Influence of thermal and current stress on high-temperature performance of AlGaAs/GaAs/AlGaAs DHBT, in *Proc. 2nd Int. High Temperature Electronics Conference, Charlotte NC, 5–10 June*, pp. P69–P74.

10. Pettenpaul, E., Heidenreich, W., Huber, J. and Flossmann, W. (1985) A high-temperature sensor based on monolithic GaAs Hall IC, in *Proc. GaAs IC Symposium, April*, pp. 169–172.

11. Katz, A., Abernathy, C.R. and Pearton, S.J. (1990) Pt/Ti ohmic contacts to ultrahigh carbon-doped p-GaAs formed by rapid thermal processing. *Appl. Phys. Lett.*, **56**(11), 1028–1030.

12. Katz, A., Nakahara, S., Savin, W. and Weir, B.E. (1990) Microstructure and contact resistance temperature dependence of Pt/Ti ohmic contact to Zn-doped GaAs. *J. Appl. Phys.*, **68**(8), 4133–4140.

13. Wang, L.C., Li, Y.Z., Kappes, M., Lau, S.S., Hwang, D.M., Schwarz, S.A. and Sands, T. (1992) The Si/Pd(Si, Ge) ohmic contact on n-GaAs. *Appl. Phys. Lett.* **60**(24), 3016–3018.

14. Wang, L.C., Wang, X.Z., Hsu, S.N., Lau, S.S., Lin, P.S.D., Sands, T., Schwarz, S.A., Plumton, D.L. and Kuech, T.F. (1991) An investigation of the Pd–In–Ge nonspiking ohmic contact to n-GaAs using transmission line measurement, Kelvin, and Cox and Strack structures. *J. Appl. Phys.*, **69**(8), 4364–4372.

15. Christou, A. (ed.) (1991) *Reliability of Gallium Arsenide MMICs*, Wiley, New York.

16. Williams, R.E. (1985) *Gallium Arsenide Processing Technologies*, 2nd edn, Artech House, Boston, MA.

17. Shen, T.C., Gao, G.B. and Morkoc, H. (1992) Recent developments in ohmic contacts for III–V compound semiconductors. *J. Vac. Sci. Technol. B*, **10**(5), 2113–2132.

18. Hartnagel, H.L. (1993) Concepts of ultrastable metal contacts and their evaluation. *Materials Science and Engineering B* **20**, 141–143.

19. Würfl, J. (1989) *Herstellung und Untersuchung zuverlässiger Metallkontakte auf GaAs zur Entwicklung von hochtemperaturstabilen Halbeiterbauelementen*, VDI Verlag, Düsseldorf.

20. Rosenberg, R. (1985) Inhibition of electromigration damage in current-stressed Al-gates as used for GaAs-MESFETs. *J. Phys. D: Appl. Phys*, **18**, 263–270.

21. Sethi, B.R. and Hartnagel, H.L. (1986) Characterization of electromigration in metal conductors on GaAs surfaces. *Int. J. Electronics*, **61**, 27–31.

22. Berger, H.H. (1972) Models for contacts to planar devices. *Solid-State Electronics*, **15**(2A), 145–158.

23. Williams, R. (1990) *Modern GaAs Processing Methods*, Artech House, Boston.

24. Shih, K.K. and Blum, J.M. (1972) Contact resistances of Au–Ge–Ni, Au–Zn and Al to III–V compounds. *Solid-State Electronics*, **15**, 1177–1180.

25. Han, W.Y., Lu, Y., Lee, H.S., Cole, M.W., Schauer, S.N., Moerkirk, R.P. and Jones, K.A. (1992) Annealing effects on heavily carbon-doped GaAs. *Appl. Phys. Lett.*, **61**, 87–89.

26. Bruce, R.A. and Piercy, G.R. (1987) An improved Au–Ge–Ni ohmic contact to n-type GaAs. *Solid-State Electronics*, **30**(7), 729–737.

27. Gupta, R.P. and Khokle, W.S. (1985) Gallium-vacancy-dependent diffusion model of ohmic contacts to GaAs. *Solid-State Electronics*, **28**(8), 823–830.

28. Zanoni, E., Callegari, A., Canali, C., Fantini, F., Hartnagel, H.L., Magistrali, F. and Paccagnella, A. (1990) Metal–GaAs interaction and contact degradation in microwave MESFETs. *Quality and Reliability Engineering International*, **6**, 29–46.

29. Sands, T. (1989) Compound semiconductor contact metallurgy. *Materials Science and Engineering B*, **1**, 289–312.

30. Fricke, K., Lee, W.Y., Krozer, V., Würfl, J., Bialas, S. and Hartnagel, H.L. (1992) Microwave characterization and comparison of performance of GaAs based MESFETs, HEMTs and HBTs operating at high ambient temperature, in *Proc. GAAS'92, Noordwijk, Netherlands, 27–29 April*.

31. Wittmer, M. (1984) Barrier layers: principles and application in microelectronics. *J. Vac. Sci. Technol. A*, **2**(2), 273–280.

32. Shih, Y.-C., Murakami, M., Wilkie, E.L. and Callegari, C. (1987) Effect of interfacial microstructure on uniformity and thermal stability of AuGeNi ohmic contact to n-type GaAs. *J. Appl. Phys.*, **62**(2), 582–590.

33. Böttner, T., Fricke, K., Goldhorn, A., Hartnagel, H.L., Rappl, A., Ritter, S. and Würfl, J. (1991) Technology and performance of a high temperature stable (up to 300 °C) operational amplifier on GaAs, in *Proc. 1st Int. High Temperature Electronics Conference, Albuquerque NM, 16–20 June*, pp. 77–82.

34. Würfl, J., Fricke, K. and Hartnagel, H.L. (1990) Nonalloyed, high temperature stable ohmic contacts to GaAs based on LaB_6 diffusion barriers, in *Proc. Int. Symp. GaAs and Related Compounds, Jersey, UK, 24–27. September*, pp. 239–244.

35. Lakhani, A.A. (1984) The role of compound formation and heteroepitaxy in indium-based ohmic contacts to GaAs. *J. Appl. Phys*, **56**, 1988.

36. Ding, J., Washburn, J., Sands, T. and Keramidas, V.G. (1986) In/GaAs reaction: effect of intervening oxide layer. *Appl. Phys. Lett.*, **49**, 818–820.

37. Richter, R. (1991) *Mikrocharakterisierung und -strukturierung von GaAs(100)-Oberflächen und deren Metallisierungen*. VDI Verlag, Düsseldorf.

38. Allen, L.H., Hung, L.S., Kavanagh, K.L., Phillips, J.R., Yu, A.J. and Mayer, J.W. (1987) Ohmic contacts to n-type using In/Pd metallization. *Appl. Phys. Lett.*, **51**, 326–327.

39. Marvin, D.C., Ives, N.A. and Leung, M.S. (1985) In/Pt ohmic contacts to GaAs. *J. Appl. Phys.*, **58**, 2659.

40. Murakami, M., Price, W.H., Shih, Y.C., Braslau, N., Childs, K.D. and Parks, C.C. (1987) Thermally stable ohmic contacts to n-type GaAs. *J. Appl. Phys.*, **62**, 3295–3303.

41. Han, C.C., Wang, X.Z., Lau, S.S., Potemski, R.M., Tischler, M.A. and Kuech, T.F. (1991) Thermally stable and nonspiking Pd/Sb(Mn) ohmic contact to p-GaAs. *Appl. Phys. Lett.*, **58**(15), 1617–1619.

42. Yano, K. and Katoda, T. (1991) Raman spectra and electric resistance of thermally treated In/GaAs structures. *J. Appl. Phys.* **70**(11), 7036–7041.

43. Pirling, T., Fricke, K., Schüßler, M., Lee, W.Y., Fueß, H. and Hartnagel, H.L. (1995) Investigations of Pd/In based high temperature ohmic contacts on GaAs by x-ray reflectometry and diffractometry. *Materials Science and Engineering B*, **29**, 70–73.

44. Fricke, K., Hartnagel, H.L., Lee, W.-Y., Pierling, T., Schüßler, M. and Würfl, J. (1994) A highly reliable ohmic contact system on GaAs based on Pd/In, in *Proc. 2nd Int. High Temperature Electronics Conference, Charlotte NC, 5–10 June*, P197–P202.

45. Wang, L.C., Wang, X.Z., Lau, S.S., Sands, T., Chan, W.K. and Kuech, T.F. (1990) Stable and shallow PdIn ohmic contacts to n-GaAs. *Appl. Phys. Lett.*, **56**(21), 2129–2131.

46. Prasad, K., Faraone, L. and Nassibian, A.G. (1991) Thermal stability of Pd–In ohmic contacts to n-GaAs formed by scanned electron beam and rapid thermal annealing. *Electronics Lett.*, **27**(2), 149–151.

47. Murakami, M., Price, W.H., Greiner, J.H., Feder, J.D. and Parks, C.C. (1989) Thermally stable ohmic contacts to n-type GaAs. V. Metal–semiconductor field-effect transistors with NiInW ohmic contacts. *J. Appl. Phys.*, **65**(9), 3546–3551.

48. Hugon, M.C., Agius, B., Varniere, F., Dubon-Chevallier, C., Bresse, J.F. and Froment, M. (1991) Thermally stable low resistance ohmic contacts to n-type gallium arsenide: Magnetron cathodic sputter-deposited NiInW contacts. *Appl. Phys. Lett.* **58**(24), 2773–2775.

49. Hallali, P.-E., Murakami, M., Price, W.H. and Norcott, M.H. (1991) Thermally stable ohmic contacts to p-type GaAs. IX. NiInW and NiIn(Mn)W contact metals. *J. Appl. Phys.*, **70**(12), 7443–7448.

50. Fricke, K., Hartnagel, H.L., Lee, W.Y. and Schüßler, M. (1994) AlGaAs/GaAs/AlGaAs DHBT operational amplifier for high temperature application. *IEEE Electron Device Letters*, **15**(3), 88–90.

51. Tsuchimoto, J., Shikata, S. and Hayashi, H. (1991) Thermally stable Pd/Ge ohmic contacts to n-type GaAs. *J. Appl. Phys.*, **69**, 6556–6563.

52. Chen, C.L., Mahoney, L.J., Finn, M.C., Brooks, R.C., Chu, A. and Mavroides, J.G. (1986) Low resistance Pd/Ge/Au and Ge/Pd/Au ohmic contacts to n-type GaAs. *Appl. Phys. Lett.*, **47**, 535.

53. Schweeger, G. (1992) *Technologie und Charakterisierung des Bipolar-Mode-Feldeffekt-transistors auf Galliumarsenid*. Verlag Shaker, Aachen.

54. Han, C.C., Wang, X.Z., Wang, L.C., Marshall, E.D., Lau, S.S., Schwarz, S.A., Palmstrom, C.J., Harbison, J.P., Florez, L.T., Potemski, R.M., Tischler, M.A. and Kuech, T.F. (1990)

Nonspiking ohmic contact to p-GaAs by solid-phase regrowth. *J. Appl. Phys.*, **68**(11), 5714–5718.

55. Allen, L.H., Hung, L.S., Kavanagh, K.L., Phillips, J.R., Yu, A.J. and Mayer, J.W. (1987) Ohmic contacts to n-GaAs using In/Pd metallization. *Appl. Phys. Lett.* **51**(5), 326–327.

56. Brillson, L.J. (ed.) (1993) *Contacts to Semiconductors*, Noyes Publications, Park Ridge NJ.

57. Lustig, N., Murakami, M., Norcott, M. and McGann, K. (1991) Low Au content thermally stable NiGe(Au)W ohmic contacts to n-type GaAs. *Appl. Phys. Lett.* **58**(19), 2093–2095.

58. Takatani, S., Uchida, Y., Yokotsuka, T. and Nakashima, H. (1987) GaAs MESFETs with a thermally stable LaB_6 self-aligned gate. *Jpn. J. Appl. Phys.*, **26**(11), L1770–L1773.

59. Würfl, J., Merkl, B. and Nold, U. (1989) Suitability of GaAs Schottky metallizations for continuous device operation at elevated temperatures up to 300 °C: a comparative study. *Int. J. Electronics*, **66**(3), 437–444.

60. Han, C.C., Wang, X.Z., Lau, S.S., Potemski, R.M., Tischler, M.A. and Kuech, T.F. (1991) The temperature dependence of the contact resistivity of the Si/Ni(Mg) nonspiking contact scheme on p-GaAs. *J. Appl. Phys.*, **69**(5), 3124–3129.

61. Kalkur, T.S. and Lu, Y.C. (1989) Preliminary studies on Mo–In–Mn based ohmic contacts to p-GaAs. *J. Electrochem. Soc.*, **136**(11), 3549–3550.

62. Han, W.Y., Lu, Y., Lee, H.S., Casas, M.W., DeAnni, A., Jones, K.A. and Yang, L.W. (1993) shallow ohmic contact to both n- and p-GaAs. *J. Appl. Phys.* **74**(1), 754–756.

63. Roedel, R.L., Davito, D., West, W. and Adams, R. (1993) Ohmic contacts to p- and n-type GaAs made with Al–Sn–Ni. *J. Electrochem. Soc.*, **140**(5), 1450–1452.

64. Uchida, Y., Yokotsuka, T., Nakashima, H. and Takatani, S. (1987) Electrical properties of thermally stable LaB_6/GaAs Schottky diodes. *Appl. Phys. Lett.*, **50**(11), 670–672.

65. Würfl, J., Singh, J.K. and Hartnagel, H.L. (1990) Reliability aspects of thermally stable LaB_6–Au Schottky contacts to GaAs, in *Proc. IEEE International Reliability Physics Symposium, New Orleans, 28–29 March*, pp. 87–93.

66. So, F.C.T., Kolawa, E., Tandon, J. and Nicolet, M.-A. (1987) Solid-phase ohmic contact to p-GaAs with W and W–N diffusion barriers. *J. Electrochem. Soc.*, **134**(7), 1755–1758.

67. Fricke, K., Hartnagel, H.L., Schütz, R., Schweeger, G. and Würfl, J. (1989) A new GaAs technology for stable FETs at 300 °C. *IEEE Electron Device Letters*, **10**(12), 577–579.

68. Saraswat, K.C. (1981) WSi_2 interconnections for very-large-scale integrated circuits. *Thin Solid Films*, **83**, 143–144.

69. Remba, R.D., Suni, I. and Nicolet, M.-A. (1985) Use of a TiN barrier to improve GaAs FET ohmic contact reliability. *IEEE Electron Device Letters*, **6**(8), 437–438.

70. Shappirio, J.R., Finnegan, J.J. and Lux, R.A. (1986) Diboride diffusion barriers in silicon and GaAs technology. *J. Vac. Sci. Technol.*, **4**(6), 1409–1415.

71. Choi, C.S., Wang, Q., Osburn, C.M., Ruggles, G.A. and Shah, A.S. (1992) Electrical characteristics of TiB_2 for ULSI applications. *IEEE Trans. Electron Devices*, **39**(10), 2341–2345.

72. Würfl, J., Singh, J.K. and Hartnagel, H.L. (1990) Reliability aspects of thermally stable LaB_6–Au Schottky contacts to GaAs, in *Proc. IEEE International Reliability Physics Symposium, New Orleans, 28–29 March*, pp. 87–93.

73. Canali, C., Castaldo, F., Fantini, F., Ogliari, D., Umena, L. and Zanoni, E. (1986) Gate metallization 'sinking' into the active channel in Ti/W/Au metallized power MESFETs. *IEEE Electron Device Letters*, **7**(3), 185–187.

74. Honig, R.E. and Kramer, D.A. (1969) Vapor pressure data for the solid and liquid elements. *RCA Review*, **30**(2), 285–305.

K. Fricke, V. Krozer and M. Schüßler

8.1 INTRODUCTION

GaAs is a very attractive material for high temperature electronics because it offers the following advantages:

- a high bandgap, especially if the $GaAs/Al_xGa_{1-x}As$ system is considered;
- low intrinsic carrier concentration;
- semiinsulating material for device isolation;
- an inherent radiation hardness [1];
- light-emitting devices are possible;
- excellent high frequency performance;
- heterostructure devices with lattice-matched materials and variable bandgaps.

GaAs high temperature circuits may find application in fields where some of these advantages are important. Heterostructure devices with performance capabilities tailored for a particular application offer a large variety of circuit realizations such as high speed, high temperature ICs, satellite power amplifiers, integration of sensors and signal processing circuitry.

For applications with higher input impedances, field effect transistors can be realized on GaAs, especially metal–semiconductor FET (MESFET), high electron mobility transistor (HEMT) and the junction field effect transistor. MOSFETs suffer from a lack of high quality natural oxide on GaAs. An alternative is the heterostructure isolated gate FET (HIGFET). In this device the gate electrode is located on an $Al_xGa_{1-x}As$ layer with an Al mole fraction of about 0.75 for a separation from the active channel. The different devices are compared in Table 8.1.

Not only is the type of transistor important for the electrical performance of a device but also the properties of the semiconductor material. For high

High Temperature Electronics. Edited by M. Willander and H.L. Hartnagel. Published in 1997 by Chapman & Hall, London. ISBN 0 412 62510 5.

Table 8.1 Typical performance of MESFET, JFET, HEMT and HBT devices

	MESFET	JFET	HEMT	HBT
Voltage gain	low	low	high	high
Current gain	low	low	low	high
High frequency noise	low	low	very low	low
$1/f$ noise	high	high	very high	very low
Low frequency input impedance	high	high	high	low
Complexity of technology	low	low	high	very high
Maturity of technology	very high	very high	high	low

temperature devices it is important to know how the material properties of GaAs are changing with temperature. A summary of the temperature-dependent properties of GaAs, AlGaAs and InGaP is provided in Table 8.2. Several collections of material data for GaAs and AlGaAs are found in the literature [2–12].

8.2 FIELD EFFECT TRANSISTORS

8.2.1 High temperature MESFET

This chapter considers the technology for a GaAs MESFET a field of considerable maturity. MESFETs are used for microwave applications, satellite power amplifiers and optical communications [13]. Some papers cover the high temperature properties of GaAs MESFETs [14–18].

The GaAs HEMT has to be considered for applications at increased temperatures and high frequency; it has better high frequency gain and noise figure especially at high frequencies [19]. Power amplifiers at high frequencies usually exhibit better performance if they are equipped with HEMTs [16,17].

Technology for high temperature MESFET

The technology outlined here for high temperature MESFETs can be used with little changes for AlGaAs/GaAs HEMTs. The technology is nearly identical to the commonly used technology for low temperature MESFETs, except for the metallizations. A conventional process with 'recessed gate' technology [20] and wet-etched mesa isolation is shown in Fig. 8.1.

A MESFET process with seven masks is described as follows:

- Metallorganic vapor phase epitaxy (MOVPE) material with an n^+-contact layer and an active n-doped channel. The material should include a heterostructure buffer for substrate leakage suppression.

Table 8.2 Temperature-dependent material properties of GaAs, AlGaAs, InGaP[a]

Parameter	Formula
	GaAs
Bandgap Γ valley (eV)	$E_{g\Gamma} = 1.517 - 5.5 \times 10^{-4} \dfrac{T^2}{T+225} - 2 \times 10^{-10} N_D^{1/3}$
Effective electron mass, m_n^*/m_0	$m_{n;eff} = \left(7.51\left[\dfrac{2}{E_{g\Gamma}} + \dfrac{1}{E_{g\Gamma}+0.341}\right] + 1\right)^{-1}$
Effective hole mass, m_h^*/m_0	$m_{001light} = 0.094$
	$m_{001heavy} = 0.3399$
	$m_{p;001} = (m_{001light}^{3/2} + m_{001heavy}^{3/2})^{2/3} = 0.3721$
	$m_{111light} = 0.08163$
	$m_{111heavy} = 0.75188$
	$m_{p;111} = (m_{111light}^{3/2} + m_{111heavy}^{3/2})^{2/3} = 0.7697$
Relative permittivity	$\varepsilon_{r;static} = 12.73(1 + 1.2 \times 10^{-4} T)$
	$\varepsilon_{r;infrared} = 10.6(1 + 9 \times 10^{-5} T)$
Effective density of states in conduction band (m^{-3})	$N_{c\Gamma} = 2\left(\dfrac{2\pi m_0 m_{n;eff} kT}{h^2}\right)^{3/2}$
Effective density of states in valence band (m^{-3})	$N_v = 2\left(\dfrac{2\pi m_{p;001} m_0 kT}{h^2}\right)^{3/2}$

Carrier lifetime (s)

$$\tau = 0.34 \times 10^{-9} \exp\left(\frac{V_T}{24\,\text{mV}}\right)$$

Electron mobility ($\text{m}^2\,\text{V}^{-1}\,\text{s}^{-1}$)

$$\mu_{n0}(N_D) = 0.6 + 0.3742\log_e\left(\sqrt{[0.5\log_{10}(N_D \times 10^{-22})]^2 + 1}\right.$$
$$\left. - 0.5\log_{10}(N_D \times 10^{-22})\right)\left(\frac{N_D}{N_D + 1\times 10^{20}}\right)^{0.5215}$$

$$\mu_{n0} = \mu_{n0}(N_D) \times 2.135\,T^{1.11}\exp\left(-\frac{T}{31.1}\right) + 1.08\left(\frac{333.3}{T}\right)^{1.6} 10^{-45.45/T}$$

Hole mobility ($\text{m}^2\,\text{V}^{-1}\,\text{s}^{-1}$)

$$\mu_p(N_A) = 0.0380 + 0.0169\log_e\left(\sqrt{[\log_{10}(N_A\times 10^{-23})]^2 + 1}\right.$$
$$\left. - \log_{10}(N_A\times 10^{-23})\right)\left[\frac{N_A}{N_A + 7.8\times 10^{22}}\right]^{0.6}$$

$$\mu_p = \mu_p(N_A)\left(0.004\,735\,47\,T^{2.41126}\exp\left(-\frac{T}{19.4652}\right)\right.$$
$$\left. + 3.46\left[\frac{52.7858}{T}\right]^{2.33642} 10^{-38.6307/T}\right)$$

Electron saturation velocity ($\text{m\,s}^{-1}$)

$$v_{n;sat} = 0.731\,25 \times 10^5(1.6915 - 2.868\times 10^{-3}T + 1.8944\times 10^{-6}T^2)$$

Hole saturation velocity ($\text{m\,s}^{-1}$)

$$v_{p;sat} = 14.1\times 10^5\left(\frac{T}{5+T}\right)^{16.91} \times \exp\left(-\sqrt{\frac{T}{68.6}}\right)^{1.1}$$

Threshold field ($\text{V\,m}^{-1}$)

$$E_{th} = [3.99\times 10^{-3} + 8.27\log_{10}(1.1777\,T) + 10.3\log_{10}(0.88\,T)^3]$$
$$\times [3.986\,36 + 13.7\times 10^{-3}\log_{10}(0.485\,217\,N_D)$$
$$+ 15.7635\times 10^{-3}(\log_{10}(1.002\,57\,N_D))^3]$$

Table 8.2 (*Continued*)

Parameter	Formula
	$Al_xGa_{1-x}As$
Bandgap Γ valley (eV)	$E_{g\Gamma} = 1.5435 + 1.1895x + 0.37x^2$
	$\quad - (3.95 + 1.15x)\, T \times 10^{-4} - 2 \times 10^{-11} N_D^{1/3}$
Bandgap L valley (eV)	$E_{gL} = 1.734 + 0.574x + 0.055x^2$
Bandgap X valley (eV)	$E_{gX} = 1.911 + 0.005x + 0.245x^2$
Electron affinity (eV)	$\chi = 4.07 - 0.62[E_{g\Gamma}(x) - E_{g\Gamma}(0)]$
Effective electron mass, m_n^*/m_0	$m_n = 0.06$
Effective hole mass, m_h^*/m_0	$m_p = \left(\dfrac{x}{0.85} + \dfrac{1-x}{0.68} \right)^{-1}$
Relative permittivity	$\varepsilon_r = 12.9 - 2.9x$
Effective density of states in conduction band (m^{-3})	$N_{c\Gamma} = 4.827 \times 10^{21} [0.067 + 0.083x]\, T]^{3/2}$
	$N_{cL} = 4.827 \times 10^{21} [(0.55 + 0.12x)\, T]^{3/2} \exp\left(\dfrac{E_{g\Gamma} - E_{gL}}{V_T} \right)$
	$N_{cX} = 4.827 \times 10^{21} [(0.85 + 0.07x)\, T]^{3/2} \exp\left(\dfrac{E_{g\Gamma} - E_{gX}}{V_T} \right)$
	$N_C = N_{c\Gamma} + N_{cL} + N_{cX}$

Effective density of states in valence band (m^{-3})	$N_{vl} = 4.827 \times 10^{21}[(0.068 + 0.192x)\,T]^{3/2}$ $N_{vh} = 4.827 \times 10^{21}(0.5\,T)^{3/2}$ $N_v = N_{vl} + N_{vh}$
Carrier lifetime (s)	$\tau_{300\,\text{K}} = (N_D \times 10^{-16} + (N_D \times 0.25 \times 10^{-20})^2)^{-1}$ $\tau = \tau_{300\,\text{K}} \times 0.34 \exp\left(\dfrac{V_T}{24\,\text{mV}}\right)$
Electron mobility (m^2 V^{-1} s^{-1})	$\mu_{n0} = 0.83\left(\dfrac{300}{T}\right)\left[1 + \dfrac{N_D}{3.98 \times 10^{15} + N_D/641}\right]^{-1/3}$
Hole mobility (m^2 V^{-1} s^{-1})	$\mu_p = \left(0.005 + 0.033\left[1 + \left(\left(\dfrac{N_D}{3.232 \times 10^{17}}\right)^{0.4956}\right)^{-1}\right]\right)\left(\dfrac{300}{T}\right)^{2.1}$
	$\text{Ga}_x\text{In}_{1-x}\text{P}$
Bandgap Γ valley (eV)	$E_{g\Gamma} = 1.1987 + 1.34x - 5.5 \times 10^{-4}\dfrac{T^2}{T + 225}$ $\qquad - 1.6 \times 10^{-10} N_D^{1/3}$
Bandgap L valley (eV)	$E_{gL} = 1.3237 + 1.34x - 7.2 \times 10^{-4}\dfrac{T^2}{T + 205}$
Bandgap X valley (eV)	$E_{gX} = 1.4157 + 1.34x - 6.05 \times 10^{-4}\dfrac{T^2}{T + 204}$
Conduction band discontinuity against GaAs (eV)	$\Delta E_c = 0.108$

Table 8.2 (*Continued*)

Parameter	Formula
Valence band discontinuity against GaAs (eV)	$\Delta E_v = 0.3$
Effective electron mass	$m_{n\Gamma} = 0.172\,357(1-x) + 0.036\,6429x$ $m_{nL} = 0.242$ $m_{nX} = 0.61$
Effective hole mass	$m_{h;light} = 0.14$ $m_{h;heavy} = 0.48$
Effective hole mass, m_h^*/m_0	$m_p = (m_{h;light}^{3/2} + m_{h;heavy}^{3/2})^{2/3}$
Relative permittivity	$\varepsilon = \dfrac{2a(x)+1}{1-a(x)} \qquad a(x) = \dfrac{11.3x}{14.3} + \dfrac{9.75(1-x)}{12.75}$
Effective density of states in conduction band (m^{-3})	$N_{ci} = 2\left(\dfrac{2\pi m_0 m_{ni} kT}{h^2}\right)^{3/2} \qquad i = \Gamma, L, X$ $N_c = N_{c\Gamma} + N_{cL} + N_{cX}$
Effective density of states in valence band (m^{-3})	$N_{vl} = \left(\dfrac{m_0 m_{h;light} kT}{2\pi\hbar^2}\right)^{3/2}$ $N_{vh} = \left(\dfrac{m_0 m_{h;heavy} kT}{2\pi\hbar^2}\right)^{3/2}$ $N_v = N_{vl} + N_{vh}$

Parameter	Value
SHR electron recombination lifetime (s)	$\tau_{n;SHR} = 2.6 \times 10^{-11}$
SHR hole recombination lifetime (s)	$\tau_{h;SHR} = 6 \times 10^{-11}$
Radiative electron recombination lifetime (s)	$\tau_{n;rad} = 1 \times 10^{-10}$
Radiative hole recombination lifetime (s)	$\tau_{h;SHR} = 1 \times 10^{-10}$
Electron mobility ($m^2 V^{-1} s^{-1}$)	$\mu_{n0}(N_D) = 4 - 3.7 \tanh\left(\dfrac{N_D \times 10^{-23} - 0.1}{N_D \times 10^{-23} + 0.7}\right)$ $+ 0.5 \tanh\left(\dfrac{0.01 - N_D \times 10^{-23}}{N_D \times 10^{-23} + 0.1}\right)$
	$\mu_{n0} = \mu_{n0}(N_D) \times \left[0.08 x^{1.26} \exp\left(\dfrac{-x}{55.73}\right) + 0.7 \tanh\left(\dfrac{x - 110}{210}\right)\right]$
Electron saturation velocity ($m\,s^{-1}$)	$v_{n;sat} = 0.62 \times 10^5$
Threshold field ($V\,m^{-1}$)	$E_{th} = 7.404 \times 10^5$

---

[a] In the formulas T is in K and N_D, N_A are in m^{-3}, k is the Boltzmann constant, m_0 is the electron rest mass, h is Planck's constant, and $V_T = kT/q$ is the thermal voltage.

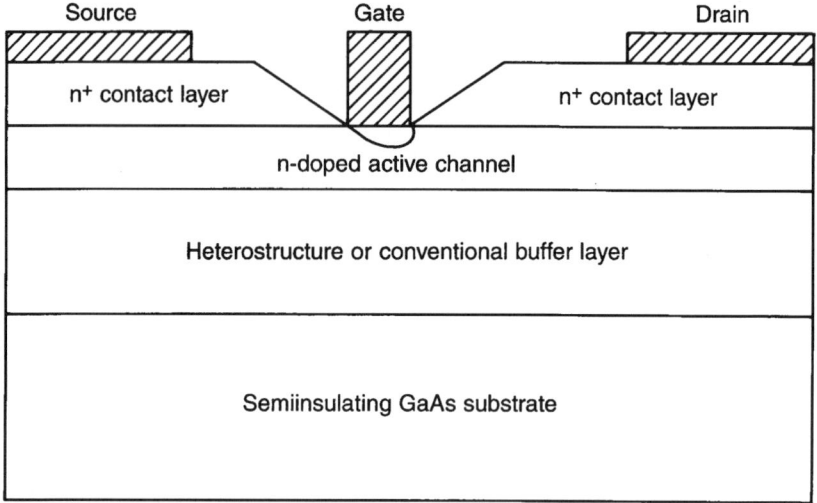

Figure 8.1 Cross section of a MESFET.

- **Mask 1** The process starts as usual for low temperature devices with wet chemical etching of the mesa structure. We use an $NH_4OH:H_2O_2:H_2O$ etchant with the composition 2:1:300 and an etch rate of approximately $80\,nm\,min^{-1}$.

- **Mask 2** The ohmic contacts are then structured with mask 2. A special high temperature stable metallization has to be used. A diffusion barrier can be evaporated on top on the ohmic contact in order to prevent interdiffusion between the contact and a top metallization (commonly Au). This process step is finalized with the annealing of the ohmic contact. Ohmic contact formation for high temperature devices may also be achieved using an InGaAs top layer to provide excellent ohmic contact resistance and high reliability.

- **Mask 3** The n^+-layer or alternatively the InGaAs top layer is removed with the help of mask 3. It is important that the ohmic contacts are not underetched in this step, if they contain Au. Such underetching may cause serious reliability problems.

- The recess may be wet etched or dry etched to achieve the required pinchoff voltage. Ti–Pt–Au may be used for the gate metal up to 523 K. LaB_6 or WSi metallizations are preferable at higher temperatures [21–23].

- In the following step a plasma-enhanced chemical vapor deposition (PECVD) SiN film is deposited. The SiN has to be optimized with respect to the pinhole density, the step coverage and the stress level.

- **Mask 4** Windows have to be etched in the SiN using a mask.
- **Masks 5 and 6** Now the second metallization can be evaporated and the plating can be performed. Two additional masks, masks 5 and 6, are used for air bridges.
- For improved reliability one may use a second SiN layer in order to prevent the second metallization from suffering environmental degradation.

Using this technology with InPd ohmic contacts, LaB_6 Schottky metallization and an optimized SiN passivation, a MESFET has been operated for 1000 h at 673 K without degradation of the current–voltage characteristic [24].

DC characteristics of high-temperature MESFETs

A GaAs MESFET with a channel doping of $1 \times 10^{17} \, \text{cm}^{-3}$ and a gate length of 5 μm exhibits a reasonable DC characteristic up to 573 K, as shown in the subthreshold characteristic of Fig. 8.2 and the output characteristic in Fig. 8.3. In this case the ambient temperature is given as a parameter, not the junction temperature, which is even higher for continuous wave (CW) operation. The MESFET used for the experiments did not have a special heat sink. It is estimated that, for this particular design, the junction temperature is about 80 K above the ambient temperature for CW operation. Pulse measurements were performed to avoid this problem.

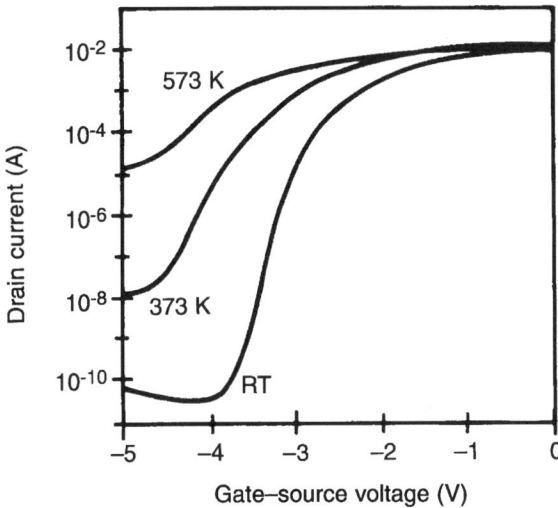

Figure 8.2 Drain current I_D of a GaAs MESFET with a gate length of 5 μm versus the gate source voltage V_{GS} with the ambient temperature as a parameter ($V_{DS} = 2 \, \text{V}$).

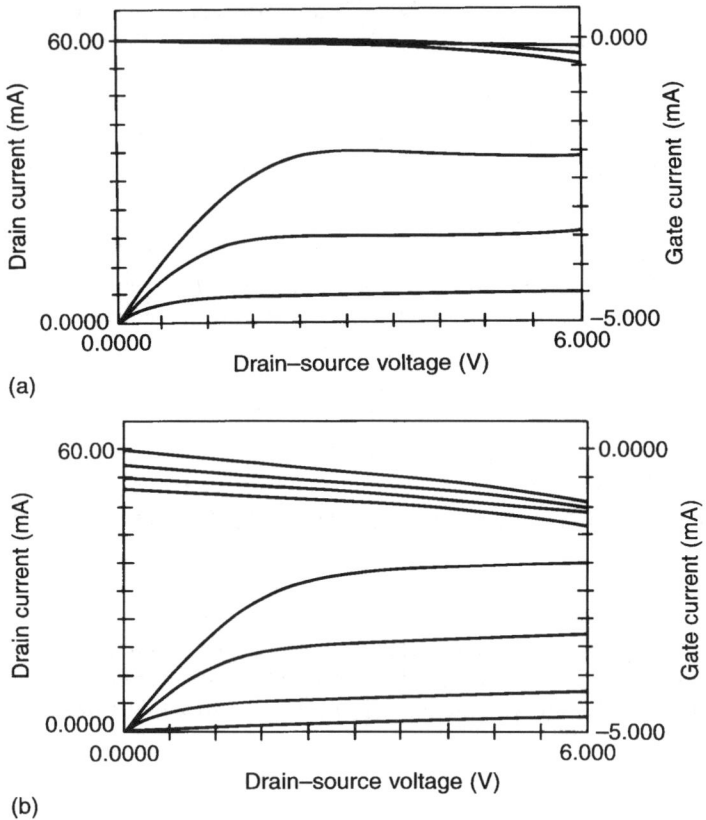

Figure 8.3 Drain current I_D of a GaAs MESFET with a gate length of $5\,\mu m$ versus the drain source voltage V_{DS} with the gate source voltage V_{GS} as a parameter at (a) 300 K and (b) 573 K.

One identifies the following problems in Figs 8.2 and 8.3:

1. The gate current increases with increasing temperature; it increases according to the voltage–current equation of the Schottky diode, depending on the barrier height. The quality of the Schottky diode is therefore an important factor. Ideally, the Schottky barrier should be temperature independent, but due to crystal imperfections, surface pinning of GaAs and interfacial layers between the metal and the crystal, the Schottky barrier drops with increasing ambient temperature. This situation can be seen in Fig. 8.4.

2. At 573 K the MESFET cannot be pinched off completely (Fig. 8.2); the ratio of the drain current in the on-state to the drain current in the off-state decreases with increasing temperature. Two reasons are responsible for this effect. Firstly, the gate current is added to the drain current.

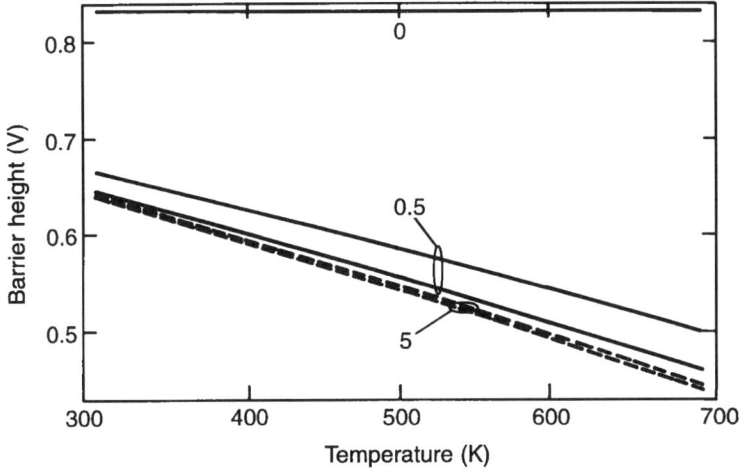

Figure 8.4 Schottky barrier lowering versus temperature with the interfacial layer thickness and interfacial density of states as a parameter. Schottky metal is Ti, layer doping $N_d = 10^{23}\,\mathrm{m}^{-3}$; surface states = $1 \times 10^{18}\,\mathrm{m}^{-2}$, oxide thickness = 2 nm, 5 nm.

Secondly, electrons are injected into the buffer or electron–hole pairs are thermally generated in the buffer layer and the depletion region, making it impossible to achieve complete pinchoff. The two currents reduce the signal-to-noise ratio (SNR) of digital circuits and reduce the efficiency of analog circuits.

3. The drain saturation current and the transconductance decrease with increasing temperature.

Figure 8.5 shows the drain current I_D in the subthreshold region plotted against the reciprocal temperature. For a semiconductor with a high bandgap, the reverse current through a diode is given by the generation current [25] as follows:

$$J_s = \frac{q n_i w}{\tau} \tag{8.1}$$

Figure 8.5 demonstrates, that the slope of the drain current I_D matches very well with the intrinsic carrier concentration divided by the minority carrier lifetime n_i/τ. This supports the fact that a large fraction of the subthreshold current of a MESFET is caused by generation currents [26]. The minority carrier lifetime τ can be expressed by the experimentally fitted equation

$$\tau = 0.34 \tau_{300\mathrm{K}} \exp(V_T/24\,\mathrm{mV}) \tag{8.2}$$

where V_T is the thermal voltage.

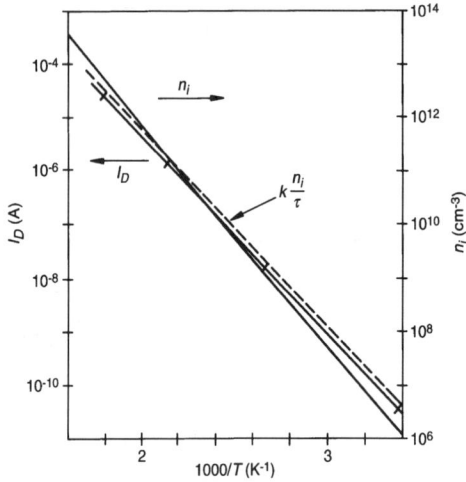

Figure 8.5 Drain current I_D of a GaAs MESFET with a gate length of $5\,\mu m$, intrinsic carrier concentration n_i and intrinsic carrier concentration divided by the minority carrier lifetime n_i/τ versus the reciprocal temperature.

This expression was obtained from analyzing temperature-dependent operation of an AlGaAs/GaAs HBT drift diffusion model with the minority carrier lifetime as the only fitting parameter. Diffusion currents do not play a significant role. Diffusion currents are proportional to n_i^2 and would exhibit a slope proportional to T^{-2}. This would cause a change in the slope of the drain current at increased temperatures.

The problem of an increased generation current can be reduced with an $Al_xGa_{1-x}As$ heterostructure buffer or a GaAs–AlAs multiquantum well structure. These layers prevent electrons in the channel from being injected into the buffer by an enlarged barrier. In principle, the barrier can also be increased by a higher doping of the channel. The disadvantage of this scheme is the fact that an increased channel doping decreases the quality of the Schottky barrier, decreases the electron mobility and increases the pinchoff voltage.

Figure 8.6 shows the DC output characteristic of a MESFET with a gate length of $5\,\mu m$, an $Al_{0.15}Ga_{0.85}As$ channel and an $Al_{0.42}Ga_{0.58}As$ buffer at 300 K and 750 K. The layer sequence is given in Table 8.3.

The MESFET with heterostructure buffer also exhibits drain leakage currents but only at temperatures beyond 750 K. The drain leakage at 750 K, $V_{DS} = 0$ and $V_{GS} = -2$ V is caused only by the gate leakage current. The leakage current through the buffer is effectively suppressed. Figure 8.6 shows that the MESFET can be operated at temperatures up to 750 K. Note that the Schottky barrier is still effective at such high temperatures and the

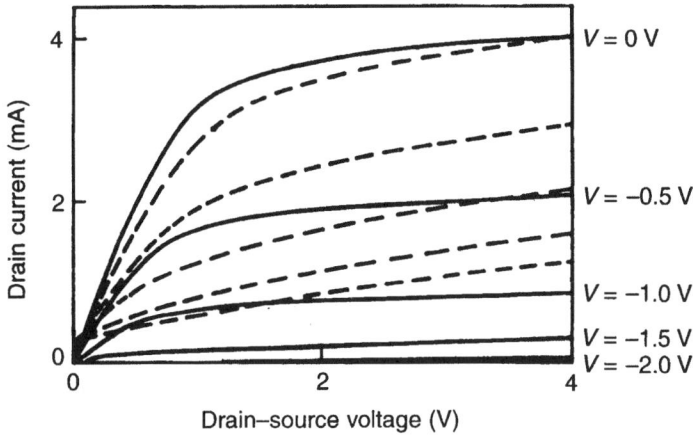

Figure 8.6 DC output characteristic of a MESFET with heterostructure buffer: drain current I_D versus the drain–source voltage V_{DS} with the gate–source voltage V_{GS} as a parameter at (—) 300 K and (---) 723 K; gate length 5 μm.

device exhibits considerable gain. And note that the saturation voltage increases with rising temperature, partly due to the deterioration of the mobility in the channel.

The increased gate leakage currents at high ambient temperatures can be reduced by a suitable choice for the material of the Schottky diode. The Schottky barrier of evaporated LaB_6 is 0.9 eV in contrast to 0.85 eV for Ti–Pt–Au Schottky diodes. These results can only be obtained with a suitable pretreatment of the GaAs surface prior to evaporation. Alternatively a thin $Al_xGa_{1-x}As$ heterostructure can be introduced under the gate. Due

Table 8.3 Sequence of the MOVPE grown layers for the MESFET with heterostructure buffer

Layer	Al concentration (%)	Thickness	Doping
Contact layer	0	200 nm	$N_d = 5.5 \times 10^{17} \, \text{cm}^{-3}$
Grading	$0.15 \rightarrow 0$	150 nm	$N_d = 1 \times 10^{17} \, \text{cm}^{-3}$
Channel	0.15	400 nm	$N_d = 1 \times 10^{17} \, \text{cm}^{-3}$
Grading	$0.45 \rightarrow 0.15$	20 nm	$N_d = 1 \times 10^{17} \, \text{cm}^{-3}$
Buffer	0.45	1000 nm	undoped
Substrate	GaAs	600 μm	$\rho \approx 10^7 \, \Omega \, \text{cm}$

to the higher bandgap of this layer, a higher Schottky barrier can be realized. This is only possible if the technological problems with the strong affinity to oxygen of AlGaAs have been solved. Another possibility could be the incorporation of an InGaP layer which is lattice-matched with GaAs. InGaP material has a larger bandgap than AlGaAs. However, both methods yield poor Schottky contacts because the thin layer is fully depleted and acts as an isolation layer between the Schottky metal and the active channel. Therefore, the ideality factor is expected to increase on this material. In an AlGaAs/GaAs HEMT the gate is located on the AlGaAs electron supply layer, so they too have this advantage. However, the improvement of the Schottky barrier height is limited. An additional problem is the decrease of the Schottky barrier height at increased temperatures. Krozer *et al.* [27] demonstrated that the Schottky barrier height decreases at high temperatures mainly due to the decrease of the bandgap, as indicated in Fig. 8.4. Figure 8.4 shows the calculated barrier height for a Ti contact with an insulating layer of 2 nm and 5 nm, respectively, and the sheet carrier density as parameter. These results are qualitatively equivalent to the measured results presented in [18].

High frequency measurements of high temperature MESFET

The high frequency performance of a GaAs MESFET with $2\,\mu m$ gate length have been measured with a heatable microwave test fixture [28]. The scattering parameters were measured up to 6 GHz with an HP-8510 network analyzer.

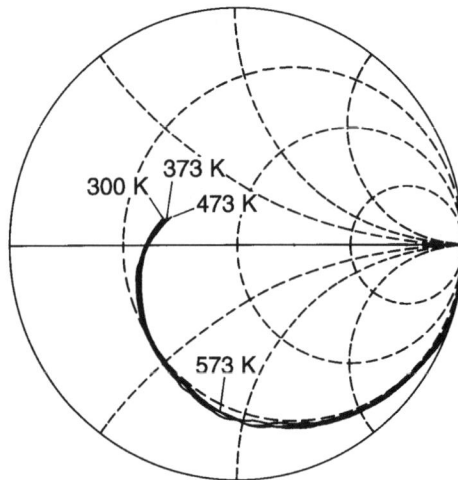

Figure 8.7 Input reflection coefficient S_{11} of a MESFET with $2\,\mu m$ gate length for different ambient temperatures; frequency range 0.045–5.5 GHz.

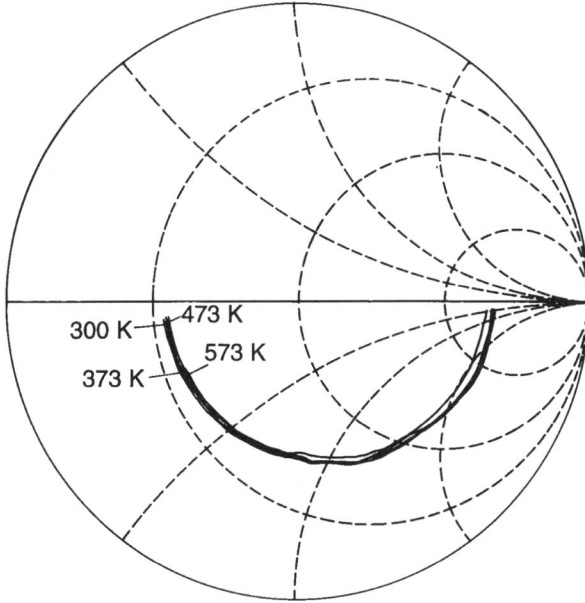

Figure 8.8 Output reflection coefficient S_{22} of a MESFET with $2\,\mu m$ gate length for different ambient temperatures; frequency range 0.045–5.5 GHz.

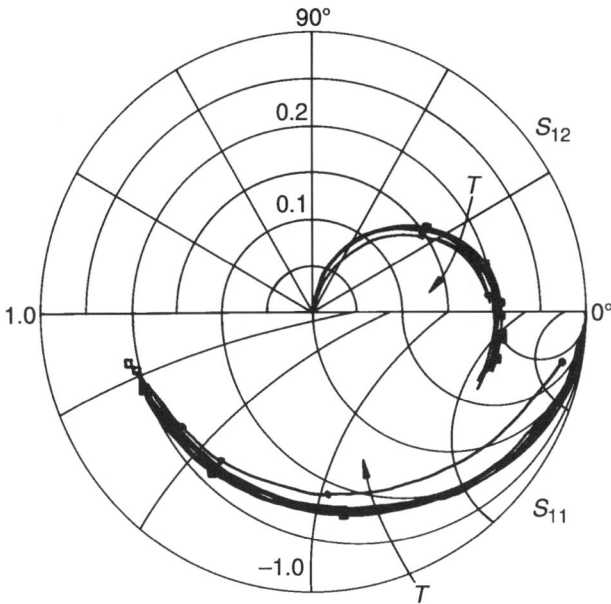

Figure 8.9 Input reflection coefficient S_{11} and reverse transconductance S_{12} of a MESFET with $0.5\,\mu m$ gate length; the ambient temperature ranges from room temperature up to 523 K.

The input and output reflection coefficients S_{11} and S_{22} are shown in Figs 8.7 and 8.8 for ambient temperatures up to 600 K. On-wafer measurements at elevated temperatures have also been reported [29] for MESFET devices with 0.5 m gate length and $5 \times 10^{17}\,cm^{-3}$ active channel doping concentration. Figures 8.9 and 8.10 illustrate the temperature dependence of all S-parameters for a larger frequency range of 1–50 GHz. The temperature was varied from room temperature up to 550 K in this experiment. The figures demonstrate that the input and output impedance of the MESFET do not change significantly with ambient temperature. This is important for the design of microwave and high speed circuits. Input and output matching networks can therefore be realized by disregarding the operating temperature of the devices. This facilitates a relatively simple matching procedure. On the other hand, S_{21} drops considerably with increasing temperature. Typically, the transconductance is halved for a temperature increase of 250 K.

The deterioration of the transconductance also manifests itself in the maximum available gain (MAG) of a GaAs MESFET, which can be determined using the measured scattering parameters. Figure 8.11 shows the MAG for different ambient temperatures with the typical 6 dB per octave decrease. The frequency f_{max} is where the MAG approaches 0 dB. Figure 8.12 compares the temperature dependence of f_{max} normalized to f_{max} at

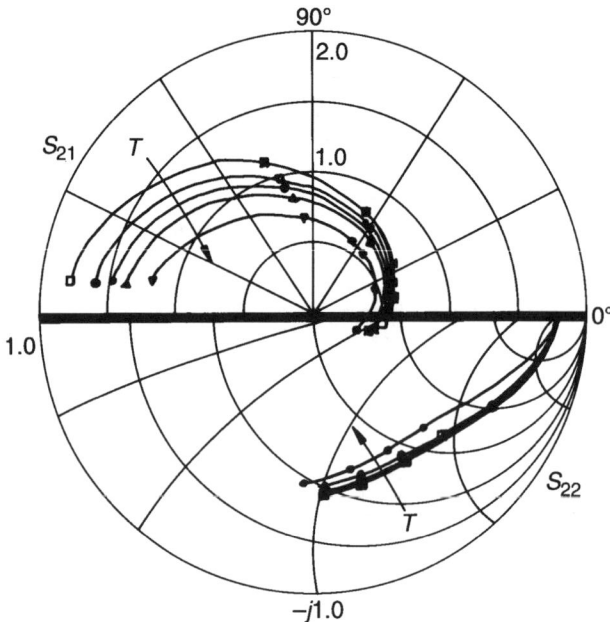

Figure 8.10 Output reflection coefficient S_{22} and transconductance S_{21} of a MESFET with 0.5 μm gate length; the ambient temperature ranges from room temperature up to 532 K.

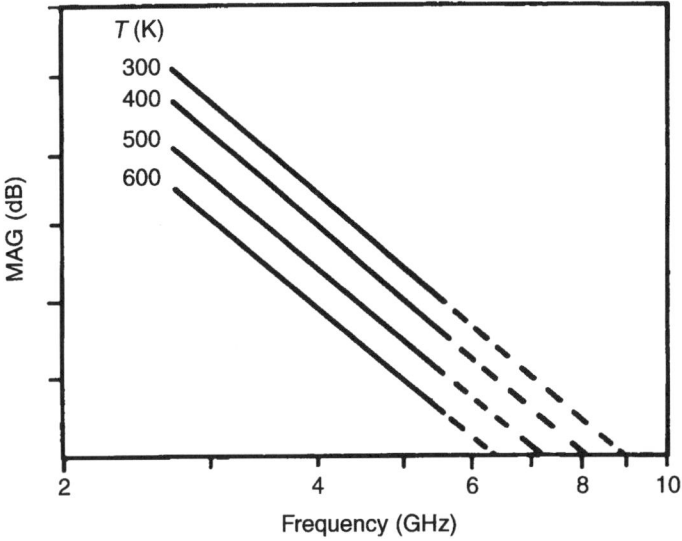

Figure 8.11 Maximum available gain (MAG) versus frequency of a GaAs MESFET with $2\,\mu m$ gate length.

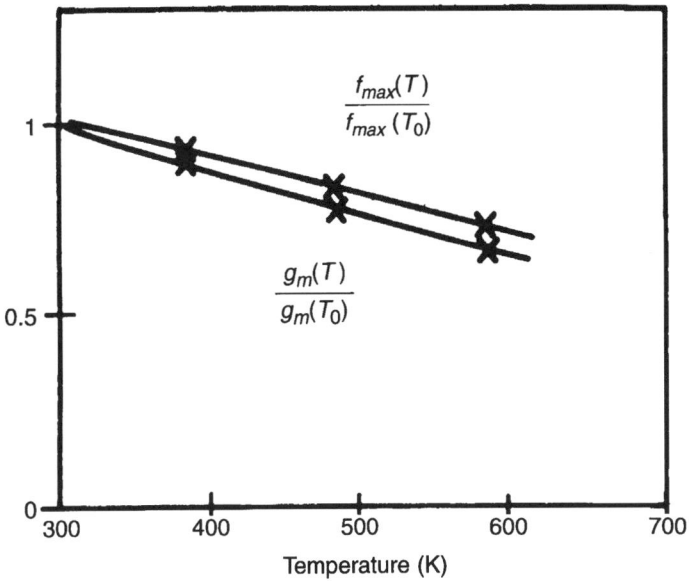

Figure 8.12 Values of f_{max} normalized to f_{max} at room temperature and transconductance normalized to the transconductance at room temperature as a function of the temperature; gate length $2\,\mu m$.

room temperature with the temperature dependence of the transconductance normalized to the transconductance at room temperature. Both curves for the transconductance and f_{max} yield nearly identical temperature characteristics, which implies that the decrease of the microwave amplification is caused mainly by the decrease of the transconductance. Since the transconductance for a MESFET with a gate length of $2 \, \mu m$ is proportional to the electron mobility, MAG is dominated by the temperature dependence of the mobility.

8.2.2 High temperature HEMT

The high electron mobility transistor (HEMT) has been developed to separate the epitaxial layer for the doping (electron supply layer) from the undoped channel in which the electrons propagate. A nearly triangular quantum well is formed, in which the electrons are distributed quasi two-dimensionally. Since the channel is not to be doped, the mobility of the

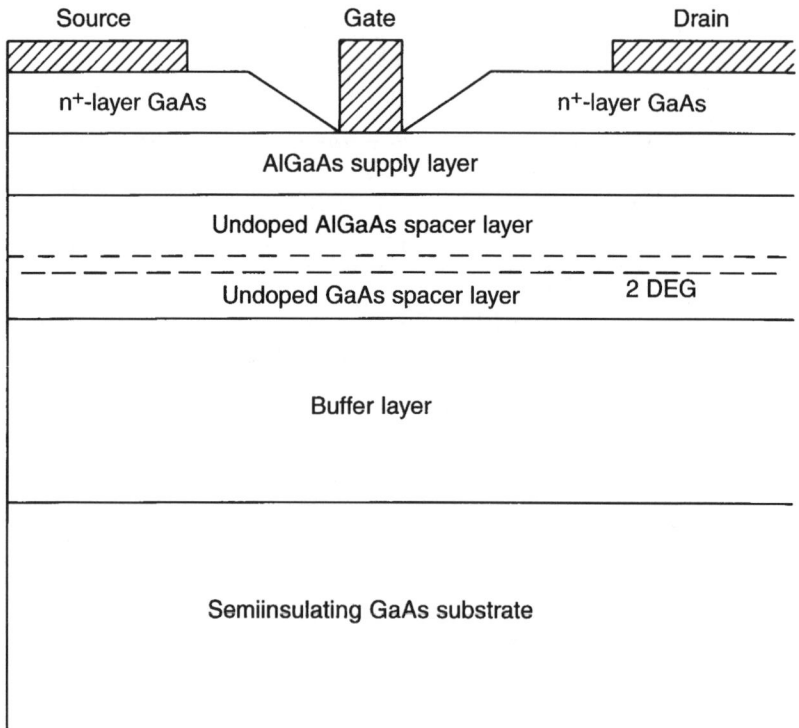

Figure 8.13 Schematic cross section of an AlGaAs/GaAs HEMT.

Table 8.4 Sequence of the layers for a high temperature HEMT

Layer	Material	Thickness	Doping
Contact layer	GaAs	30 nm	$N_d = 3 \times 10^{18}\,\mathrm{cm}^{-3}$
Electron supply layer	$Al_{22}Ga_{78}As$	40 nm	$N_d = 1 \times 10^{18}\,\mathrm{cm}^{-3}$
Spacer	$Al_{22}Ga_{78}As$	2.5 nm	$N_d \approx 1 \times 10^{14}\,\mathrm{cm}^{-3}$
Buffer	GaAs	800 nm	$N_d \approx 1 \times 10^{14}\,\mathrm{cm}^{-3}$
Substrate	GaAs	490 μm	$\rho \approx 10^7\,\Omega\,\mathrm{cm}$

electrons is higher than in MESFET devices. The mobility is further improved at low temperatures because the phonon scattering decreases. Pseudomorphic HEMTs on GaAs, which use an InGaAs layer as channel and an AlGaAs layer for the electron supply layer, have especially good performance at millimeter wavelengths. Furthermore, the performance of HEMT devices is generally superior to MESFETs at higher operating frequencies.

It is also interesting to look at the high temperature performance of this device. Even at increased temperatures there remains an advantage in electron mobility for the HEMT in comparison with the MESFET. The higher bandgap of the AlGaAs heterostructure layer between the active channel and the Schottky metal improves the barrier height of the gate Schottky diode.

The AlGaAs/GaAs HEMT (Fig. 8.13) was fabricated on the MOVPE material in Table 8.4 [28, 30]. The technology was similar to the MESFET technology described above. Instead of the PECVD passivation, an ultra-violet chemical vapor deposition (UVCVD) was used because it creates less surface damage.

The DC characteristics of this AlGaAs/GaAs HEMT are shown in Fig. 8.14 for different ambient temperatures. Its transconductance decreases with increasing temperature, similar to the MESFET. At temperatures above $\sim 470\,\mathrm{K}$ the electrons are no longer confined in the two-dimensional electron gas (2DEG). This causes a drop in the effective electron mobility and an increase in parasitic resistance. Up to 700 K the HEMT shows a relatively good pinchoff. A parallel conduction to the channel can be observed at higher temperatures. Another probable cause is degradation of the confinement for the 2DEG at higher temperatures. This deficiency could partly be overcome by incorporation of an additional barrier towards the substrate, a barrier which would inhibit the scattering of electrons into the undoped regions outside the 2DEG.

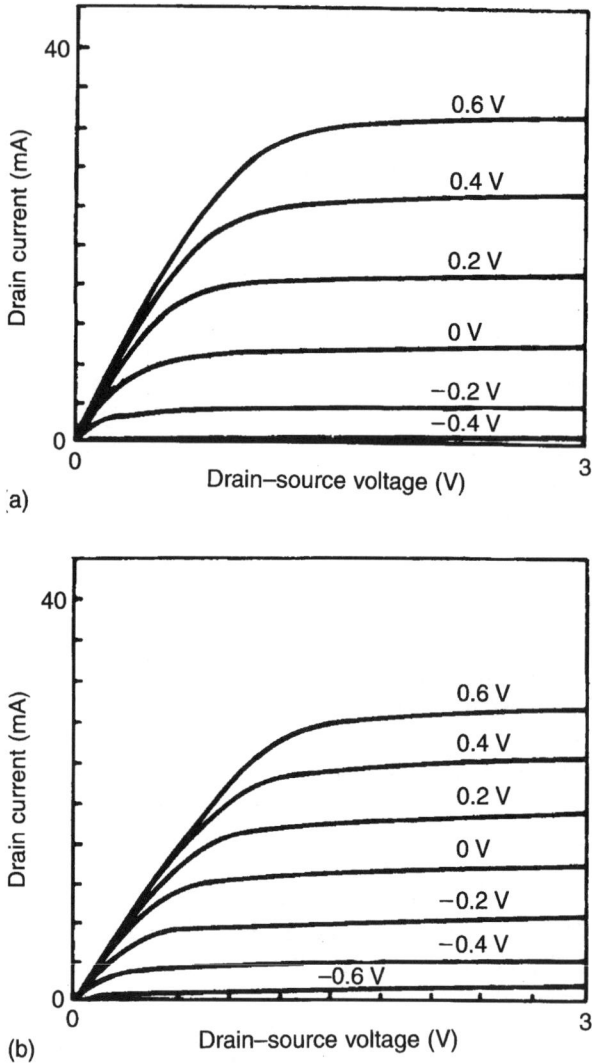

Figure 8.14 Drain current I_D of an AlGaAs/GaAs HEMT versus the drain–source voltage V_{DS} with the gate–source voltage V_{GS} as a parameter at (a) 300 K, (b) 473 K, (c) 573 K and (d) 623 K; gate length 1 μm.

8.2.3 Comparison between MESFET and HEMT

The same geometry was used when comparing HEMTs with MESFETs [28]; the gate length was 1 μm. In both cases the passivation was performed with UVCVD. The transconductances of both devices are presented in Fig. 8.15. In some temperature ranges the curves decrease in proportion to the

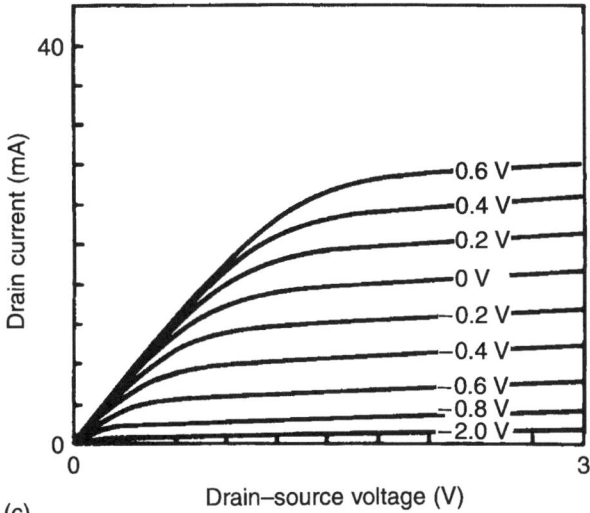

(c)

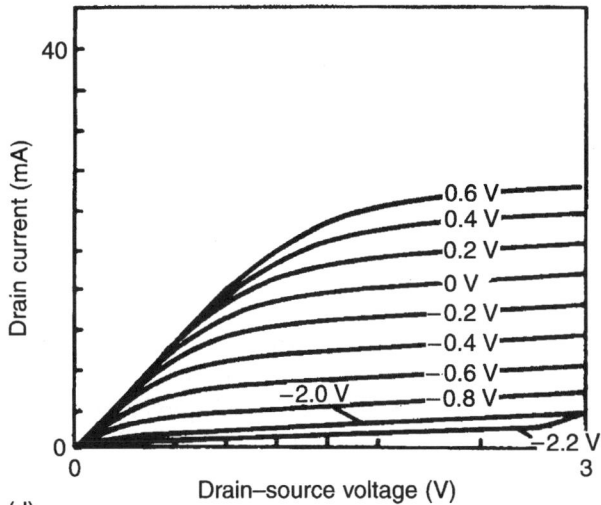

(d)

Figure 8.14 (*Continued*).

inverse temperature. This can be explained by the temperature dependence of the mobility of the electrons, which is also proportional to T^{-1} in GaAs [25]. However, the transconductance of the HEMT decreases at higher temperatures faster than T^{-1}. This is also caused by degradation of the electron gas confinement. In the whole temperature range the transconductance of the HEMT is better than for the MESFET.

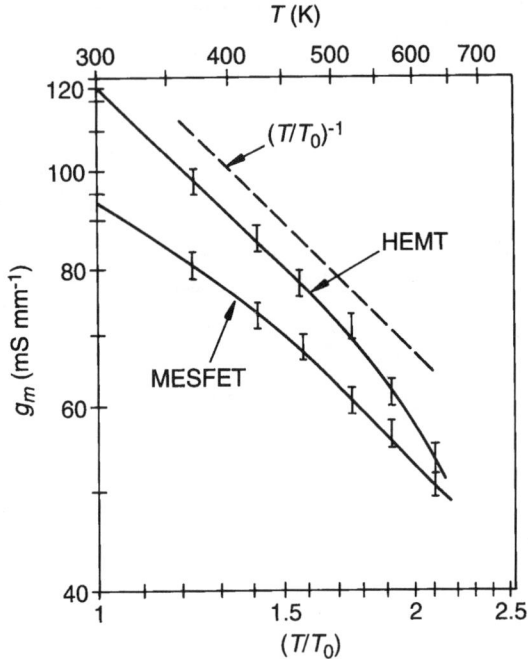

Figure 8.15 Transconductance g_m versus temperature of HEMT and MESFET with the same geometry; gate length 1 μm. The slope of T^{-1} is shown for comparison.

One difference between the output characteristics of MESFET and HEMT, as shown in Figs 8.3 and 8.14, is the more pronounced slope for the MESFET. The conductance g_s parallel to the channel is used to model this effect (Fig. 8.16). The advantages of the HEMT are due to the larger barrier for the electrons to enter the buffer. This is also confirmed by the simulations performed for AlGaAs/GaAs HEMT devices [16] for the subthreshold current of a MESFET and a HEMT. The simulation results are shown in Fig. 8.17.

The gate currents of MESFET and HEMT are plotted as a function of T^{-1} in Fig. 8.18. The HEMT exhibits lower gate currents. Since the gate Schottky diode of the HEMT is located on top of an AlGaAs layer after etching the recess, it is possible to realize Schottky contacts with greater barrier height. Using the method of activation energy, a barrier height of about 1 eV was measured for this HEMT. The Schottky diode was structured with a Ti–Pt–Au metallization. Better values may be obtained with an LaB$_6$ metallization.

The drain saturation currents of HEMT and MESFET as a function of temperature are plotted in Fig. 8.19. The apparent difference is that the drain

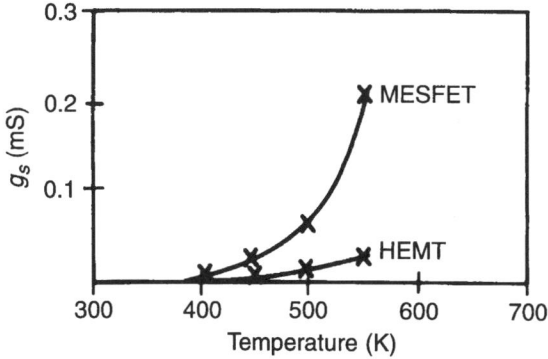

Figure 8.16 Parallel conduction g_s versus temperature of HEMT and MESFET with the same geometry; gate length 1 μm.

saturation current of the MESFET has a negative temperature coefficient. This is caused by the temperature dependence of the electron mobility. The thermal generation of carriers, which may cause the opposite effect, is dominated by the higher channel doping at the temperatures discussed here. On the other hand, the drain saturation current of the HEMT has a positive

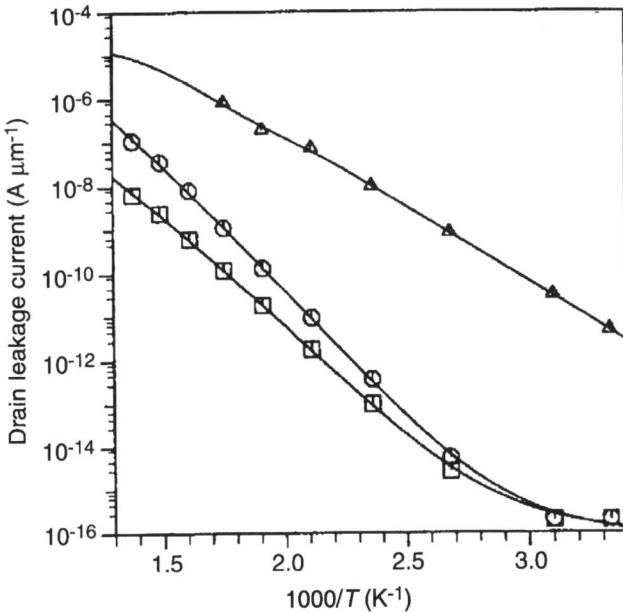

Figure 8.17 Comparison of the drain subthreshold currents for an AlGaAs/GaAs HEMT and a GaAs MESFET as a function of the inverse temperature; gate length 0.5 μm. ($\square$) HEMT constant barrier, ($\bigcirc$) HEMT variable barrier, ($\triangle$) MESFET.

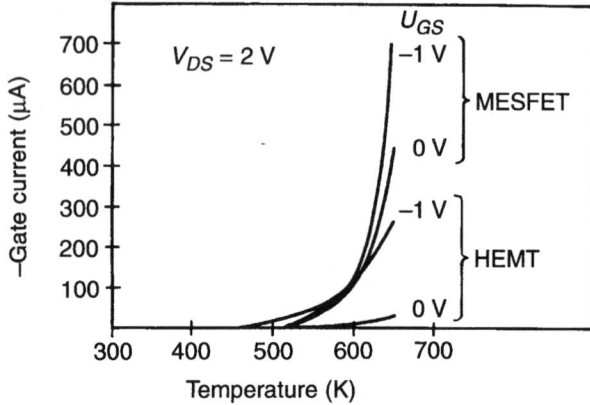

Figure 8.18 Gate currents I_G versus temperature of HEMT and MESFET both having a gate length of 1 μm.

temperature coefficient at temperatures higher than 400 K. This is because all the carriers thermally generated in the undoped GaAs increase the number of electrons in the 2DEG and contribute to the conduction.

An important issue for the fabrication of circuits is the temperature-compensated operation point (ZTC) of MESFET and HEMT (Fig. 8.20) [31]; both devices exhibit this phenomenon even at increased temperatures.

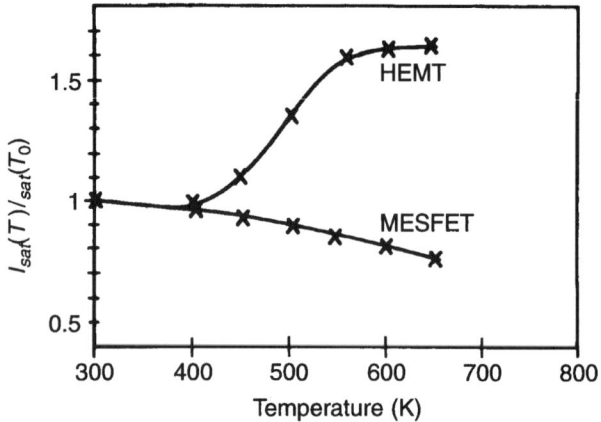

Figure 8.19 Drain saturation currents I_{sat} of HEMT and MESFET as a function of temperature; gate length 1 μm.

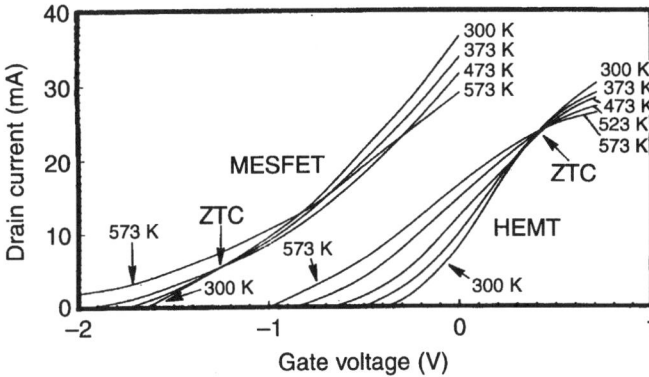

Figure 8.20 Temperature-compensated operating points of HEMT and MESFET: drain current I_D as a function of the gate–source voltage V_{GS} for different temperatures ($V_{DS} = 2\,\text{V}$).

8.3 HETEROJUNCTION BIPOLAR TRANSISTORS ON GaAs

High temperature electronics with compound semiconductors have the advantage of heterostructuring and tailoring of material properties for a particular application. The heterojunction bipolar transistor (HBT) is an example of operational device improvement by proper material choice.

8.3.1 Design of high temperature heterojunction bipolar transistors on GaAs

The AlGaAs/GaAs heterostructure bipolar transistor is a relatively new device, so the technology is not as advanced as for MESFET and HEMT devices. Even at room temperature the question of long-term stability is not yet fully established.

The advantage of an HBT in comparison with a conventional bipolar transistor (BJT) is based on its high bandgap emitter. In the case of the npn transistor, generally employed for high frequency and high speed applications, the valence band offset prevents the injection of holes from the base into the emitter; this allows a high current gain. In a BJT only when the emitter is more heavily doped than the base can there be a current gain. However, an HBT has high base doping exceeding that of the emitter in order to achieve a low base resistance and consequently a high maximum frequency of operation f_{max}.

GaAs/AlGaAs HBTs are ideal for power amplification in the microwave range [32]. Rezazadeh [33] showed that HBTs are also well suited for power amplification at increased ambient temperatures. He used HBTs with an Al content of 30% in the emitter. However, the devices were not optimized for high ambient temperatures. An AlGaAs/AlGaAs HBT for high temperature application was presented by Frost [34]. The emitter had an Al content of 45%, base and collector had 25% Al content. The current gain was of the order of 12 at room temperature and dropped to 4 at 800 K. The reason for the low current gain was the low valence band offset in the emitter–base junction. Alternative designs were presented by Zipperian [35]; he used the GaP/AlGaP system for the fabrication of high temperature HBTs.

We now present a technology for the fabrication of high temperature stable HBTs as used by the Technische Hochschule, Darmstadt, Germany. A classical wet etched mesa structure on MOVPE material was used; a self-aligned process is also described. In order to suppress the hole injection into the emitter, even at elevated temperatures, a high Al content of 40% was chosen for the emitter. This Al content is a compromise between maximum valence band offset and the achievable crystal and material quality. Conventional HBT devices as well as double-heterostructure bipolar transistors with an AlGaAs collector have been realized in order to avoid a high collector leakage current due to thermally generated electron–hole pairs in the collector–base space charge layer. The advantage will be obvious in the comparison between the single-HBT (SHBT) with the double-HBT (DHBT).

Table 8.5 MOVPE layer structure for the heterostructure bipolar transistors (SHBT and DHBT)

	SHBT			DHBT		
	Thickness (μm)	Doping (cm^{-3})	Al content $(\%)$	Thickness (μm)	Doping (cm^{-3})	Al content $(\%)$
GaAs emitter	0.1	1×10^{18}	0	0.15	5×10^{18}	0
				0.5	4×10^{17}	0
AlGaAs emitter	0.05		$0 \rightarrow 45$	0.02	4×10^{17}	$0 \rightarrow 42$
	0.15	4×10^{17}	45	0.03	4×10^{17}	42
	0.01		$45 \rightarrow 0$	0		
Spacer	0.01	–	0	0	–	–
Base	0.15	5×10^{18}	0	0.1	2×10^{19}	0
Collector	0.5	4×10^{16}	0	0.03	4×10^{16}	$0 \rightarrow 20$
				0.5	4×10^{16}	20
Subcollector	0.5	1×10^{18}	0	0.5	5×10^{18}	0
Substrate	600	–	0	400	–	0

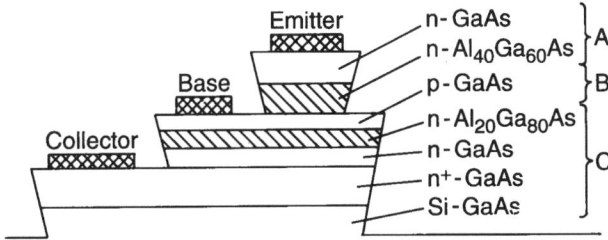

Figure 8.21 Schematic cross section of a typical DHBT.

As an example, two typical epitaxial structures are presented in Table 8.5. In both cases a relatively thin base was used to achieve good high frequency performance. The schematic cross section of a DHBT is shown in Fig. 8.21. In this particular SHBT only the $Al_{0.2}Ga_{0.8}As$ layer in the collector is missing. The design of the epitaxial layers is now discussed for a high temperature SHBT and a DHBT. The layer structure of the DHBT is suitable for a self-aligned technology:

- The thicknesses of the emitter layers of SHBT and DHBT are designed according to the different technologies used for the fabrication. The DHBT has a relatively thick GaAs emitter to separate the emitter and the base metallization in the self-aligned technology. The GaAs emitter improves the ohmic contact of the emitter in the SHBT. A highly doped InGaAs layer may further improve the emitter contact resistance.
- The base of the DHBT was doped with carbon, a very stable dopant, so a spacer layer was not required [36, 37]. Carbon has a much lower diffusion constant in GaAs than Zn and is therefore much more suitable for base doping. In the case of carbon as a dopant, a substitutional diffusion has to be considered because of the absence of vacancies in the lattice. Substitutional diffusion is much slower than other diffusion mechanisms, e.g. interstitial diffusion at a metal–semiconductor interface. Be exhibits nearly the same stability to carbon [38]. Carbon-base doping has a particular advantage when used in conjunction with InGaP/GaAs HBT devices because C can hardly be activated in the InGaP layer and therefore remains electrically neutral.
- A lower Al content was chosen for the DHBT emitter in order to achieve a better material quality. The $Al_{0.45}Ga_{0.55}As$ emitter grown for the SHBT had a poorer material quality.
- The base of the DHBT is thinner than for the SHBT and is therefore doped more heavily ($2 \times 10^{19}\,cm^{-3}$).
- The SHBT has heterojunctions with a grading on both sides to avoid spikes in the conduction band. As shown in the design of the DHBT, it is possible to omit the grading in the emitter at the emitter–base junction

and the grading in the collector at the collector–base junction without changing the electrical performance.

- The two transistors also differ in the design of the collector. The breakdown performance of the DHBT is improved because of the higher breakdown voltage of its high bandgap material. The doping in the collector and the collector layer thickness are chosen for maximum breakdown voltage. A thick collector would improve the collector–emitter breakdown voltage but the transit time would increase correspondingly. A low doping in the collector increases the parasitic collector resistance at high operating temperatures.

An important aspect in the design of an HBT is the control of the breakdown voltage. Two effects may degrade the breakdown voltage:

- The avalanche effect causes electron–hole pairs which increase the drain current at high base–collector voltages.
- Especially at increased temperatures, an additional current occurs due to the thermal generation of carriers in the collector space charge region. Because the collector–base voltage modulates the base–collector space charge layer, this thermal current is also voltage dependent.

Both effects cause a hole current from the collector into the base; this contributes to the current amplification of the transistor. The hole current is added to the external base current. If the emitter current is kept constant, the external base current changes its sign at a certain collector–base voltage because of these additional currents [39,40]. The effect causes an increase of the current gain at increased temperatures [41,42].

The temperature-dependent coefficients of the avalanche effect have been measured with HBT in common base mode by Fricke *et al.* [40]. By measuring the base current with fixed emitter current, the temperature-dependent multiplication factor M has been obtained as a function of the collector–base voltage. The results show that the avalanche effect has a positive temperature coefficient. Therefore the danger of voltage breakdown is relaxed at increased temperatures [43,44].

Due to the field dependence of the avalanche effect, a lower doping of the collector produces a high breakdown voltage. The disadvantage of this design is the high parasitic resistance of such a collector. It is advantageous to achieve the higher breakdown voltage with a wide bandgap collector.

8.3.2 Technology for high temperature stable HBTs

A typical fabrication sequence for high temperature stable HBTs is as follows:

- The mesa structure for the base is etched with a number of selective and nonselective etchants [45–47]. Selective etching of the GaAs emitter from the AlGaAs is accomplished using $NH_4OH:H_2O_2 = 3:500$ at $3\,^\circ C$ [47]; selective etching of the AlGaAs is accomplished using $NH_4OH:H_2O_2$ at room temperature and a pH of 5.2 [45]. The nonselective etchant required to etch the grading is: $NH_4OH:H_2O_2:H_2O = 3:1:150$ at $3\,^\circ C$. A 50 nm thick AlGaAs layer is left on top of the base, which is fully depleted and suppresses the surface recombination.

- Prior to the evaporation of the base metallization, the AlGaAs layer is removed using photoresist as a mask.
- The base and the collector mesas are formed with nonselective etchants until the subcollector is revealed.
- An additional mesa is etched for device isolation.
- The emitter and collector metallizations are evaporated in one step. The contact systems $Ni-Au-Ge-Ni-W_5Si_2-Ti-W_5Si_2-Au$ [48] and Pd–In–Pd–W–Au system may be used for high temperature stable operation.
- For the ohmic contact of the base, either Ti–Pt–Au or Pd–In–Pd–W–Au may be used [49, 50]. The metallization systems of base and emitter are annealed simultaneously if they have the same formation temperature. For different formation temperatures, the contact system with the higher formation temperature is evaporated and annealed first.
- A 100 nm thick PECVD $Si_xN_yO_z$ layer is sufficient for passivation.
- Windows are etched into the $Si_xN_yO_z$ using dry etching or buffered HF.
- The connections to emitter, base and collector are formed with plated gold air bridges for good microwave performance.

The technology for self-aligned HBTs [51] differs from the previous technology in several steps. First the collector mesa is etched, then the collector and emitter contacts are evaporated. The base mesa is etched using the emitter metallization as a mask. The emitter mesa is also used as a mask for the evaporation of the base metallization. Different solutions exist for this technology:

- If the epitaxial layer system has an InGaAs layer on top, it can be structured with the shape of the emitter. Later it can be used for etching the emitter mesa (using selective etchants). The overhang formed during the etching yields very good separation of base and emitter metallization. Ti–Pt–Au may be used for the metallization on p-GaAs ohmic contacts.
- The emitter metallization may also be used as a mask for the emitter etching. The emitter metallization should be very homogeneous in this case and should not be attacked by the etching solution. It is also advantageous if no catalytic elements like Au or Pt are present in the metallization in order to avoid increased etch rates at the metal–semiconductor interface – an advantage of this Pd–In–Pd–W–Au system. A W–Au or LaB_6–Au top layer is evaporated after the etching step and evaporation of the base metallization.

Parallel conduction between the bond pads of an HBT can be detected at increased temperatures. This is important for temperatures of about 700 K and higher. An additional output resistance r_{ce} is due to the semiinsulating substrate. The resistance between closely spaced (30 μm) bond pads (100 μm × 100 μm) is at 600 K to $r_{ce} \approx 10^4 \, \Omega$ and at 700 K to $r_{ce} \approx 500 \, \Omega$. These values for r_{ce} are less than or equal to the base–collector reverse resistance. A solution to this problem is either a large separation of the pads or the realization of conducting pads on top of an SiN passivation layer for isolation purposes.

If HBTs should be used at temperatures up to 700 K, the bond pads have to be isolated. Lee [51, 52] solved this problem by evaporating an Al_2O_3 layer between the semiconductor and the metal for the bond pads.

8.3.3 Performance of high temperature HBTs

The DC output characteristic of an SHBT is shown in Fig. 8.22 for different ambient temperatures; the results of a numerical drift diffusion model are also shown for comparison. The model does not include the avalanche effect and the conduction between the bond pads.

The measurements demonstrate the suitability of the HBT for high temperature operation. However, a parallel conduction can be observed in this figure, which is not predicted by the model. This parallel conduction is most probably due to surface conduction or pad-to-pad conduction, because the effect of carrier generation is inherently included in the numerical analysis. The Gummel plot is presented in Fig. 8.23. The agreement of measurement and model is very good; almost ideal behavior for the HBT operation is demonstrated in the figure.

Comparison of the DHBT and the SHBT in Fig. 8.24 shows the improved breakdown performance of the DHBT. The current gain of the DHBT at increased temperatures is lower since the generation of carriers in the collector space charge region is suppressed. The improved breakdown characteristics of HBTs at high temperatures is a consequence of the increased scattering rates due to phonons in the space charge region. This is demonstrated in Fig. 8.25, where the DC $I–V$ characteristics at room temperature and at 523 K are directly compared. For low base currents, only minor differences exist in the collector current at room temperature and 523 K. Thermal generation or variation in the current gain cannot therefore explain the improvement in breakdown voltage. For the design of circuits it is important to note that the collector emitter resistance r_{ce} of the DHBT remains very high, even at high ambient temperatures.

The current gain of the SHBT and the DHBT are compared in Fig. 8.26. The current gain of the SHBT increases at higher temperatures. This is due to the extra current generated in the collector space charge region and added to the external base current, thus improving the current gain. This

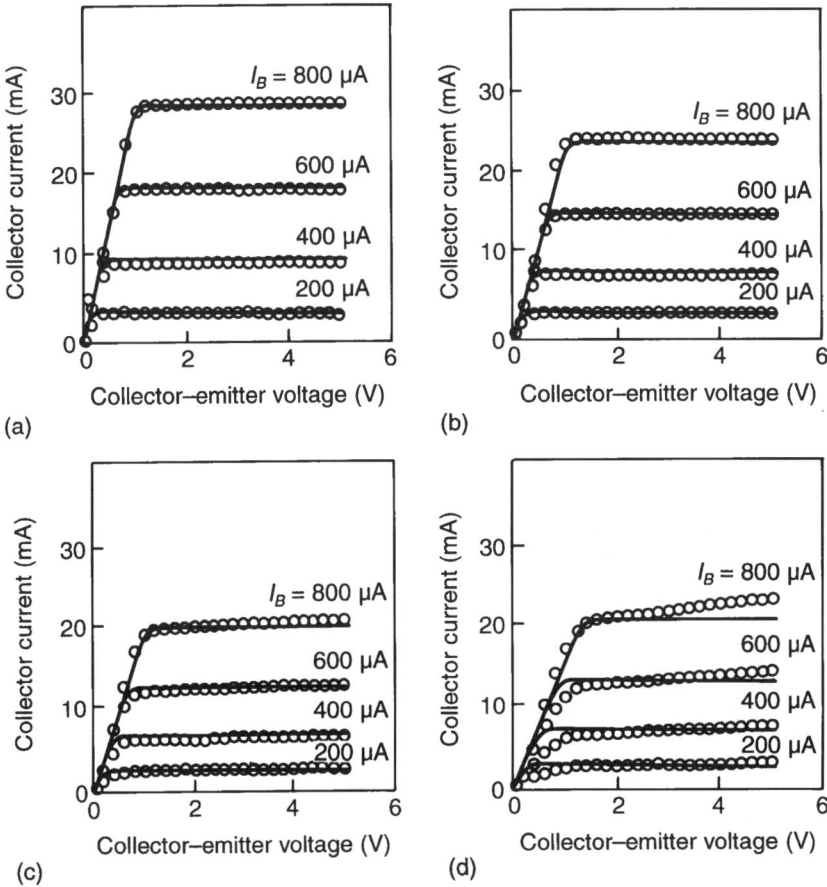

Figure 8.22 Comparison of measured and simulated DC characteristics of an SHBT: collector current I_C over the collector–emitter voltage V_{CE} for a range of base currents I_B at ambient temperatures of (a) 300 K, (b) 373 K, (c) 473 K and (d) 573 K: (—) simulated and (O) measured.

current is effectively suppressed in the DHBT by the wide bandgap collector. The different layer structure of the two transistors means that, compared with the SHBT, the current gain of the DHBT decreases faster with increasing temperature.

Figure 8.27 shows the DC output characteristic and the Gummel plot of a DHBT with isolated bond pads at 673 K. The epitaxial layer structure corresponds to the DHBT in Table 8.5. Besides a reasonable current amplification of 18, Fig. 8.27 shows that no current between the bond pads can be detected. Otherwise a dependence of the collector current on the collector–emitter voltage would be visible.

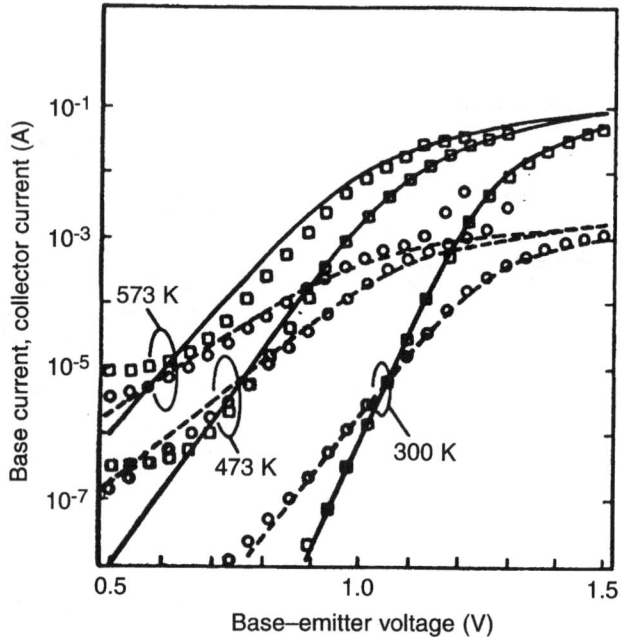

Figure 8.23 Comparison of measured and simulated Gummel plot of an SHBT: collector current I_C and base current I_B versus the base–emitter voltage V_{BE} at different ambient temperatures: (—) simulated I_C, (---) simulated I_B, ($\square$) measured I_C and ($\bigcirc$) measured I_B.

An interesting feature of HBT device operation is the possibility of constant collector bias. The deterioration of the transconductance with increasing temperature (Fig. 8.28) can be partly compensated by biasing the device for constant collector current as compared to constant base current. The parameter S_{21} decreases by a factor of approximately 2 for a constant base current bias condition when the temperature increases from room temperature up to 523 K. This is comparable with MESFET and HEMT high temperature device operation. It is also possible to bias the device for constant collector current. In this case S_{21} remains nearly constant (Fig. 8.28). This feature can be efficiently utilized in circuit design. Figure 8.29 illustrates all S-parameters for different operating temperatures of an InGaP/GaAs/InGaP DHBT. When using a DHBT the collector leakage current is suppressed and the output reflection coefficient S_{22} remains essentially unchanged at increased ambient temperatures. Conversely, the input reflection coefficient S_{11} decreases with increasing ambient temperature.

The variation of the transmission scattering parameter S_{21} and small-signal current gain h_{21} with temperature can be seen in Figs 8.30 and 8.31 for AlGaAs/GaAs and InGaP/GaAs HBT single-finger devices. The small-

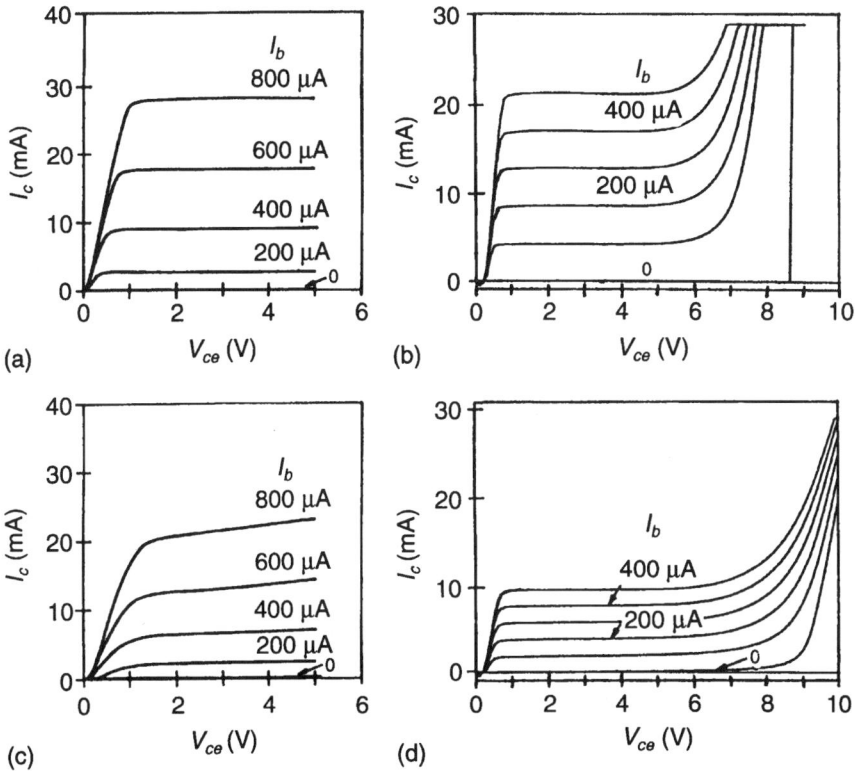

Figure 8.24 Collector current versus the emitter–collector voltage for different base currents (I_b) at 300 K and 573 K ambient temperature: (a) SHBT, 300 K, 200 μA per step; (b) DHBT, 300 K, 100 μA per step; (c) SHBT, 573 K, 200 μA per step; (d) DHBT, 573 K, 100 μA per step.

signal current gain h_{21} shows larger variation with temperature as compared with S_{21} for both technologies. However, HBT devices with AlGaAs/GaAs base–emitter heterojunction exhibit larger temperature-dependent variation as compared to the InGaP/GaAs HBT devices. These results support the argument that the InGaP-based heterojunction more effectively inhibits the hole current from the base to the emitter.

On the other hand, Figs 8.32 and 8.33 demonstrate that only minor differences exist with varying temperature between the extracted transit frequency f_T of AlGaAs/GaAs HBTs and InGaP/GaAs HBTs, respectively. The same holds true for the maximum frequency of oscillation f_{max}.

8.3.4 Lifetime investigations

$C–V$ and $V–I$ measurements have been performed at increased temperatures to evaluate the lifetime of HBTs [44]. The measurements were made using

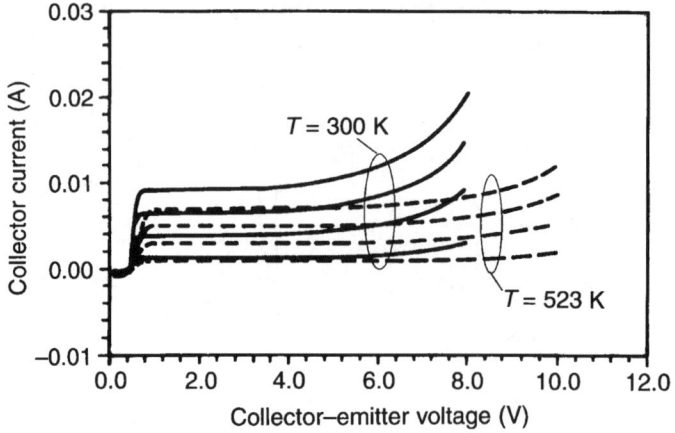

Figure 8.25 Output $I-V$ characteristic of an HBT device at room temperature (300 K) and 523 K: $I_b = 50 \, \mu A$, $150 \, \mu A$, $250 \, \mu A$, $350 \, \mu A$.

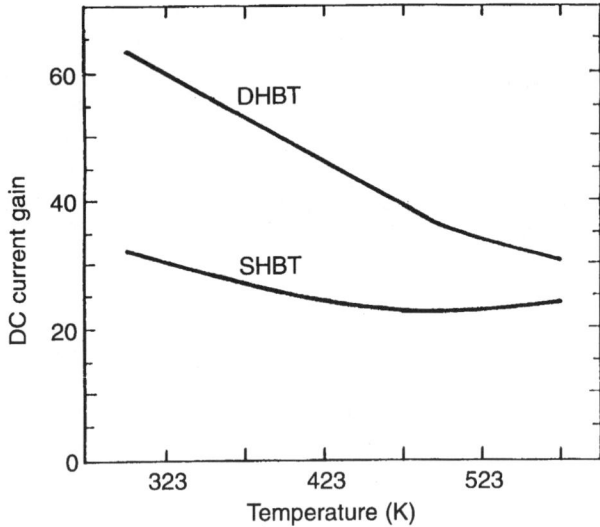

Figure 8.26 Current gain of DHBT and SHBT versus the ambient temperature.

DHBTs fabricated according to the technology described earlier. The results can be summarized as follows:

- After an unbiased lifetime test at 700 K for 250 h, no changes of the current gain β and the breakdown voltage could be detected. Small

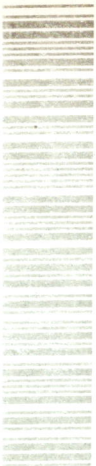

Dear Customer, please take a moment to provide contact information.

1. Contact Name _____
(A person not travelling with you today)

2. Contact Phone Number _____
(Include country code, area code and number)

3. Are you a U.S. Citizen?   Yes ☐   No ☐

4. I decline to provide this information. ☐

This information is only retained for 24 hours and will remain confidential. Thank you.

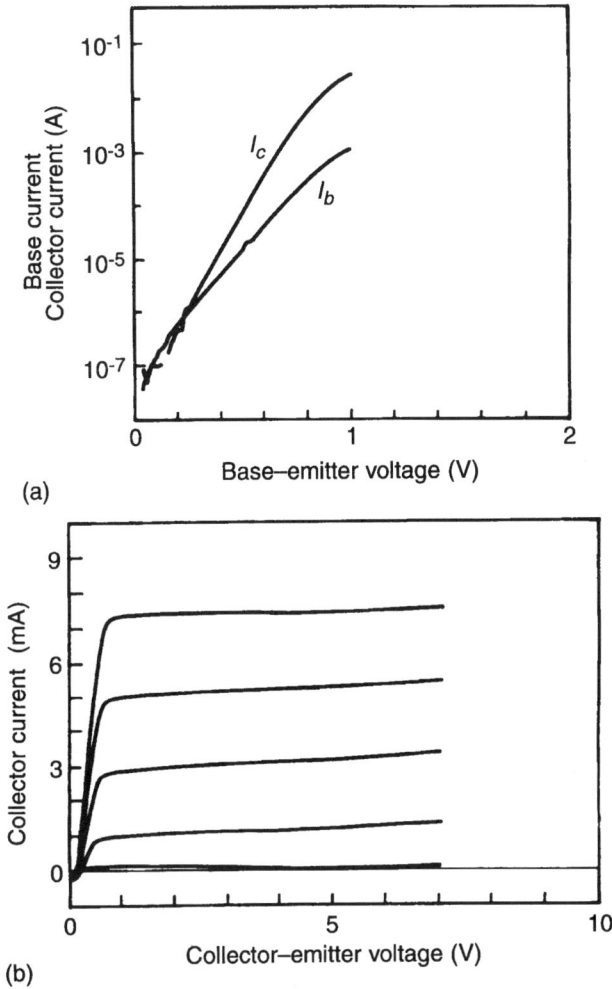

Figure 8.27 DHBT with isolated bond pads: (a) Gummel plot at 673 K; (b) collector current I_C versus collector–emitter voltage V_{CE} for different base currents I_B.

changes in the base current have been observed in the lower current range [27].

- After a 50 h stress test at 700 K and at a collector current density of $6 \times 10^3 \, \text{A cm}^{-2}$, no changes compared with the lifetime without bias have been detected.

- C–V measurements of the emitter–base diode by Lee and Schüßler [27] indicated no measurable change of the diffusion voltage after the stress-test under bias. This demonstrates that no degradation of the doping

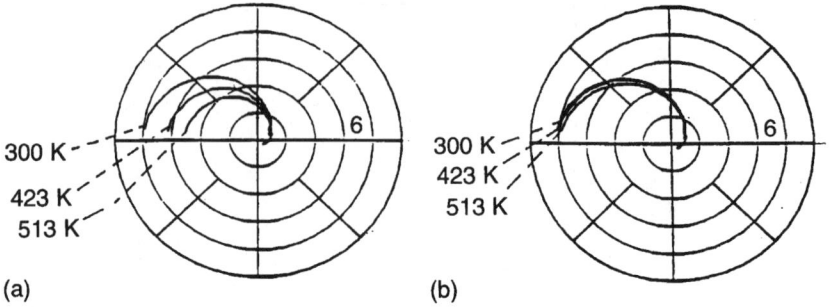

Figure 8.28 Measured S_{21} for an AlGaAs/GaAs HBT at different ambient temperatures: (a) constant base current bias condition $I_B = 370\,\mu A$, $V_{CE} = 4\,V$; (b) constant collector current bias condition $I_C = 10\,mA$, $V_{CE} = 4\,V$; frequency range $45\,MH_2$ to $10\,GHz$.

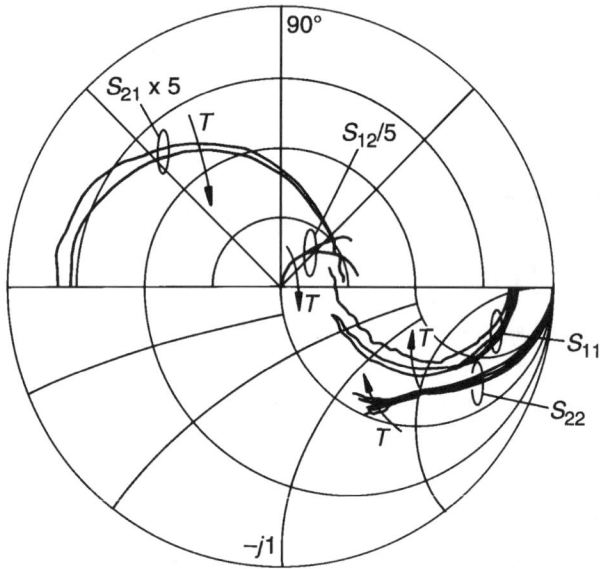

Figure 8.29 Measured S-parameters for an InGaP/GaAs/InGaP HBT at different ambient temperatures; frequency range $45\,MHz$ to $10\,GHz$.

profile of the base had taken place. However, the capacitance of the emitter–base diode changed marginally. It was shown that the change of the capacitance can be modeled by an increase of the surface charge at the emitter–base interface.

The results demonstrate that a reasonable reliability has been obtained with the special high temperature technology for HBTs. Further investigations

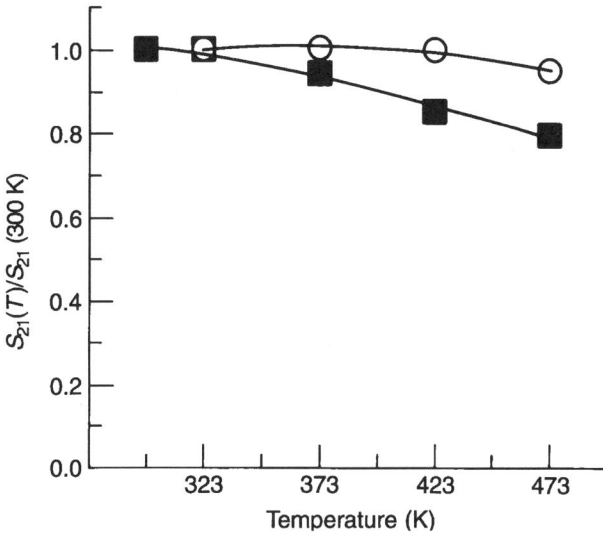

Figure 8.30 Measured normalized $S_{21}(T)/S_{21}(RT)$ for (■) AlGaAs/GaAs and (○) InGaP/GaAs HBTs as a function of temperature.

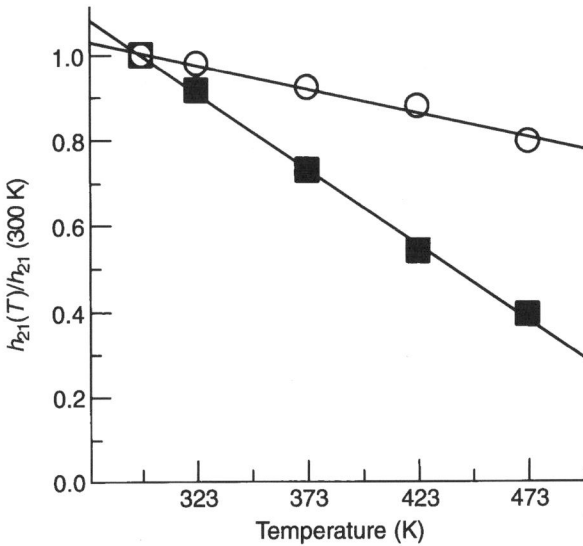

Figure 8.31 Measured normalized small-signal current gain $h_{21}(T)/h_{21}(RT)$ for (■) AlGaAs GaAs and (○) InGaP/GaAs HBTs as a function of temperature.

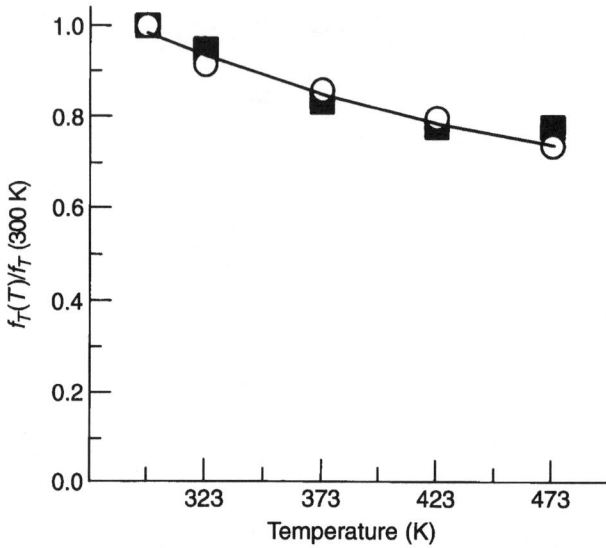

Figure 8.32 Extracted normalized transit frequency $f_T(T)/f_T(RT)$ from measured values for (■) AlGaAs/GaAs and (○) InGaP/GaAs HBTs versus temperature.

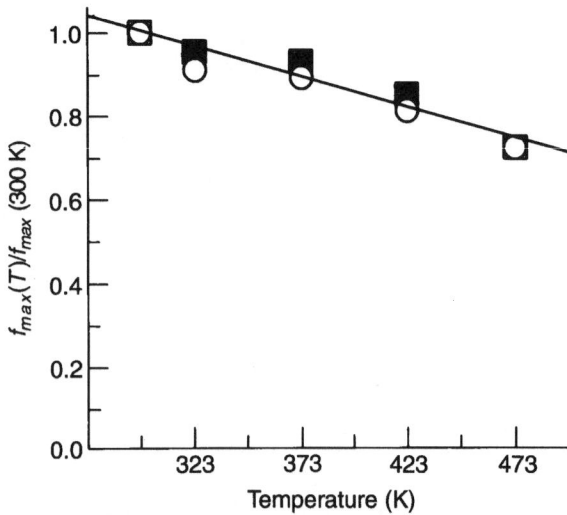

Figure 8.33 Extracted normalized maximum frequency $f_{max}(T)/f_{max}(RT)$ from measured values for (■) AlGaAs/GaAs and (○) InGaP/GaAs HBTs versus temperature.

have to be undertaken to understand the reliability issues of devices operating at high ambient temperatures. Accelerated lifetime testing is critical for devices designed to operate at temperatures around 600–700 K. This is because an additional temperature increase approaches the temperatures used in fabrication procedure of the devices. Other tests for accelerated lifetime testing should therefore be designed.

8.4   MODELING OF HIGH TEMPERATURE DEVICE OPERATION

The design of GaAs circuits and the optimization of high temperature device performance requires models capable of simulating the high temperature performance of individual devices. Models for GaAs-based devices have been developed for microwave circuits; they are widely employed in commercial software packages. Simulations of power devices have focused on temperature-dependent modeling. However, for accurate modeling of the operation at increased ambient temperatures, it is essential to examine the temperature behavior of all device parameters. Physically based modeling is best suited for this purpose. Temperature-dependent material parameters for GaAs, AlGaAs and InGaP are given in Table 8.2.

8.4.1   Modeling of MESFET and HEMT

There are many publications concerned with self-heating effects in power devices, but only a few about modeling and characterization of high temperature MESFETs or HEMTs based on GaAs [16–18, 26–28, 53, 54].

Some research [18, 26] focuses on modification of existing empirical models for high temperature simulation purposes. Physically based models for MESFET and HEMT devices have been developed in analytical form by Krozer *et al.* [27, 55] and using a numerical approach by Wilson *et al.* [16, 17]. Both models take into account the impact of the gate reverse current on the device operation and the Schottky barrier lowering at high ambient temperatures. Gate currents are crucial for high temperature simulation because they decrease the depletion depth under the gate and hence the drain–source current. Similar effects can be observed in HEMT devices. The substrate current can be modeled by the thermal generation of carriers. A simple relation for the thermal generation current can be given according to the formulas developed for the current flow between two metal stripes on a two-layer material:

$$J_{sub;th} = V_{ds} \frac{W}{L} \frac{r_{s1}r_{s2}}{r_{s1} + r_{s2}} \tag{8.3}$$

where W is the device gate width, L is the device gate length and the resistances r_{s1}, r_{s2} are given as follows:

$$r_{s1} = \frac{1}{qN_c v_{sat} dL_{DS}} \qquad r_{s2} = \frac{1}{qn_{i0} V_{ds} \max(L_D, L_S)}$$

with N_c the active channel doping concentration, n_i the intrinsic concentration, v_{sat} the saturation velocity, subscript 0 the electron mobility, d the active channel thickness, L_{DS} the drain–source spacing, and L_S and L_D the source and drain length, respectively.

The physical analytical model includes the following features:

- temperature- and field-dependent Schottky barrier;
- thermal generation of the carriers for reverse gate–source bias;
- analytical two-dimensional solution of Poisson equation under the gate following the approach presented in [55].
- nonabrupt transition between the active channel and the depletion region;
- continuous field and velocity characteristic versus electric field and temperature;
- nonhomogeneous RC transmission line between gate and source;
- extension of the depletion region towards the drain.

A comparison of the measured and simulated I–V characteristics of a $0.5\,\mu m$ GaAs MESFET are shown in Fig. 8.34. Simulation results for GaAs

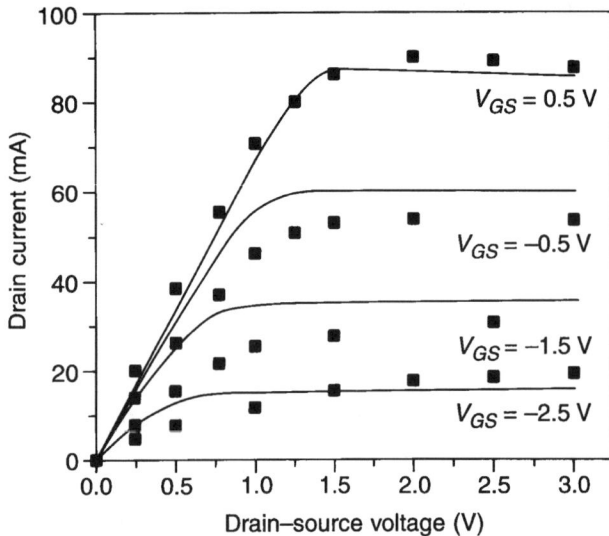

Figure 8.34 Comparison of (—) simulated and (■) measured I–V characteristics of a $0.5\,\mu m$ GaAs MESFET at room temperature.

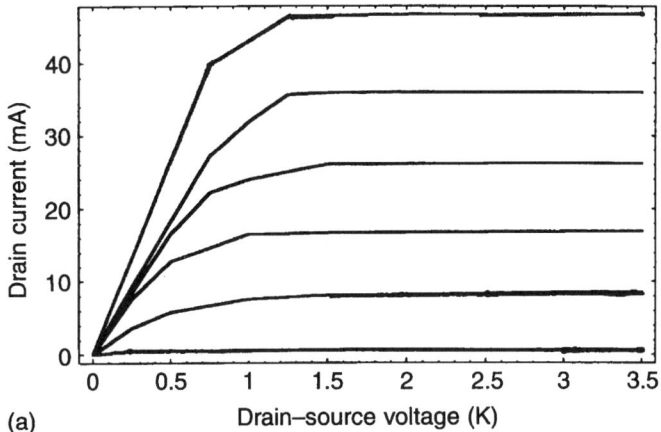

(a)

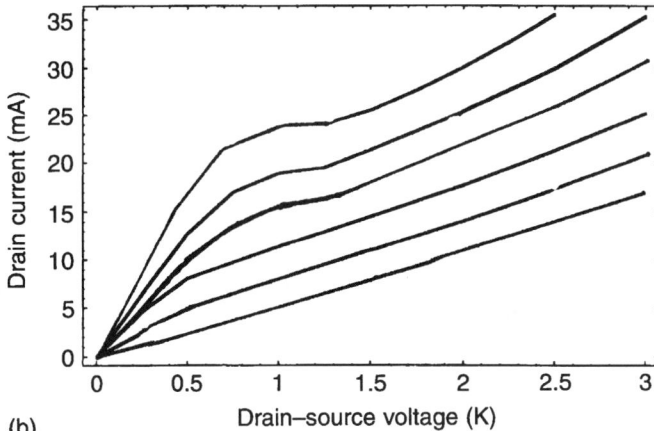

(b)

Figure 8.35 Simulated $I-V$ characteristic for a 0.5 μm GaAs MESFET at (a) room temperature (300 K) and (b) 523 K.

MESFET devices with 0.5 μm gate length are presented in Fig. 8.35 for room temperature and 523 K. In the calculation of the high temperature $I-V$ characteristic it was assumed that no barrier existed between the channel and the buffer layer and that the substrate current was due to thermally generated carriers. At voltages exceeding the saturation voltage, the drain current is dominated by the substrate current in the case assumed. These simulation results stress the importance of an effective buffer layer between the active channel and the substrate. Simulation of the high frequency performance at high temperatures (Fig. 8.36) confirm the measured results presented earlier. Variation of S_{22} with temperature is due to the strong contribution of the substrate layer.

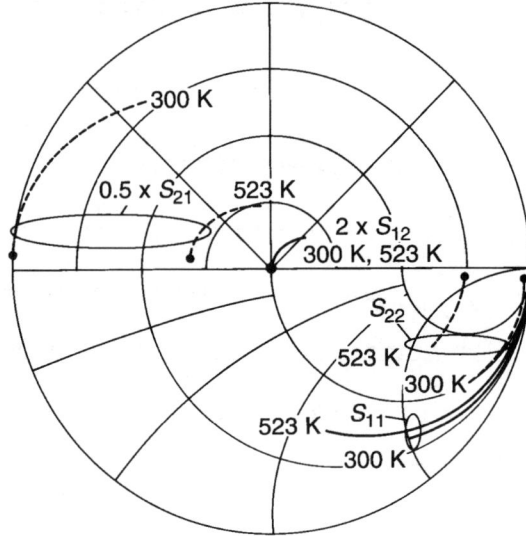

Figure 8.36 Simulated S-parameters for a $0.5\,\mu$m GaAs MESFET for room temperature (300 K) and 523 K.

Simulations carried out for HEMT devices [16] show improved current leakage characteristics compared to the GaAs MESFET; this has already been discussed and is illustrated in Fig. 8.18.

8.4.2 Modeling of HBT

Temperature-dependent modeling of HBT devices has received much attention in the recent literature [27, 56–59]. Most contributions focus on the thermal properties of power HBT devices. This interest is due to the strong self-heating effects which cause increased junction temperatures of HBT power devices of up to 600 K. Another effect is the collector current collapse due to excessive junction temperature. It has recently been demonstrated that the thermal performance of an HBT device is the limiting factor in power applications.

A physically based DC and RF model has been developed to understand the physical origin and the effects occurring in high power or high temperature operation. It includes the following features:

- tunneling and thermionic emission at heterojunctions;
- avalanche breakdown at large reverse voltages;
- self-heating effects due to power dissipation;

- revised boundary conditions at the base–collector junction with large reverse bias;
- accurate determination of the depletion width for forward-biased junctions;
- interface and surface recombination currents;
- current spreading from the emitter towards the collector;
- field-dependent emitter and collector resistances at high current levels.

The following equations are used for the collector and base currents versus base–collector and base–emitter voltages, respectively:

$$
\begin{pmatrix} I_C \\ I_B \end{pmatrix}^T = \left\{ \begin{pmatrix} J_{21} & J_{22} \\ J_{11} - J_{21} \dfrac{A_E}{A_C} & J_{12} - J_{22} \end{pmatrix} \begin{pmatrix} A_E \left[\exp\left(\dfrac{U_{BE}}{U_T}\right) - 1 \right] \\ A_C \left[\exp\left(\dfrac{U_{BC}}{U_T}\right) - 1 \right] \end{pmatrix} \right.
$$

$$
\left. + \begin{pmatrix} -J_{p_{nn^+}} A_{Cintr.} \\ J_{p_{nn^+}} A_{Cintr.} \end{pmatrix} + \begin{pmatrix} -1 & 0 \\ 1 & 1 \end{pmatrix} \begin{pmatrix} (J_{rcI} + J_{rcB}) A_C \\ (J_{reI} + J_{reB}) A_E \end{pmatrix} \right\}^I
$$

$$
\times \begin{pmatrix} M_{aval} & \left(\dfrac{1}{2\cosh(W_B/L_{eB})} \dfrac{A_E}{A_C} - 1 \right)\left(1 - \dfrac{1}{M_{aval}} \right) \\ 0 & 1 \end{pmatrix} \tag{8.4}
$$

The individual components in the current relations have the following meaning:

$$
J_{11} = J_{nE} + J_{pE} \tag{8.5}
$$

$$
J_{12} = -(J_{nC} + J_{genC})\frac{1}{\cosh(W_B/L_{eB})} \tag{8.6}
$$

$$
J_{21} = J_{nE}\frac{1}{\cosh(W_B/L_{eB})} \tag{8.7}
$$

$$
J_{22} = -J_{nC} - 2J_{genC} - \begin{cases} J_{pC} & \text{if } U_{BC} > -5\,U_T \\ 0 & \text{otherwise} \end{cases} \tag{8.8}
$$

With the following additional symbols:

A_E = base–emitter junction area
$A_{Cintr.}$ = intrinsic base–collector junction area corrected for current spreading
A_C = collector–base junction area
$J_{p_{nn^+}}$ = collector hole current density calculated using improved boundary conditions at the collector–subcollector junction (forward biased base–collector junction indicates $J_{p_{nn^+}} = 0$)

$J_{rcI}, J_{rcB},$

J_{reI}, J_{reB} = base–collector (base–emitter) space charge region interface
and bulk recombination current densities

J_{genC} = base–collector thermionic generation current density

M_{aval} = avalanche multiplication factor

The other terms have their usual meaning.

Thermal coupling of multifinger devices is accomplished using an equivalent thermal resistance for one finger and connecting several fingers in parallel according to the design for the device. The temperature of the individual HBT finger is determined by the surrounding temperature coupled from the neighboring fingers and the additional heat generated by the finger itself. The coupling coefficients can be determined from the electrical equivalent of the thermal problem [60].

Figure 8.37 illustrates the agreement between measured and calculated results for the $I-V$ characteristics at room temperature and 473 K. The only

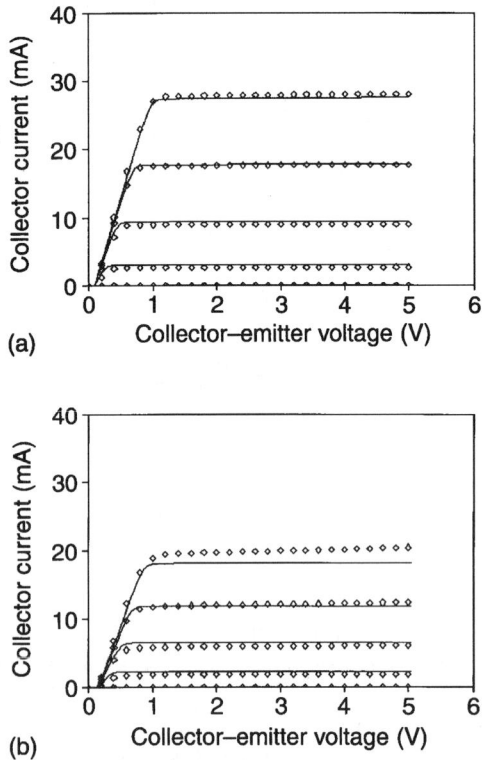

Figure 8.37 Comparison of (—) simulated and (◇) measured $I-V$ characteristics of an AlGaAs/GaAs HBT for (a) room temperature (300 K) and (b) 473 K.

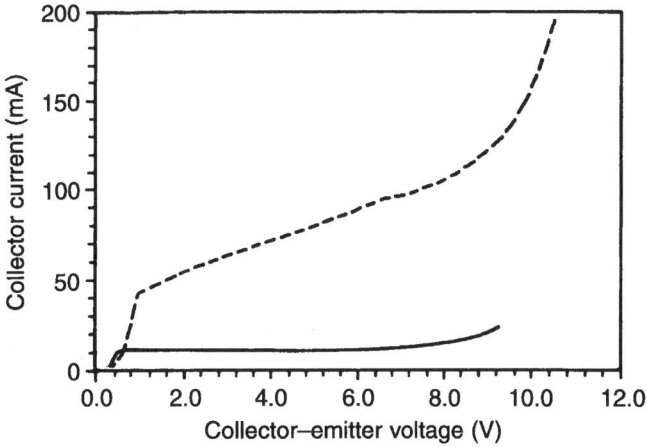

Figure 8.38 Simulated collector current of an AlGaAs/GaAs HBT versus collector–emitter voltage for 1 nA base current for (—) room temperature (300 K) and (---) 523 K.

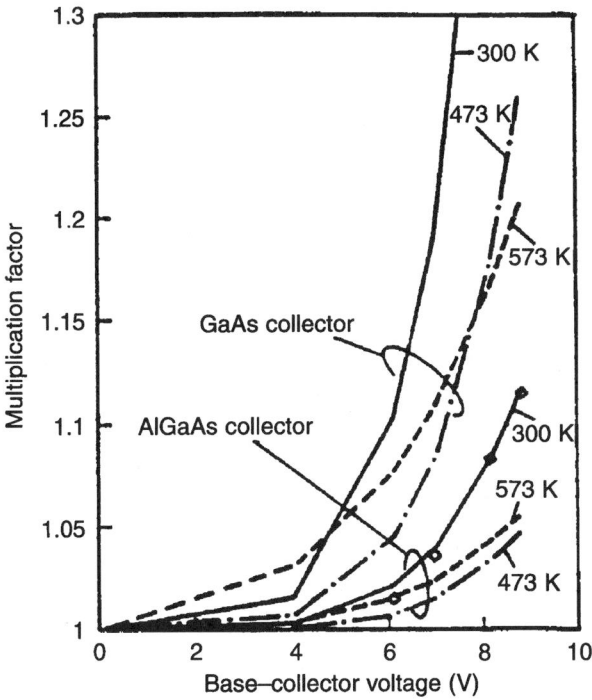

Figure 8.39 Comparison of simulated and measured multiplication factor for an AlGaAs/GaAs single heterojunction bipolar transistor at three different temperatures: 300 K, 373 K and 473 K.

parameter which has been varied between room temperature and high temperature operation is the effective interfacial recombination factor. The HBT in Fig. 8.37 has a graded emitter–base heterojunction. The graded layer is modeled as a number of successive steps in the bandgap as the Al content is increased from the emitter side of the junction. Each step can be characterized by an interfacial carrier density at the interface. The increase in the interfacial carrier density at the interface at high temperatures is partly due to the increase of emitter doping in the AlGaAs layer at high temperatures and partly due to activation of additional carriers at the interface.

Thermal carrier generation is important in the base–collector junction. Figure 8.38 shows the calculated base–collector currents at a low base current of 1 nA at room temperature and 523 K. There is a remarkable increase of collector current at high ambient temperatures. Note that the holes generated in the base–collector depletion region are swept away

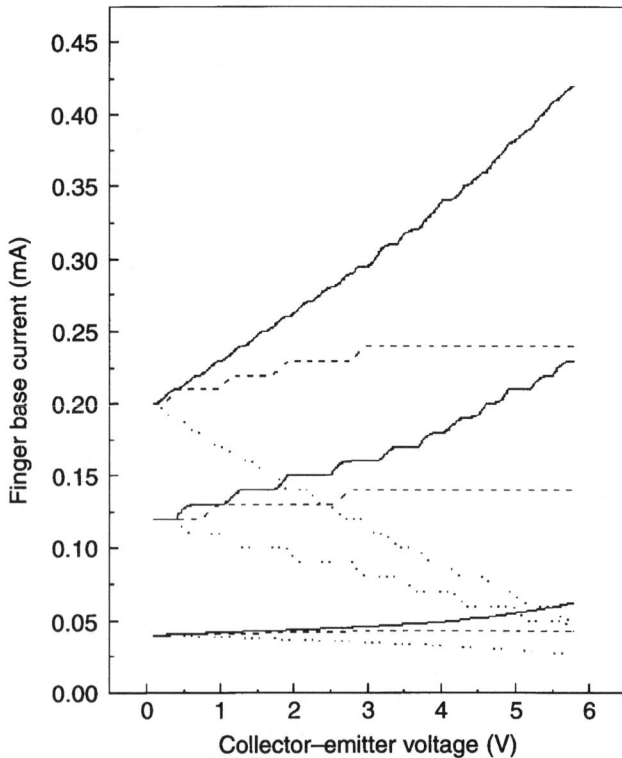

Figure 8.40 Calculated base current of different emitter fingers of an AlGaAs/GaAs five-finger HBT: (····) outer finger, (---) middle finger and (—) inner finger.

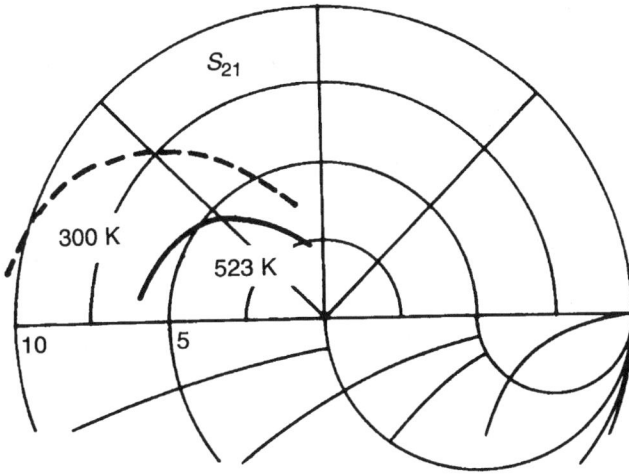

(a)

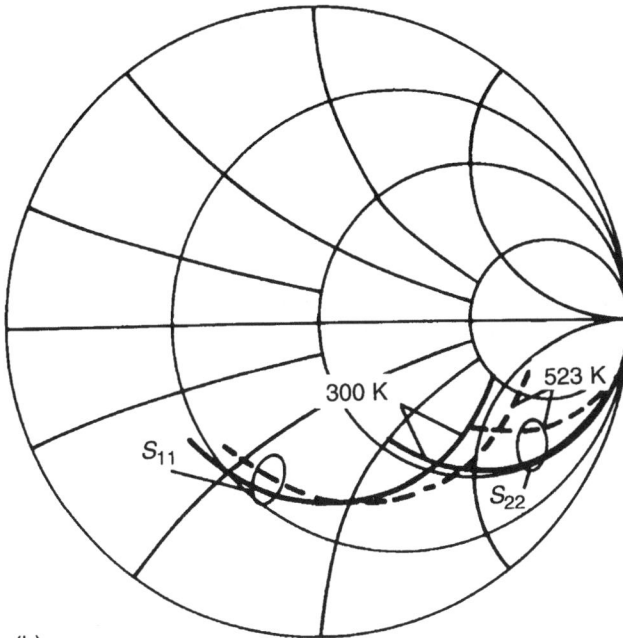

(b)

Figure 8.41 Simulated S-parameters for two temperatures; frequency range is approximately 1–18 GHz: (a) S_{21}, (b) S_{11} and S_{22}.

towards the base and are added to the overall base current. This effect is important in the accurate determination of the multiplication factor in the base–collector junction. A comparison of measured and simulated results of the multiplication factor for an AlGaAs/GaAs SHBT are given in Fig. 8.39 for three different ambient temperatures. The breakdown voltage increases with temperature.

The operation of multifinger HBT devices utilizing the higher breakdown voltage at increased ambient temperatures is limited by the effect of thermal shutdown of individual emitter fingers due to excessive power dissipation. Figure 8.40 illustrates the base current distribution among the individual fingers of a five-finger AlGaAs/GaAs HBT. The hottest finger is in the middle of the device and drains the most current. The other HBT fingers are concurrently shut down and do not contribute to the gain of the device. This effect continues until the hottest finger collapses from thermal breakdown.

The temperature-dependent RF properties are more complicated in HBT devices. As indicated in Fig. 8.41, both reflection coefficients change with increasing temperature. The real part of the output reflection coefficient decreases because of the base–collector leakage current. On the other hand, the real part of the input reflection coefficient is decreasing because of the increasing base current in order to maintain a constant collector current. The calculated S_{21} is also given in Fig. 8.41 for the case of constant base current. There exists a large variation of S_{21} with temperature.

8.5 HIGH TEMPERATURE CIRCUITS ON GaAs

There are several publications about high temperature stable circuits on GaAs [62–64]. A high temperature stable Hall sensor with integrated electronics [61] and a simple differential amplifier [73], which showed amplification at temperatures up to 700 K, are two examples with MESFET technology. An integrated multivibrator [74] with MESFETs could be operated up to 600 K.

A digital circuit has also been published [75]. An integrated GaAs MESFET n-channel NOR gate was realized. The circuit was operational up to 500 K. The delay time was 20–60 ns with the higher values at the higher temperatures. The risetime was a temperature-independent 140 ns. These values correspond to state-of-the-art values for the 5 μm gate technology used in this experiment. The experiment shows that the output voltage swing decreases at temperatures higher than 500 K because of the subthreshold currents of the MESFETs. The n-channel MESFET technology suffers mainly from this drawback. Complementary devices would be more advantageous for digital circuits. A very promising approach for this application is the complementary technology proposed by Baier et al. [65]. HBTs may be used as an alternative. And operational amplifiers based on HBTs have recently been published [28, 66]. The next section looks at two operational amplifiers, one with MESFETs and the other with HBTs.

8.5.1 Technology for high temperature circuits on GaAs

The fabrication of high temperature circuits requires a special technology which provides stable devices at increased temperatures and maintains the electrical specifications for the circuit. The technological processes for MESFETs, HEMTs and HBTs has already been described in sections 8.2.1 and 8.3.2. The technology of ohmic and Schottky contacts is discussed in Chapter 7. The following sections outline the important technological aspects of device passivation and isolation.

Passivation

The passivation layer of high temperature GaAs devices has to prevent the outdiffusion of As from the GaAs substrate. A problem is the relatively high vapor pressure of As which reaches values as high as 8×10^2 Pa at 700 K. An Si_3N_4 passivation is known to be a very good barrier against the outdiffusion of As. Si_3N_4 is used for the annealing of implantation damage up to temperatures of 1150 K for up to 30 min [67]; only marginal degradation of the GaAs surface can be detected. PECVD is suitable for HBT and MESFET devices since it offers a very dense SiN layer.

Alternatively SiO_2 and phosphorus silicate glass (PSG) have been studied by some authors for the passivation of MESFETs. SiO_2 can either be evaporated on to GaAs or deposited by CVD. For the fabrication of PSG, phosphorus is evaporated and mixed with the process gases (NO_2 and SiH_4) [68].

For HEMTs in particular, passivation has to start with a low damage process as remote plasma-CVD or photoenhanced CVD deposition in order to avoid an ion bombardment of the relatively thin HEMT layers. UV light is used for the photoenhanced CVD process. A disadvantage of the photo process is the minor adhesion of the SiN film to the GaAs surface. The pressure for this deposition is 1000 Pa, much higher than for PECVD, in order to achieve a reasonable deposition rate. Even with this pressure, a 1 h deposition yields an SiN layer of only 70–80 nm thickness. The substrate temperature is 423 K in this case. The layers exhibit very low damage because the saturation current of HEMT does not drop after the deposition of the passivation layer. The photoluminescence of a sample with photoenhanced CVD deposited SiN layer is about twice that of a sample deposited by PECVD. This indicates a high quality GaAs surface after the deposition.

Separation of the devices
Separation and isolation of individual devices is crucial for high temperature
operation. Poor isolation causes high leakage currents and degrades the
circuit performance by parasitic signal coupling and voltage shifting.

Isolation by the semiinsulating substrate
In conventional GaAs circuits the individual devices are separated by the
semiinsulating substrate or by an isolation implantation. The isolation
resistance drops exponentially with increasing temperature because of the
thermally generated increase of intrinsic concentration n_i. Figure 8.42 shows
this effect for a Cr-compensated substrate. Due to the high temperature of
operation, effective isolation can be achieved by layers with high bandgap,
e.g. AlGaAs superlattices.

Isolation with an ion implantation
Ion implantation is a widely used method for the isolation of the individual
devices. It offers the advantage of a planar surface which facilitates structur-
ization in the submicron range. One may distinguish between three pro-

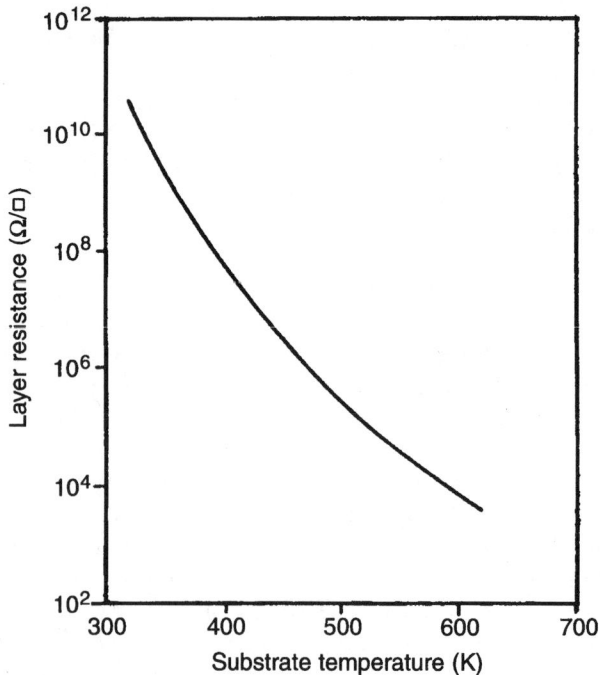

Figure 8.42 Specific resistance of a Cr-compensated GaAs substrate as a function of the
ambient temperature.

cesses. The first uses implantation of protons like boron or nitrogen. The damage of the implanted layers yields high resistivity layers. The second process relies on implantation of oxygen. The isolation is due to the traps which are formed by the incorporation of oxygen. This method cannot be used for optoelectronic circuits. The last implantation process utilizes nitrogen and an annealing step for high isolation.

Boron implantation is not suitable for high temperature circuits because it anneals out in a heat treatment at $600\,K$ for $1\,h$ [69]. But nitrogen-implanted layers are generally annealed for high resistivity at around $800\,K$ [70]. This heat treatment yields very high isolation of up to $10^9\,\Omega\,cm$ [71,72]. The isolation degrades with a second annealing at temperatures above $950\,K$, which could prove satisfactory for high temperature devices on GaAs. The implantation of oxygen is also believed to give stable results at high temperatures.

Isolation by an AlGaAs buffer
An alternative approach to decrease the currents between neighboring devices is the use of an AlGaAs buffer layer or an AlAs/GaAs superlattice. Used for the fabrication of a MESFET circuit, it gives slightly better results than isolation by the semiinsulating substrate on its own.

Isolation by shielding
A metal shield at a fixed potential can be used to suppress surface currents. The method reduces the surface currents efficiently. A disadvantage is the increase in the capacitance to ground.

Isolation by a p–n junction and epitaxial liftoff
This method is used in silicon technology. It gives little advantage for GaAs circuits because the process is complicated. The epitaxial liftoff process (ELO) requires the removal of the overall substrate. After substrate removal, the individual devices or circuits are placed on highly insulating substrates such as quartz, diamond, AlN or others. The process of ELO is used in power amplifiers where excessive heat generation occurs due to high power dissipation. However, this technique is not suitable for high volume fabrication because of the complicated process.

8.5.2 Degradation of doping profiles

Epitaxial doping profiles may degrade at increased temperatures. Appropriate dopants and controlled growth conditions are important for solving this problem. The most severe problem is the high base doping in HBTs.

Problems with the degradation of doping profiles are less severe than the degradation of ohmic contacts. Diffusion during contact degradation occurs within a lattice containing many dislocations, so it is much faster than the interstitial diffusion in the base of an HBT. The doping concentrations in the emitter increase with temperature due to increased activation of dopants. In an AlGaAs emitter the doping concentration increased by 10% at 523 K from its nominal value of $5 \times 10^{17} \, \text{cm}^{-3}$.

8.5.3 Resistors

Resistors fabricated by flash-evaporated NiCr layers have good stability and a low temperature coefficient [73]. A stabilization anneal is needed at 600 K for 3 h. The oxygen content in the chamber during the flash evaporation has to be controlled for reproducible temperature coefficients. The substrate conductivity in the higher temperature range means that very long resistors have to be placed on top on the Si_3N_4 passivation. A second passivation is needed to prevent the NiCr resistors from suffering environmental degradation.

A second method is to use mesa resistors, particularly for MESFET ICs. Resistors using the active layer of the MESFET with a doping concentration of $N_D \simeq 1 \times 10^{17} \, \text{cm}^{-3}$ have the same temperature dependence as the electron mobility. At 600 K the resistance therefore increases to about twice its value at room temperature. Mesa resistors cannot be placed on top of the Si_3N_4 passivation, so crosstalk with nearby devices has to be considered.

Depending on the application, either temperature-independent NiCr resistors or temperature-dependent mesa resistors may be preferred for circuit design. The temperature dependence of the transconductance and the temperature dependence of a mesa resistor are equivalent; both of them are caused by the carrier mobility. This effect can be utilized in dedicated high temperature circuit design.

8.5.4 Capacitors

Capacitors may be realized with the Si_3N_4 passivation as a dielectric. Schweeger *et al.* [74] used reference capacitors with an Si_3N_4 layer of 100 nm thickness. The relative dielectric constant was $\varepsilon_r = 2.5$. The capacitors were used in conjunction with NiCr resistors for the frequency control of a multivibrator circuit [74]. No change of the resonant frequency could be measured after operation at 573 K for 750 h, so the capacitors appear to have reasonable long-term stability [75]. Problems may arise due to electrostatic discharge (ESD) [76] at the corners of the metallization of the capacitor [75].

Interdigital capacitors, as used in the microwave range, should be located on top of the SiN passivation. Although the capacitance decreases, a better isolation of the electrodes and consequently a higher Q-value can be obtained, especially at higher operating temperatures.

8.5.5 Interconnects

Interconnect lines cause a serious problem at increased temperatures. Firstly, electromigration is increased at increased ambient temperatures [77,78]. The chosen maximum current density in the conductors must therefore be lower than at room temperature. The Arrhenius law is used to calculate the required cross section.

The conductivity of the substrate increases at higher ambient temperatures causing crosstalk between the interconnect lines. A possible solution is to locate the interconnect on top of the Si_3N_4 passivation. A second passivation on top of the interconnect lines increases the lifetime of the interconnect. Other solutions are presented in section 8.5.1.

Via-holes [79] for the connection to the back of the substrate should not be filled with Au completely otherwise the different expansion coefficients of GaAs and Au will cause reliability problems.

8.5.6 Packages

The packaging has to be suitable for operation at high ambient temperatures. Ceramic circuit boards are well suited because of their high chemical inertness. The expansion coefficients of the components need to be matched carefully. Conductors on ceramics may be fabricated with thick film techniques or an Au thin film system. Bonding with Au wire is a convenient way to connect the chip to the chip carrier. To avoid stress, the GaAs chips may be glued to the substrate by silver epoxy.

Electrolyte capacitors are not available for operation at increased temperatures, but magnetic materials [80,81] may be used. Large inductors may be used to stabilize a current.

Different packaging concepts are presented in the literature [82,83]. Commercial ceramic packages may be used. In order to keep the space charge temperature of the active devices as low as possible, especially at increased ambient temperatures, it is important to keep a low thermal resistance to the heat sink [84–86].

8.5.7 Operational amplifiers on GaAs

Several circuits have been realized on GaAs, e.g. differential amplifiers [73], multivibrators [74] and digital circuits [75]. Two operational amplifiers have been realized, one equipped with MESFETs [87] the other with HBTs [66]. The circuit diagrams (Figs 8.43 and 8.44) show that both have two-stage amplification. Although the circuits look quite similar, the technology for the HBT amplifier is much more complicated. The first stage is optimized in both amplifiers for maximum gain in order to improve noise performance and to reduce the offset voltage. The first stage is followed by a level shifter with a current source and a resistor.

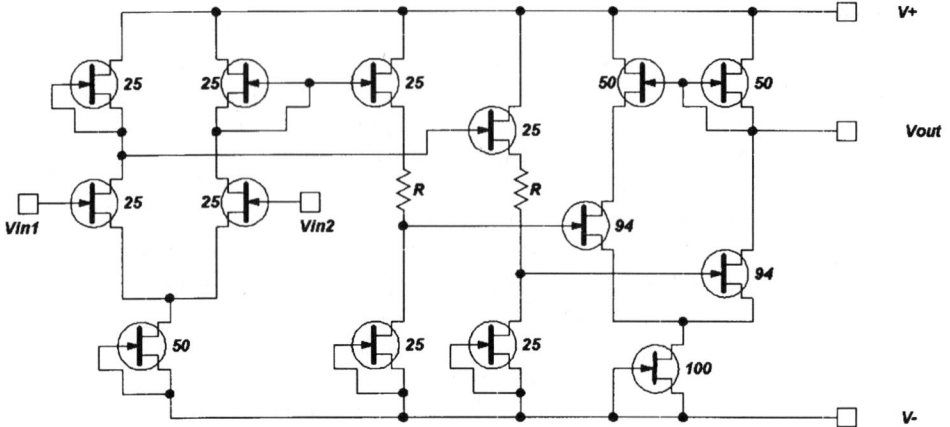

Figure 8.43 Circuit diagram of the MESFET operational amplifier; the numbers beside the MESFETs are the gate widths in micrometers.

The third stage of the MESFET operational amplifier is made symmetrical by using the bootstrap technique [88], which gives a better gain compared to a conventional differential amplifier with active load. The HBT operational amplifier has an output stage with a quasi complementary transistor pair to improve the output voltage swing. The bias point is stabilized by feedback.

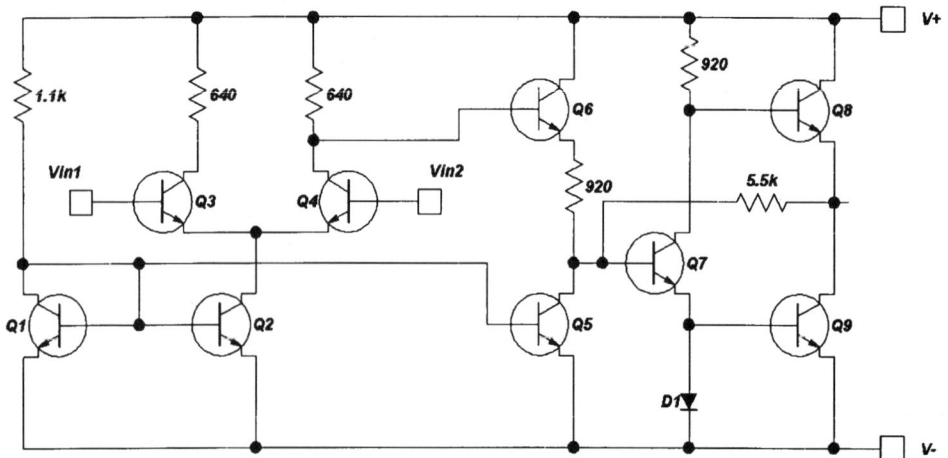

Figure 8.44 Circuit diagram of the HBT operational amplifier.

Both circuits use resistors with flash-evaporated NiCr. The transistors of the MESFET operational amplifier have a gate length of 5 μm indicated on the circuit diagram. The ohmic contacts are based on Ni–Au–Ge–Ni–W$_5$Si$_2$–Ti–W$_5$Si$_2$–Au, the gates are metallized with LaB$_6$–Au.

The active elements of the HBT operational amplifier are the DHBTs described earlier in the chapter. Pd–In–Pd–W–Au is used for the n-type ohmic contacts and the Ti–Pt–Au system is used for the p-type ohmic

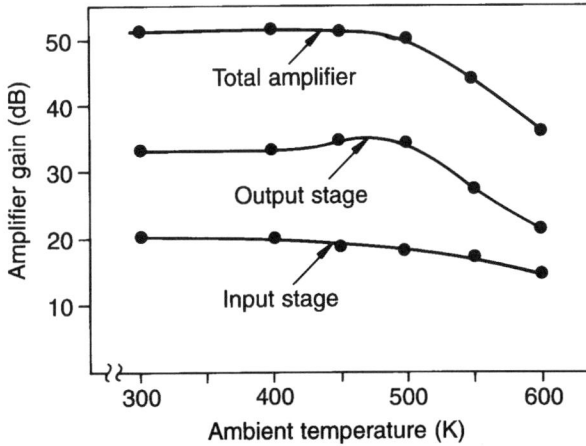

Figure 8.45 Open-loop voltage gain versus ambient temperature for the MESFET operational amplifier, the input stage on its own and the output stage on its own.

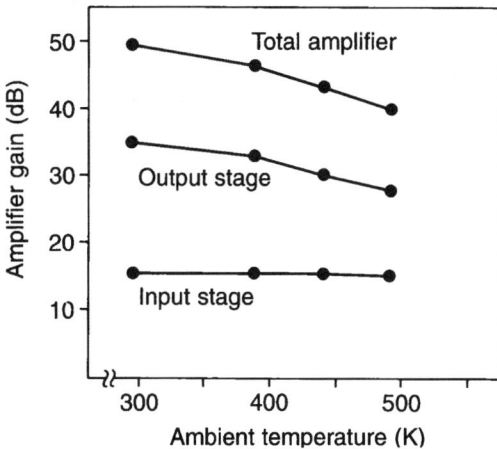

Figure 8.46 Open-loop voltage gain versus ambient temperature for the HBT operational amplifier, the input stage on its own and the output stage on its own.

contacts. The interconnect lines were plated with $1\,\mu$m Au. An air bridge technology was used for crossover points and for connections to emitter, collector and base.

Performance of the operational amplifiers

The open-loop voltage gain of the two amplifiers is shown in Figs 8.45 and 8.46. The MESFET operational amplifier exhibits slightly better gain and a better temperature stability. The HBT operational amplifier could only be operated up to 573 K because the bias current increased excessively at

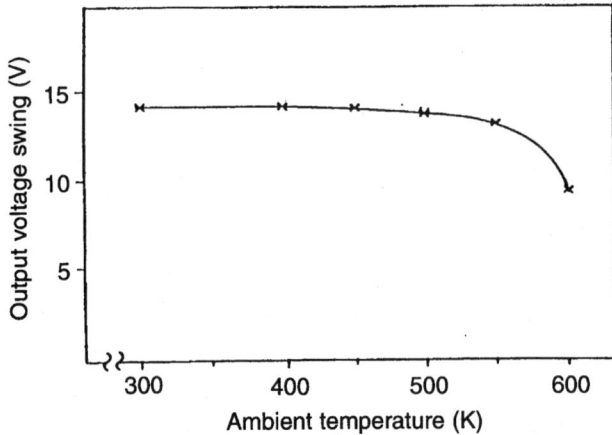

Figure 8.47 Output voltage swing of the MESFET operational amplifier versus the ambient temperature.

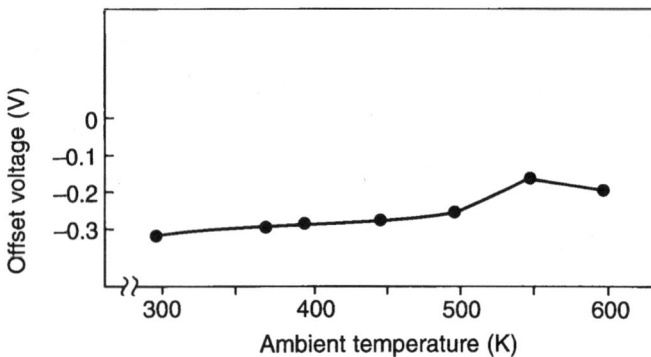

Figure 8.48 Offset voltage of the MESFET operational amplifier versus the ambient temperature.

higher temperatures. Figure 8.47 shows the output voltage swing of the MESFET operational amplifier as a function of the ambient temperature.

The temperature dependence of the offset voltage is relatively stable (Fig. 8.48) even though its value is high. This is because of the difficulty in achieving a homogeneous threshold voltage of the MESFETs over a wafer.

8.6 CONCLUSION

The results for the MESFET devices may be concluded as follows:

- At high ambient temperatures, MESFET and HEMT exhibit a high parallel conductance to the channel. The leakage current is proportional to n_i and has the same temperature dependence as n_i. This effect can be minimized by the introduction of AlGaAs heterostructure buffers or by an AlAs/GaAs superlattice.
- The leakage current of the gate Schottky diode increases exponentially at high temperatures. Even with improved gate metallization and high barrier height, the leakage current of the gate remains a problem.
- The $1/f$ noise is relatively temperature independent; measurements are not presented here.
- The microwave performance of a MESFET is mainly governed by the temperature dependence of S_{21}. S_{11} and S_{22} are relatively temperature independent. MAG and g_m drop in proportion to the inverse temperature.

The results for the HBT can be summarized as follows:

- The HBT is well suited for operation up to 700 K.
- Thermal generation of carriers in the collector space charge region is a problem. This problem can be circumvented by the introduction of a wide bandgap material into the collector (e.g. $Al_{20}Ga_{80}As$).
- A substantial current flows between the contacting pads at high operating temperatures above 523 K. The contacting pads must therefore be placed on top of the passivation or isolation layers. This effect also limits the packing density of the individual transistors.
- The HBT does not suffer from excessive barrier leakage current compared with Schottky gate leakage in MESFETs and HEMTs.
- The current gain of HBTs decreases with temperature. The breakdown voltage increases with ambient temperature.
- Lifetime investigations have been presented but have to be extended.
- The microwave performance of an HBT is determined by the deterioration of the current gain with increasing temperature. However, S_{21} can be kept constant with temperature when biased for constant collector current instead of constant base current. The input and output reflection coefficients change with temperature, complicating the circuit design.

REFERENCES

1. Zulegg, R. (1989) Radiation effects in GaAs FET devices. *Proc. IEEE*, **77**(3), 389–407.
2. Lee, H.J., Juravel, L.Y., Wooley, J.C. and Sring-Thorpe, A.J. (1980) Electron transport and band structure of $Ga_xAl_{1-x}As$ alloys. *Phys. Rev. B*, **21**, 659–669.
3. Tiwari, S., Wright, S.L. and Frank, D.J. (1990) Compound semiconductor heterostructure bipolar transistors. *IBM J. Res. Dev.*, **34**(4), 550–566.
4. Watanabe, M.O., Yoshida, J., Mashita, M., Naknishi, T. and Hojo, A. (1985) Band discontinuity for GaAs/AlGaAs heterojunction determined by CV profiling technique. *J. Appl. Phys.*, **57**, 5340–5344.
5. Casey, H.C. and Panish, M.B. (1978) *Heterostructure Lasers*, Academic Press, New York.
6. Walton, A.K. and Mishra, U.K. (1968) Light and heavy hole masses in GaAs and GaSb. *J. Phys. C*, **1**, 533–538.
7. Dumke, W.P., Lorenz, M.R. and Petit, G.D. (1972) Enhanced indirect optical absorption in AlAs and GaP. *Phys. Rev. B*, **5**, 2978–2985.
8. Chang, C.S. and Fetterman, H.R. (1986) Electron drift velocity versus electric field in GaAs. *Solid-State Electronics*, **29**(12), 1295–1296.
9. Kurata, M. and Yoshida, J. (1984) Modelling and characterization for high speed GaAlAs–GaAs heterojunction bipolar transistors. *IEEE Trans. Electron Devices*, **31**, 467–473.
10. Blakemore, J.S. (1982) Semiconducting and other major properties of gallium arsenide. *J. Appl. Phys.*, **53**(10), R123–R181.
11. Beyzavi, K., Lee, K., Kim, D.M., Nathan, M.I., Wrenner, K. and Wright, S.L. (1991) Temperature dependence of minority-carrier mobility and recombination time in p-type GaAs. *Appl. Phys. Lett.*, **58**, 1268.
12. Nathan, M.I., Dumke, W.P., Wrenner, K., Tiwari, S., Wright, S.L. and Jenkins, K.A. (1988) Electron mobility in p-type GaAs. *Appl. Phys. Lett.*, **52**, 654.
13. Fricke, K., Krozer, V. and Hartnagel, H.L. (1989) A new GaAs power MESFET structure for improved power capabilities. *IEEE Trans. Microwave Theory and Tech.*, **37**(9), 1334–1339.
14. Sadwick, L.P., McDonald, R.M., Crofts, R.J., Koniak, J. and Hwu, R.J. (1994) 350 °C GaAs MESFET-based electronic technology, in *Proc. 2nd Int. High Temperature Electronics Conference, Charlotte NC, 5–10 June*, pp. V27–V32.
15. Papanicolaou, N.A., Anderson, W.T., Katzer, D.S., Jones, S.H. and Jones, J.R. (1994) All-refractory GaAs FET for high temperature applications, in *Proc. 2nd Int. High Temperature Electronics Conference, Charlotte NC, 5–10 June*, pp. V9–V14.
16. Wilson, C.D. and O'Neill, A.G. (1994) GaAs heterojunction devices for integrated circuit technology at elevated temperatures, in *Proc. 2nd HITEC'94, High Temperature Electronics Symposium, Charlotte NC, 23–27 May*.
17. Wilson, C.D., O'Neill, A.G., Baier, S. and Novaha, J. (1995) A complementary III–V heterostructure field-effect transistor technology for high temperature integrated circuits. *Materials Science and Engineering B*, **29**(1–3), 54–58.
18. Shoucair, F.S. (1994) Device modeling for high-temperature integrated circuits: silicon (CMOS), gallium arsenide (MESFETs), and silicon carbide (MOSFETs), in *Proc. 2nd Int. High Temperature Electronics Conference, Charlotte NC, 5–10 June*, short course notes.
19. Kraemer, B., Basset, R., Chye, P., Day, D. and Wei, J. (1994) Power PHEMT module delivers 12 watts, 40% PAE over the 8.5 to 10.5 GHz band, in *Proc. IEEE MTT-S Symposium, San Diego CA, 23–27 May*, pp. 683–686.
20. Williams, R.E. (1985) *Gallium Arsenide Processing Technologies*, 2nd edn, Artech House, Boston, MA.
21. Uchida, Y., Yokotsuka, T., Nakashima, H. and Takatani, S. (1987) Electrical properties of thermally stable LaB_6/GaAs Schottky diodes. *Appl. Phys. Lett.*, **50**(11), 670–672.
22. Würfl, J., Singh, J.K. and Hartnagel, H.L. (1990) Reliability aspects of thermally stable LaB_6–Au Schottky contacts to GaAs, in *Proc. IEEE International Reliability Physics Symposium, New Orleans, 28–29 March*, pp. 87–93.

23. Würfl, J., Fricke, K. and Hartnagel, H.L. (1990) Nonalloyed, high temperature stable ohmic contacts to GaAs based on LaB$_6$ diffusion barriers, in *Proc. Int. Symp. GaAs and Related Compounds, Jersey, UK, 24–27 September*, pp. 239–244.

24. Fricke, K., Hartnagel, H.L., Schütz, R., Schweeger, G. and Würfl, J. (1989) A new GaAs technology for stable FETs at 300 °C). *IEEE Electron Device Letters*, **10**(12), 577–579.

25. Sze, S.M. (1981) *Physics of Semiconductor Devices*, 2nd edn, Wiley Eastern, New Dehli.

26. Shoucair, F.S. and Ojala, P.K. (1992) High temperature electrical characteristics of GaAs MESFETs (25–400 °C). *IEEE Trans. Electron Devices*, **39**(7), 1551–1557.

27. Krozer, V., Ruppert, M., Miao, J.M., Lee, W.Y., Schüßler, M., Fricke, K. and Hartnagel, H.L. (1994) Modelling of high temperature performance of MESFETs and HBTs, in *Proc. 2nd Int. High Temperature Electronics Conference, Charlotte NC, 5–10 June*, pp. IV9–IV14.

28. Fricke, K., Lee, W.Y., Krozer, V., Würfl, J., Białas, S. and Hartnagel, H.L. (1992) Microwave characterization and comparison of performance of GaAs based MESFETs, HEMTs and HBTs operating at high ambient temperatures, in *Proc. GAAS'92, European Gallium Arsenide and Related III–V Compounds Application Symposium, Noordwijk, Netherlands, 27–29 April*.

29. Würfl, J., Janke, B., Rooch, K.-H. and Thierbach, S. (1994) GaAs microwave devices operating at high ambient temperatures: technology and performance, in *Proc. 2nd Int. High Temperature Electronics Conference, Charlotte NC, 5–10 June*.

30. Steinhagen, F. (1991) Aufbau und experimentelle Untersuchung eines Heterostruktur-Feldeffekt-Transistors auf GaAs/AlGaAs. Diplomarbeit an der Technischen Hochschule Darmstadt.

31. Würfl, J. and Fricke, K. (1991) Schaltungstechniken für Hochtemperaturelektronik. *Mikroelektronik*, **5**(6), LXXXVI.

32. Fricke, K., Gatti, G., Hartnagel, H.L., Krozer, V. and Würfl, J. (1992) Performance capabilities of HBT devices and circuits for satellite communication. *IEEE Trans. Microwave Theory and Tech.*, **40**(6), 1205–1214.

33. Rezazadeh, A.A., Morgan, D.V., Mawby, P.A. and Kerr, T.M. (1987) Current transport mechanism at the emitter base junction of an npn GaAs/AlGaAs heterojunction bipolar transistor prepared by MBE. *IEEE Trans. Electron Devices*, **34**, 947–949.

34. Frost, M.S., Riches, M. and Kerr, T. (1986) A pnp AlGaAs heterojunction bipolar transistor for high temperature operation. *J. Appl. Phys.*, **60**(6), 2149–2153.

35. Zipperian, T.E. and Dawson, L.R. (1983) GaP/Al$_x$Ga$_{1-x}$P heterojunction transistors for high-temperature electronic applications. *J. Appl. Phys.*, **54**(10), 6019–6025.

36. Kren, E., Rezazadeh, A.A., Rees, P.K. and Tothill, J.N. (1993) High c-doped base InGap/GaAs HBTs with improved characteristics grown by MOCVD. *Electronics Lett.*, **29**(11), 961–963.

37. Schaper, U., Bachem, K.H., Krner, M. and Zwicknagl, P. (1993) Scaling of small-signal equivalent circuit elements for GaInP/GaAs hole–barrier bipolar transistors (HBT). *IEEE Trans. Electron Devices*, **40**(1), 222–224.

38. Streit, D.C., Oki, A.K., Umemoto, D.K., Velebir, J.R., Stolt, K.S., Yamata, F.M., Saito, Y., Hafizi, M.E., Bui, S. and Tran, L.T. (1991) High-reliability GaAs–AlGaAs HBTs by MBE with Be base doping and InGaAs emitter contacts. *IEEE Electron Device Letters*, **12**, 471–473.

39. Fricke, K., Hartnagel, H.L., Lee, W.Y. and Würfl, J. (1992) AlGaAs/GaAs HBT for high temperature application. *IEEE Trans. Electron Devices*, **38**(9), 1977–1981.

40. Fricke, K., Hartnagel, H.L., Lee, W.Y. and Würfl, J. (1993) Reply to comments on 'AlGaAs/GaAs HBT for high temperature applications'. *IEEE Trans. Electron Devices*, **40**(9), 1717.

41. Rein, H.M., Schad, T. and Zühlke, R. (1972) Der Einfluß des Basisbahnwiderstandes und der Ladungsträgermultiplikation auf das Augangskennlinienfeld von Planartransistoren. *Solid-State Electronics*, **15**, 481–500.

42. Huang, J.T. (1967) Study of transistor switching circuit stability in the avalanche region. *IEEE J. Solid-State Circuits*, **2**(1), 10–21.

43. Schüßler, M., Fricke, K., Lee, W.-Y., Krozer, V. and Hartnagel, H.L. (1993) Collector design for improved breakdown in high temperature AlGaAs/GaAs HBT, in *Proc.*

Workshop on High Performance Electronic Devices for Microwave and Optoelectronic Applications, London, UK, 18 October.

44. Fricke, K., Krozer, V., Lee, W.-Y., Schüßler, M., Schweeger, G., Sigurdardóttir, A. and Hartnagel, H.L. (1994) Design and technology of GaAs/AlGaAs HBT for high temperature circuits, in *Proc. Gallium Arsenide Applications Symposium 94, Turin, Italy, 28–29 April*, pp. 123–126.

45. Tijburg, R.P. and Van Dongen, T. (1976) Selective etching of III–V compounds with redox systems. *J. Electrochem. Soc.*, **123**(5), 687–691.

46. Kenefick, K. (1982) Selective etching characteristics of peroxide/ammonium hydroxide solutions for GaAs/Al$_{0.16}$Ga$_{0.84}$As. *J. Electrochem. Soc.*, **129**, 2380–2382.

47. LePore, J.J. (1980) An improved technique for selective etching of GaAs and Al$_x$Ga$_{1-x}$As. *J. Appl. Phys.*, **51**(12), 6441–6442.

48. Würfl, J. (1989) *Herstellung und Untersuchung zuverlässiger Metallkontakte auf GaAs zur Entwicklung von hochtemperaturstabilen Halbleiterbauelementen.* VDI Verlag, Düsseldorf.

49. Pirling, T., Fricke, K., Schüßler, M., Lee, W.Y., Fueß, H. and Hartnagel, H.L. (1995) Investigations of Pd/In based high temperature ohmic contacts on GaAs by x-ray reflectometry and diffractometry. *Materials Science and Engineering B*, **29**, 70–73.

50. Fricke, K., Hartnagel, H.L., Lee, W.-Y., Pierling, T., Schüßler, M. and Würfl, J. (1994) A highly reliable ohmic contact system on GaAs based on Pd/In, in *Proc. 2nd Int. High Temperature Electronics Conference, Charlotte NC, 5–10 June*, pp. P197–P202.

51. Lee, W.-Y. (1995) Technologieentwicklung und Charaketrisierung des Hochtemperatur-Hochfrequenz-Heterostruktur-Bipolar-Transistors auf AlGaAs/GaAs bis 673 K. PhD Thesis, TH Darmstadt.

52. Schüßler, M., Lee, W.-Y., Fricke, K. and Hartnagel, H.L. (1994) Influence of thermal and current stress on high-temperature performance of AlGaAs/GaAs/AlGaAs DHBT, in *Proc. 2nd Int. High Temperature Electronics Conference, Charlotte NC, 5–10 June*, pp. P69–P74.

53. Dikmen, C.T., Dogan, N.S. and Osman, M. (1994) Modeling and characterization of AlGaAs/GaAs heterojunction bipolar transistors for high-temperature applications, in *Proc. HITEC'94, High Temperature Electronics Symposium, Charlotte NC, 23–27 May*.

54. Rizzoli, V., Constanzo, A., Lipparini, A., Mastri, F. and Muzzarelli, G. (1994) Thermal modelling of microwave devices for the electrothermal analysis of nonlinear microwave circuits, in *Proc. 24th European Microwave Conference, Nice, France, September 1994*.

55. Krozer, V. (1991) Verfahren der Kleinsignal- und Großignal-Analyse und Charakterisierung von Mikrowellenschaltungen und Bauelementen mit Hilfe der Volterra-Reihe. PhD Thesis, TH Darmstadt.

56. Krozer, V., Ruppert, M., Lee, W.Y., Grajal, J., Goldhorn, A., Schüßler, M., Fricke, K. and Hartnagel, H.L. (1994) A physics-based temperature-dependent SPICE model for the simulation of high temperature microwave performance of HBTs and experimental results, in *1994 IEEE MTT-S, Int. Microwave Symp. Dig., San Diego CA, 23–27 May*.

57. Lee, W.Y., Fricke, K., Krozer, V., Schüßler, M. and Hartnagel, H.L. (1993) AlGaAs/GaAs HBT for microwave applications up to 240 °C, in *Proc. Workshop on High-Speed Bipolar Circuits and Devices, Ulm, Germany, 11–12 October*.

58. Liu, W. and Bayraktaroglu, B. (1993) Theoretical calculations of temperature and current profiles in multi-finger heterojunction bipolar transistors. *Solid-State Electronics*, **36**(2), 125–132.

59. Liu, W. and Khatibzadeh, A. (1994) The collapse of current gain in multi-finger heterojunction bipolar transistors: its substrate temperature dependence, instability criteria, and modeling. *IEEE Trans. Electron Devices*, **41**(10), 1698–1707.

60. Donzelli, P., Ghione, G. and Naldi, C.U. (1990) Thermal models for low- and high-power GaAs MESFET devices, in *Proc. Gallium Arsenide Applications Symposium, Rome, Italy, 19–20 April*.

61. Pettenpaul, E., Heidenreich, W., Huber, J. and Flossmann, W. (1985) A high-temperature sensor based on monolithic GaAs Hall IC, in *Proc. GaAs IC Symposium*, pp. 169–172.

62. Schweeger, G., Fricke, K., Menke, K. and Hartnagel, H.L. (1991) A GaAs integrated differential amplifier for operation up to 300 °C. *Solid-State Electronics*, **34**(7), 731–733.

63. Schweeger, G., Singh, J.K., Fricke, K., Klingelhöfer, C. and Hartnagel, H.L. (1989) Monolithic GaAs multivibrator for operation at temperatures up to 300 °C. *Electronics Lett.*, **25**(20), 1385–1386.

64. Fricke, K., Schweeger, G. and Hartnagel, H.-L. (1990) Integrated circuits on GaAs for the temperature range from room temperature up to 300 °C, in *Proc. Gallium Arsenide Applications Symposium, Rome, Italy, 19–20 April*, pp. 284–289.

65. Baier, S., Nohava, J., Jeter, R., Carlson, R. and Hanka, S. (1994) High temperature electronics using complementary heterostructure FET (CHFET) technology, in *Proc. 2nd Int. High Temperature Electronics Conference, Charlotte NC, 5–10 June*, pp. V21–V26.

66. Fricke, K., Hartnagel, H.L., Lee, W.Y. and Schüßler, M. (1994) AlGaAs/GaAs/AlGaAs DHBT operational amplifier for high temperature application. *IEEE Electron Device Letters*, **15**(3), 88–90.

67. Miao, J.-M., Hartnagel, H.L., Rück, D. and Fricke, K. (1994) The use of ion implantation for micromachining of GaAs towards sensor applications, in *Proc. Eurosensors VIII, Toulouse, France, 25–28 September*.

68. Riemenschneider, R., DasGupta, N., Schütz, R. and Hartnagel, H.L. (1993) Low temperature deposition of SiO₂ and PSG using SiH₄, N₂O and phosphorus vapour for damage-free passivation of InP-based PIN-diodes by photo and plasma-assisted LPCVD. *Applied Surface Science*, **69**(5), 277–280.

69. Singh, B.R., Fricke, K., Würfl, J. and Hartnagel, H.L. (1992) Boron implant isolation for high temperature GaAs integrated circuits. *J. Electrochem. Soc.*, **139**(5), 1470–1472.

70. Miao, J., Hartnagel, H.L., Rück, D. and Fricke, K. (1995) The use of ion implantation for micromachining GaAs for sensor applications. *Sensors & Actuators A*, **46–47**, 30–34.

71. Miao, J., Hjort, K., Hartnagel, H.L., Schweitz, J.-A., Rück, D. and Tinschert, K. (1995) Resonant sensors on thin semi-insulating GaAs membranes, in *Proc. 8th Int. Conf. on Solid-State Sensors and Actuators, and Eurosensors, Stockholm, Sweden, 25–29 June*, pp. 604–607.

72. Miao, J., Hartnagel, H.L., Weiss, B.L. and Wilson, R.J. (1995) Improved free-standing semi-insulating GaAs membranes for sensor applications. *Electronics Lett.*, **31**(13), 1047–1049.

73. Schweeger, G., Fricke, K., Menke, K. and Hartnagel, H.L. (1991) A GaAs integrated differential amplifier for operation up to 300 °C. *Solid-State Electronics*, **34**(7), 731–733.

74. Schweeger, G., Singh, J.K., Fricke, K., Klingelhöfer, C. and Hartnagel, H.L. (1989) Monolithic GaAs multivibrator for operation at temperatures up to 300 °C. *Electronics Lett.*, **25**(20), 1385–1386.

75. Fricke, K., Schweeger, G. and Hartnagel, H.-L. (1990) Integrated circuits on GaAs for the temperature range from room temperature up to 300 °C, in *Proc. Gallium Arsenide Applications Symposium, Rome, Italy, 19–20 April*, pp. 284–289.

76. Bock, K., Fricke, K. and Hartnagel, H.-L. (1990) Improved ESD-protection of GaAs-FET microwave devices by new metallization strategy, in *Proc. 12th Annual Electrical Overstress/Electrostatic Discharge Symposium, Orlando FL, 11–13 September*, pp. 193–195.

77. Rosenberg, R. (1985) Inhibition of electromigration damage in current-stressed Al-gates as used for GaAs-MESFETs. *J. Phys. D: Appl. Phys.*, **18**, 263–270.

78. Sethi, B.R. and Hartnagel, H.L. (1986) Characterization of electromigration in metal conductors on GaAs surfaces. *Int. J. Electronics*, **61**, 27–31.

79. Daga, O.P., Fricke, K. and Hartnagel, H.L. (1986) Improved via-hole etching for source grounding of microwave MESFET. *J. Electrochem. Soc.*, **123**(12), 2660–2661.

80. Schwarze, G.E., Niedra, J.M. and Wiederman, W.R. (1991) High temperature, high frequency experimental investigation of soft magnetic materials, in *Proc. 1st Int. High Temperature Electronics Conference, Albuquerque NM, 16–20 June*, pp. 364–373.

81. Kaufman, B. (1991) Miniature pulse transformers for high temperature applications, in *Proc. 1st Int. High Temperature Electronics Conference, Albuquerque NM, 16–20 June*, pp. 355–363.

82. Christou, A., Dasgupta, A., Pecht, M. and Barker, D. (1991) Design optimization of high temperature electronics packaging, in *Proc. 1st Int. High Temperature Electronics Conference, Albuquerque NM, 16–20 June*, pp. 377–383.

83. Hieber, H. (1992) Packaging-choice of materials and built-in reliability, in *Proc. Euroform Seminar on High Temperature Semiconductor Electronics, Darmstadt, Germany, 6–8 May*, pp. 6.1–7.24.

84. Fricke, K., Krozer, V. and Hartnagel, H.L. (1995) Thermodynamics of microwave semiconductor devices, in *Handbook of Microwave Technology* (ed. K. Ishii), Academic Press, San Diego CA.
85. Fricke, K., Krozer, V., Bock, K. and Hartnagel, H.-L. (1990) Thermal problems in microwave semiconductor devices, in *Proc. MICROCOLL, Budapest, Hungary, 14 September*, pp. 107–111.
86. Fallmann, W., Hartnagel, H.L. and Matur, P.C. (1971) Experiments on heat sinking of semiconductor devices. *Electronics Lett.*, **7**, 12–13.
87. Böttner, T., Fricke, K., Goldhorn, A., Hartnagel, H.L., Rappl, A., Ritter, S. and Würfl, J. (1991) Technology and performance of a high temperature stable (up to 300 °C) operational amplifier on GaAs, in *Proc. 1st Int. High Temperature Electronics Conference, Albuquerque NM, 16–20 June*, pp. 77–82.
88. Abidi, A.A. (1987) An analysis of bootstrapped gain enhancement techniques. *IEEE J. Solid-State Circuits*, **22**(6), 1200–1204.

PART FOUR

Future Materials

Silicon carbide: material and device properties | 9

Y. M. Tairov and M. Willander

9.1 INTRODUCTION

Continuous progress in high temperature electronics creates a demand for novel, frequently unique materials, processing technologies and electronic devices. Physical and chemical properties of the wide bandgap semiconducting materials such as diamond, silicon carbide, gallium nitride and aluminum nitride have recently become highly interesting and important for high temperature regimes. Various device implementations not only use the common properties (for wide bandgap semiconductors) but also the special peculiarities of different materials. Tight chemical bonding means the diamond-like semiconductors are chemically inactive at elevated temperatures; they also possess hardness, mechanical firmness and resistance to radiation. These important physical properties guarantee their wide-ranging use in high safety space, aircraft, military and automobile electronics, etc., but it is hardly possible in the near future that their processing technologies will ever be cheap and free from problems.

For certain special purposes, other definite material peculiarities may be more significant. Thus high breakdown electric field, saturated carrier drift velocity and thermal conductivity are of great importance for high power and high frequency applications. In order to ensure higher quantum efficiency of optoelectronic devices, the wide bandgap semiconductors with direct bandgap, such as gallium or aluminum nitrides and solid solutions based on them are more favorable [1–3]. Some progress has already been achieved in processing technologies, and at lower costs. These advantages in the future may determine the preferential use of one of the diamond-like materials in micromechanical applications [4].

High Temperature Electronics. Edited by M. Willander and H.L. Hartnagel. Published in 1997 by Chapman & Hall, London. ISBN 0 412 62510 5.

In order to illustrate the potential opportunities for wide bandgap semiconductors in high temperature electronics, some of the important material properties are listed in Table 9.1 and compared with silicon and gallium arsenide.

It is easy to infer from Table 9.1 that, when comparing diamond-like semiconductors with silicon, there is no competition based on physical properties alone. State-of-the-art fabrication and processing technologies may also determine and/or limit the choice of all the prospective materials available for device implementation. Bulk crystal and epitaxial growth, selective doping by both n- and p-type impurities, formation of ohmic contacts, etching and isolation techniques have also to be elaborated.

Diamond has a superior combination of physical properties, including wide bandgap, high thermal conductivity, high electron mobility and high hole mobility. Considerable progress has recently been made towards the development of a controllable technique for low pressure chemical vapor deposition (LPCVD) of thin film diamond at acceptable production cost. Unfortunately, it is only possible to obtain p-type diamond with relative ease, using boron doping; n-type diamond at room temperature has proved much more difficult to achieve because of the significantly higher ionization energy of commonly used donors.

Unlike diamond and silicon carbide, which have indirect bandgap, aluminum nitride and gallium nitride are direct bandgap semiconductors and would therefore be suitable for producing more efficient light-emitting and light-detecting devices as well as the short wavelength injection laser. They show complete miscibility with each other and with indium nitride. This is very important for optoelectronic device implementations because it allows the energy bandgap to be controlled, and thus the wavelength of the spectral characteristic maximum. The solid solubility of AlN with SiC has also been noted. Furthermore, the advantage of the group III nitrides is their piezoelectricity, which permits the production of surface acoustic wave devices. Device implementation of these materials was long held back because of the difficulties in achieving reproducible p-type material with high conductivity and in forming the abrupt p–n junction.

Silicon carbide processing technology is now the most advanced of all the wide bandgap semiconductors, so SiC has the best chance of realizing the advantages listed earlier. Relatively large single crystals and high quality epitaxial layers have been reproducibly grown; both p-type and n-type conductivity appear to be feasible; the technological operations required for device fabrication have been demonstrated; and prototypes of basic electron devices have been fabricated. The most common items for SiC electronics up to now have been the junction-gate field effect transistor (JFET), the metal–oxide–semiconductor (MOSFET), and the Schottky barrier (MESFET) semiconductors. Operating temperatures of up to 650 °C were achieved for some prototype devices [5].

Table 9.1 Properties of the wide bandgap semiconductors in comparison with silicon

Semiconductor	Si	GaAs	GaN	AlN	3C-SiC	6H-SiC	Diamond
Band transition	indirect	direct	direct	direct	indirect	indirect	indirect
Bandgap, E_g (eV)	1.1242	1.42	3.45	6.28	2.39	3.02	5.50
Carrier mobilities							
n $(cm^2 V^{-1} s^{-1})$	1450	9500	600	80	1000	400	2200
p $(cm^2 V^{-1} s^{-1})$	500	450	10	14	60	60	2100
Saturated carrier drift							
Velocity, V_{sat} $(cm\ s^{-1})$	1.0×10^7	2.0×10^7	2.5×10^7	–	2.5×10^7	2.0×10^7	2.7×10^7
Breakdown field, E_b $(V\ cm^{-1})$	3×10^5	4×10^5	20×10^5	–	20×10^5	70×10^5	100×10^5
Thermal conductivity at 300 K $(W\ cm^{-1} K^{-1})$	1.56	0.44	1.30	0.30	–	3.9	10
Debye temperature (K)	674	344	600	744	1093	907	–
Dielectric constant							
(0)	11.9	12.9	10.4 ($\parallel c$) 9.5 ($\perp c$)	8.50	9.72	10.03 ($\parallel c$) 9.66 ($\perp c$)	5.70
(∞)		10.9	5.8 ($\parallel c$) 5.35 ($\perp c$)	4.68	6.52	6.70 ($\parallel c$) 6.52 ($\perp c$)	–
Reproducibly obtained in conductivity types	n, p	n, p	n	n	n, p	n, p	p

Silicon carbide is a perfect radiation resistance material having wide bandgap, high Debye temperature and high thermal conductivity. It may be quite possible that devices in silicon carbide electronics would operate very efficiently, e.g. in control and operating systems of nuclear reactors. Improving work safety, lowering costs on cooling and radiation shielding of electronic appliances, simplifying reactor construction and thus reducing their size. All these aspects would offset the extra costs of using silicon carbide. Another field ripe for silicon carbide electronics could be sensor and actuator engineering. The chemical and thermal stability of SiC would allow the sensing elements to be integrated with amplifying and processing electronics on the same chip, leading to improved reliability and lower costs. Chemical inactivity and higher reliability would make SiC electronics attractive for medicine and ecological engineering.

We believe silicon carbide will be the first of the wide bandgap semiconductors to be used in high temperature commercial electronics. Consequently, we should focus much more attention upon it. The properties of silicon carbide, the key processing technologies and some SiC devices are discussed in the following sections. As to other wide bandgap semiconductors, we refer the interested reader to the excellent reviews [1–3,6] and the references therein.

9.2 STRUCTURAL AND ELECTRICAL PROPERTIES OF SiC

Silicon carbide exists in a wide variety of structural forms called polytypes. Polytypism is a special one-dimensional case of polymorphism. All polytypes consist of identical closely packed Si–C double-layers, whose stacking sequences differ along a certain direction. The nearest-neighbor arrangement of atoms is identical in all crystal structures. Each carbon atom is tetrahedrally surrounded by four silicon neighbors and each silicon atom is tetrahedrally bonded to four carbon atoms by the sp^3 hybrid orbitals. The ionicity of the Si–C bond is about 12%. The next nearest neighbors may be placed in two possible ways, forming either cubo-octahedron or so-called hexagonal cubo-octahedron.

Different ways of stacking the subsequent layers will form different crystal structures. Up to now, more than 140 SiC polytypes have been reported [7,8]. The simplest and most well known of them are the sphalerite or 3C (three-layer cubic) and the wurtzite or 2H (two-layer hexagonal) modifications. The 6H (six-layer hexagonal), 15R (fifteen-layer rhombohedral) and the 4H (four-layer hexagonal) are the most commonly occurring polytypes, but the multilayer structures consisting of some hundred strongly repeated layers, sometimes more, have also been observed.

To understand the special features of the more complicated structures their cross sections by the (1 1 0) plane are to be scrutinized. Using this cross

Table 9.2 Structural and chemical properties of silicon carbide polytypes

	Ramsdell notation	Zhdanov notation	Wyckoff–Jagodzinski notation	Space group	Lattice parameters a (nm)	c/n (nm)	$\Delta H^{\circ}_{f,298}$ $(kJ\,mol^{-1})$	$\Delta S^{\circ}_{f,298}$ $(J\,mol^{-1}\,K^{-1})$
β	3C	∞	c	F43m	0.308 269 0.435 98^a	0.251 708	−62.7	14.88
α	2H	(11)	h	P6$_3$mc	0.307 63	0.2524	–	–
	4H	(22)	hc	P6$_3$mc	0.307 997	0.252 035	−66.5	10.62
	15R	(23)$_3$	hchcc	R3m	0.308 043	0.252 009	−65.6	24.70
	6H	(33)	hcc	P6$_3$mc	0.308 086	0.251 955	−65.2	21.61

^{a}For cubic unit cell.

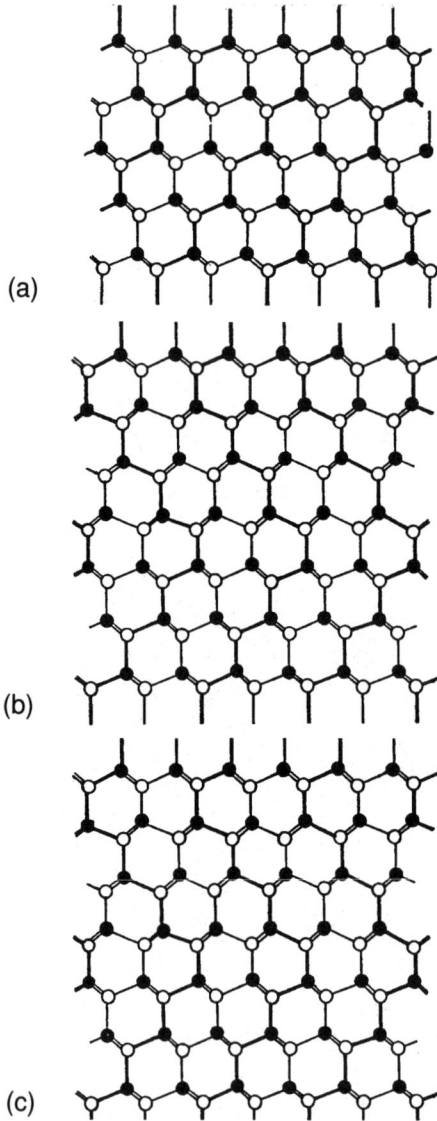

Figure 9.1 Projections of polytype structures on the (1 1 0) plane: (a) 3C-SiC, (b) 4H-SiC, (c) 6H-SiC and (d) 15R-SiC.

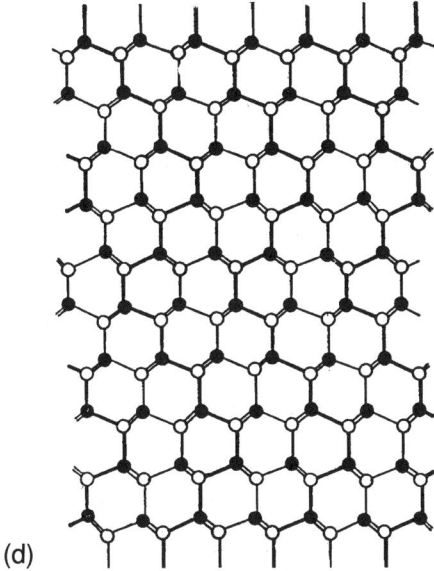

(d)

Figure 9.1 (*Continued*).

section (the most significant from the practical viewpoint), the atomic arrangement in polytypes is shown in Fig. 9.1.

The crystallographic features of several SiC polytypes (Fig. 9.1) are summarized in Table 9.2. Any notation has advantages as well as deficiencies. The Ramsdell notation includes information about the symmetry of the crystal structure. The Wyckoff–Jagodzinski notation clearly shows the number of nonequivalent sites in the lattice. The Zhdanov notation illustrates the zigzag layer consequence in the (1 1 0) plane.

The chemical bonding and thermodynamics of different polytypic modifications seem to be nearly the same. Quite the reverse is true for details of the bandgap structure, so the electronic properties differ very strongly from one polytype to another. The polytypes may be regarded as a family of semiconducting materials having almost the same lattice parameter and similar chemical properties. In this respect, the silicon carbide polytypes may be outstanding as an excellent material to construct the heterojunction devices.

On the other hand, although many hypotheses have been proposed, the origin of the polytypism is not yet well understood, and only limited success has been achieved in the controlled nucleation and growth of SiC single crystals [8]. The crystallographic aspect and most of the theories being proposed were discussed in the relatively old but up-to-date monograph of Verma and Krishna [9].

Table 9.3 Properties of silicon carbide polytypes

Polytype		E_g at 4 K (eV)	Debye temperature (K)	Carrier mobility ($cm^2\ V^{-1}s^{-1}$)	
				n	p
2H	(11)	3.330	–	–	–
4H	(22)	3.265	1297	800	60
15R	$(23)_3$	2.986	841	500	60
6H	(33)	3.023	907	360	60
21R	$(34)_3$	2.853	–	–	–
8H	(44)	2.800	866	–	–
3C	(∞)	2.390	1093	1000	60

The dislocation theory of Frank postulates that a screw dislocation with a definite Burgers vector causes the corresponding crystal structure formation during the growth, but hardly any studies have been carried out as to why just that kind of dislocation arises. The Jagodzinsky theory, with special emphasis on the vibrational entropy, establishes that the simplest polytypic modifications, at least, seem to have original thermodynamics, so the conditions favor their preferential nucleation, but that theory is only semiquantitative and cannot predict anything.

The absence of common assumptions on the nature of the polytypism phenomenon illustrates why polytypic control during the growth of SiC single crystals remains a matter for ongoing experimental investigation.

The problem of polytypic control was never an abstract curiosity of purely scientific interest. Certain properties of several SiC polytypes are listed in Table 9.3, properties significant for implementation in microelectronics. It is clearly seen that 2H-SiC has the widest bandgap of all. Unfortunately, 2H-SiC single crystals up to now have only been obtained as whiskers. 3C-SiC has higher charge carrier mobility but does not have a wide bandgap compared with other polytypes. The usual polytypic modification (obtained by the modified sublimation method) is 6H-SiC. Because 4H-SiC has a wider bandgap and a higher carrier mobility, many researchers are addressing the problem of 4H-SiC polytypic control during its growth [10–12].

Formation of each of the polytypes requires a special growth technique as discussed below.

9.3 SiC SINGLE CRYSTAL GROWTH

The most widely practiced growth techniques of SiC are sublimation techniques for growing single-crystal ingots and (hetero)epitaxy for producing monocrystalline thin films.

Single crystals of semiconducting silicon carbide are being grown in bulk at present by the modified sublimation method. At first Lely [13] developed a sublimation process for growing high purity α-SiC single crystals. A graphite crucible filled with polycrystalline or powdered SiC is heated to about 2500 °C. SiC powder sublimates in the hot part of crucible and condenses on the cooler walls of the inner graphite cavity, producing randomly sized and mostly intergrown roughly hexagonally shaped platelets of some uncontrolled polytype, most commonly 6H type.

Tairov and Tsvetkov [14, 15] have proposed putting a seed into the crucible and have thus developed the modified seeded-growth sublimation process. This approach was then adopted, later by Ziegler *et al.* [16], Koga *et al.* [17] and Carter *et al.* [18]. The modified sublimation method is a well-controlled technique and allows the growth of large single-crystal boules of the pure 6H polytype with diameters up to 40 mm [19] as well as those of 4H polytype [10, 12]. Doping of SiC crystals by donor and acceptor impurities during the growth process has been realized without considerable difficulty by introducing nitrogen or other gaseous specimens into the growth chamber.

The seeded-growth method makes it possible to achieve heavily doped crystals (up to 10^{20} cm^{-3}) as well as lightly doped crystals. But note that, so far, dislocation densities have been much greater when crystals are spontaneously grown (10^4 to 10^5 cm^{-2} and $< 10^2$ cm^{-2}, respectively). Moreover, seeded growth produces crystals containing many micropores and micropipes ($\approx 10^2$ cm^{-3}), defects completely absent in spontaneously grown crystals. In order to eliminate these defects or at least to reduce their number, the initial stage of single-crystal growth requires thorough investigation.

Several techniques have been proposed for growing high quality SiC epitaxial layers doped in a controlled way. The sublimation sandwich method [20] or liquid phase epitaxy [21] were previously used to obtain SiC thin films and p–n junctions. Fabrication possibilities have been successfully demonstrated for blue light-emitting diodes (LEDs) made of SiC [22–25]. These processes were believed to be too complicated for realizing multilayer epitaxy [26].

Chemical vapor deposition (CVD) is now the most advanced epitaxial technique and the method best suited for device implementation. Once Nishino *et al.* [27] had solved the problem of large lattice mismatch between growing SiC layer and silicon substrate (20%), the CVD heteroepitaxial growth of 3C-SiC on Si was improved significantly. Furthermore, using the off-axis wafer has made it possible to eliminate the antiphase domain boundaries that are the main type of intrinsic defect in the artificially grown layers. The details of this technique can be found in the literature [28, 29]. We only want to point out here that, despite the progress achieved in the SiC thin film growth, the as-grown epitaxial layers still contain many misfit dislocations, which limit carrier mobilities.

We believe there will soon be intensive research into CVD, substrates made out of wafers cut from SiC single-crystal boules [30, 31].

9.4 SiC PROCESSING TECHNOLOGIES

One of the key tasks in device-related technological studies is the controlled doping of SiC by donor and acceptor impurities. Group Va elements, N, P and Sb, are the usual donors in SiC technology. Group IIIa elements, B, Al, Ga and Be, are acceptors. Aluminum, gallium and most other impurities substitute for silicon atoms, whereas nitrogen and boron substitute for carbon.

Nitrogen is the most commonly used donor in SiC technology. The existence of nonequivalent states in the lattices of various polytypes complicates the electrical and optical behavior of incorporated impurity atoms. The ionization energy of nitrogen atoms in cubic-like positions is almost twice as much as the ionization energy of those occupying the hexagonal-like positions and strongly depend on the polytypic modification.

Nearly the same average value independent of polytype is observed for the ionization energy of aluminum, i.e. the shallow impurity in SiC. Boron causes deeper levels to appear in the bandgap. These data are listed in Table 9.4. Further discussion on this topic may be found in the literature [32–34].

The solubility of impurities in different SiC polytypes and their diffusion mobilities are almost completely independent of polytypic modification [32]. The average interatomic distance is small, so it is only tiny atoms which dissolve markedly in SiC. Tight chemical bonding leads to a very high energy of point defect formation in SiC, about 4 eV for simple vacancies. The

Table 9.4 Impurity levels in silicon carbide polytypes

Polytype and site	Impurity level (meV)		
	N	B	Al
3C-SiC	53.6	~730	~260
4H-SiC			
h	52.1	~655	~275
c	91.8	–	–
6H-SiC			
h	81.0	698	239
c_1	137.6	723	249
c_2	142.4	–	–

Table 9.5 Solubility and diffusion of impurities in silicon carbide

| Diffusant | Diffusion coefficient $D = D_0 \exp(-W/kT)$ | | Maximum solubility attained (cm^{-3}) |
	$D_0 (cm^2 s^{-1})$	$W (eV)$	
Si	5.0×10^2	-7.22	–
C	8.6×10^5	-7.41	–
N	–	–	8.0×10^{20}
P	–	–	4.0×10^{19}
B	50.0	-5.6	3.0×10^{21}
Al	8.0	-6.1	1.1×10^{21}
Ga	0.17	-5.5	1.2×10^{19}
Be	0.3	-3.1	5.0×10^{19}

activation barrier of their migration is also high. Therefore, most impurities diffuse very slowly in SiC even at elevated temperatures. Only boron and beryllium have more or less acceptable diffusion coefficients (Table 9.5), but surface carbonization during the annealing restricts the use of diffusion in SiC electronics.

The main problem of ion implantation is the annealing of radiative defects and the electrical activation of implanted atoms; high temperature and long-term heat treatment are therefore required. Only implantation into the hot wafer at about 1000 °C with postimplant furnace annealing at about 1400 °C makes it possible to obtain high quality p–n junctions exhibiting the reverse bias leakage current as compared with grown epitaxial p–n junctions [35, 36].

In situ doping during the CVD process has proven very effective for obtaining high quality abrupt p–n junctions as well as multilayer structures. The carrier concentration in grown layers appears to be well controlled, depending on the partial pressure of the corresponding dopant gas source [37, 38].

Standard *in situ* doping during epitaxial growth is a nonlocal method; the differently doped regions must be separated from one another. This explains why the mesa style often occurs in SiC device design. To solve this problem, plasma-assisted dry etching techniques are now sufficiently advanced and prove the most convenient [39, 40].

Device fabrication also requires an appropriate metallization technique. It is obvious that traditional Al contacts are not suitable for the special purposes of high temperature electronics. Certain metals, forming either stable silicides or carbides, have been tested in order to obtain high temperature ohmic contacts to silicon carbide and Schottky barriers too. Ni, Fe, Cr, Ti, Ta, Co, Al, Ag, Au, Pt, Pd, Tb, W and some silicides have been proved in use [41–44]. The problem is that the barrier height depends on the wafer doping level and on the special features of the annealing process as well as on polytypic modification and wafer orientation. For example, the Schottky barrier Pt contact is extremely stable when positioned on a C-terminated (0 0 1) surface of a 6H-SiC wafer but less stable when deposited on a (0 0 1) Si surface.

Deposition of Au or Pt [45, 46] seems to be the most advanced technique for Schottky contact formation. Al, Ni, Ni/Ti, Ti/Au and Ta/Au ohmic contacts have been reported to ensure the requirement of low on-resistance and stability for concrete device implementation [5, 47–49].

SiO_2 layers obtained by dry or wet oxidation of SiC (at higher temperatures than needed for silicon oxidation) not only facilitate device isolation and surface passivation but also have acceptable characteristics for fabricating MOS devices [50, 51].

So we can conclude that all the processing technologies necessary for device fabrication have been demonstrated. Their peculiarities and difficulties have a distinct influence on device design using silicon carbide electronics.

9.5 DEVICE PROPERTIES

9.5.1 Metal–insulator–semiconductor devices

Metal–oxide–semiconductor field effect transistors (MOSFETs), the key transistors in most electronic circuits, have been realized in 3C-SiC and 6H-SiC. Both the polytypes have high electron saturation velocity, but 3C-SiC has much higher electron mobility and is therefore regarded as a potential material for device implementation. However, no 3C-SiC substrates exist, so 3C-SiC is usually grown on silicon substrates. Due to the large lattice mismatch (20%) between 3C-SiC and silicon, mechanical defects can appear up to the active region of the transistor. Very good crystal quality has been demonstrated by growing thick 3C-SiC layers 4° off axis [52]. At 400 °C threshold leakage currents as low as 225 pA have been obtained [53]. For long channels (3–5 µm) the transconductance had a maximum value at 400 °C; a physical explanation is given below in the discussion for 6H-SiC. Depletion mode n-channels have been operated up to 650 °C.

Historically, 3C-SiC films grown on silicon substrates were first realized by S. Nishino et al. from Japan [26, 54] in the early 1980s. During the second half of the decade, Japanese groups successfully produced enhancement mode MOSFETs [55, 57], and later they were fabricated by Palmour et al. in the United States [5]. Depletion mode MOSFETs were also produced at the same time [5, 60]. However, 3C-SiC films have also been grown on 6H-SiC with high quality results due to the lattice matching between the two polytypes [28]. In this way improved characteristics for 3C-SiC MOSFETs have been obtained; Fig. 9.2 shows some results. Despite

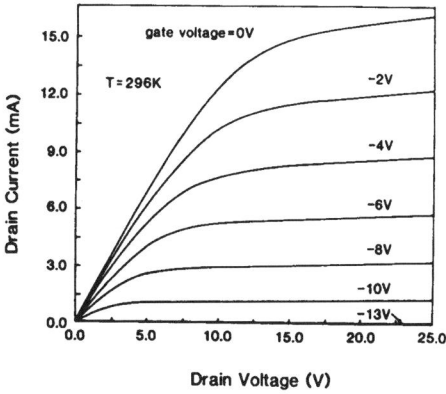

(a)

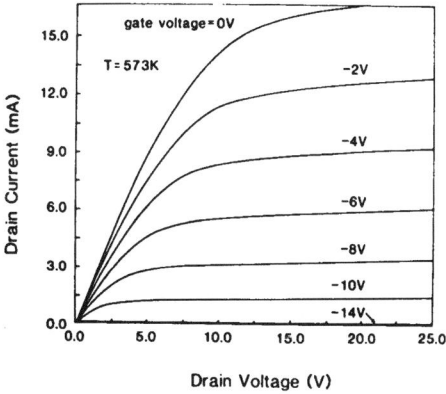

(b)

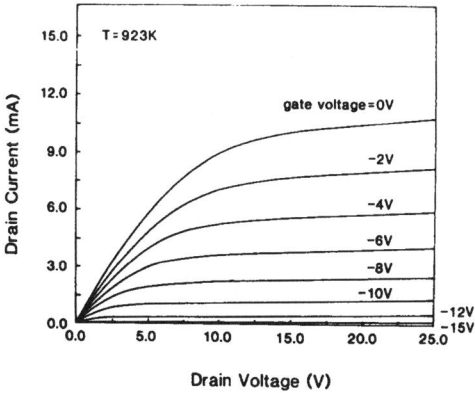

(c)

Figure 9.2 Current–voltage characteristics at different temperatures of a depletion mode n-channel MOSFET with gate length 7.2 μm and gate width 390 μm. The transistor was made of β-SiC (1 1 1) thin film grown on an α(6H)-SiC (0 0 1) substrate: (a) $T = 296\,\text{K}$, (b) $T = 573\,\text{K}$ and (c) $T = 923\,\text{K}$. (From Davis *et al.* [28])

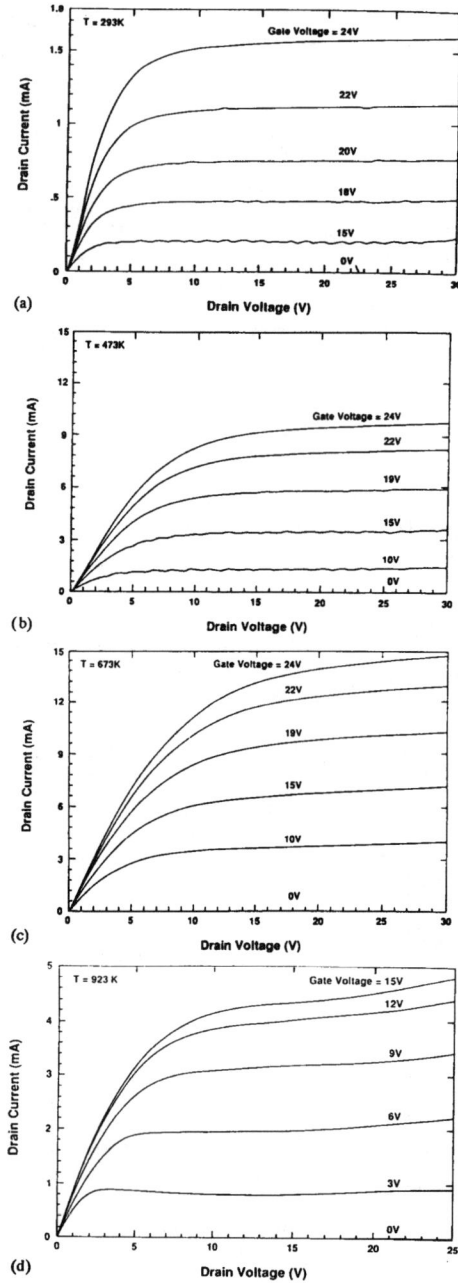

Figure 9.3 Current–voltage characteristics of a 6H-SiC n-channel inversion mode MOSFET at different temperatures: (a) $T = 293$ K, (b) $T = 473$ K, (c) $T = 673$ K and (d) $T = 923$ K (From Davis *et al.* [28])

its inferior transport properties, the larger bandgap and the ready availability of its substrates have made 6H-SiC a sought-after material for MOS applications. The first high quality 6H-SiC MOSFETs were demonstrated by Davis et al. [28] with high temperature $I-V$ characteristics shown in Fig. 9.3.

The SiC–SiO$_2$ interface is the key to understanding the high temperature operation of SiC MOSFETs. Brown et al. [59] found that phosphorus (n-type) atoms are rejected from the oxide during thermal oxidation; for aluminum (p-type) the Al atoms are incorporated into the oxide. Brown et al. came up with the hypothesis that Al at the interface between silicon carbide and oxide produces dangling bond states, slowly responding states and trapping centers. From the temperature mobility measurements on the NMOS transistor and by using the formula

$$\mu = \frac{1}{(W/L)(V_{GS} - V_T)} \frac{1}{C_{ox}} \frac{dI_{DS}}{dV_{GS}}$$

Brown et al. found an activation energy of 0.3 eV, indicating an interface trap at 0.3 eV from the conduction band, which may be related to Schoucair et al. measurement characteristics with an anomalous temperature behavior for the mobility [60]. A temperature dependence of $T^{-1.5}$ in the mobility should be expected for silicon (and for SiC). For temperatures higher than 200 °C the 6H-SiC MOSFET field effect mobility behaved as for Si MOSFETs.

However, the threshold voltage, V_T, behaved differently from Si MOSFETs. Theoretically, we can show that

$$dV_{TH}/dT \approx -dQ_f/dT + T^{-1}[E_G/(2q) - |\Phi_B|]$$
$$\times [2 + C_i^{-1}[2\varepsilon_{SiC}qN_A/(2\Phi_B + V_{SB})]^{1/2}]$$

A different type of behavior was also found experimentally [60]. For the field effect mobility, μ, and the threshold voltage, any anomalous results are probably due to interface traps. As a result of the two different kinds of behavior of V_T and μ, it was found that, for a certain V_{GS}, I_{DS} characteristics are temperature stable [60].

Gate transconductance, g_m, is proportional to $\mu(T)[V_{GS} - V_T(T)]$. When the temperature increases, mobility increases and threshold voltage decreases, so g_m increases for temperatures up to and around 200 °C. Above this temperature the transconductance decreases, so g_m peaks before 200–300 °C. The value of g_m in silicon always decreases for increasing temperatures.

A drastically increased I_{on}/I_{off} ratio (I_{off} = leakage current from source and drain and drain–substrate body junction), by more than 100 times at temperatures above 200 °C, was experimentally measured by Rebello et al. [60] for 6H-SiC MOSFETs. They found a low leakage current which was five orders of magnitude lower at 300 °C than for Si MOSFETs. There were indications that the recombination–generation current dominated below 200 °C and the diffusion current dominated above 200 °C.

Silicon carbide power MOSFETs have a major advantage over their silicon counterparts because silicon carbide has a much higher critical breakdown electric field (E_{max}) than silicon. This allows the device dimensions to be drastically reduced as well as the on-resistance of the power MOSFET. A vertical 6H-SiC MOSFET has also been produced and the transistor showed good characteristics up to 300 °C [61]. Since the electron mobility of 4H-SiC is twice that of 6H-SiC, 4H-SiC is a more promising material for high power operation. (And 6H-SiC has an unisotropic mobility.) Palmour et al. also fabricated a 4H-SiC vertical MOSFET [62] with an R_{on} stable up to 200 °C. The threshold voltage decreased from 6.4 V at room temperature to 4.3 V at 300 °C. The main problem at high temperatures (above 300 °C) was the gate oxide degradation.

Two silicon carbide circuits (Figs 9.4 and 9.5) were demonstrated in 1994 for a flame detector circuit [63]; both circuits worked up to 350 °C. The depletion mode 6H-SiC MOSFET is shown in Fig. 9.6. Note that high reliability gate oxide can be retained for the depletion mode devices. The UV detectors (flame detectors) were 6H-SiC p–n diodes and the LEDs were also 6H-SiC p–n diodes (see p–n diodes below). Finally, the resistors were made in 6H-SiC.

In 1993 the first monolithic integrated 6H-SiC analog circuits (operational amplifiers) were demonstrated by Brown et al. of General Electric [64]. The basic element was an NMOS depletion mode 6H-SiC transistor. Three problems were identified: (1) Al-doped p-type SiC produces highly nonuniform n-channel inversion mode NMOS FETs across the wafer because of the large amounts of fixed charge in the oxide and variable fast and slow interface state densities; (2) high temperature gate oxide reliability is poor with positive gate bias; (3) the quality of thermal oxides grown on

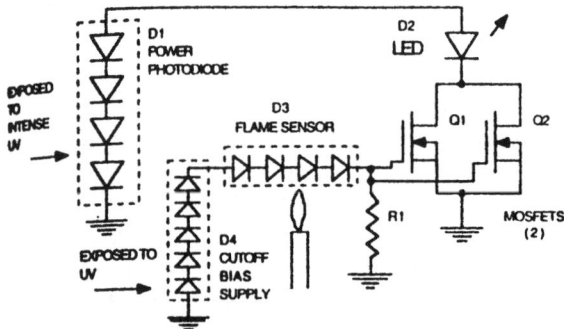

Figure 9.4 Flame detector circuit with LED optical output. (From Palmour et al. [63]).

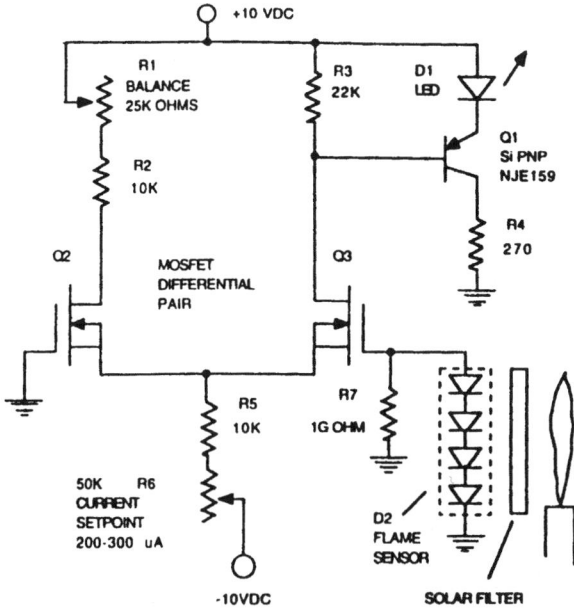

Figure 9.5 Flame detector circuit with solar filter and LED optical output. (From Palmour *et al.* [63])

N ion implanted diode regions is considerably degraded. Trench isolation was also regarded as the only practical method for isolation. Figure 9.7 shows two photomicrographs of SiC MOSFET operational amplifier chips. Table 9.6 shows the gain and bandwidth as a function of temperature. Westinghouse has also demonstrated SiC analog monolithic integrated circuits (single-stage transistor amplifiers, depletion-loaded differential pair and resistance-loaded differential pair) [65].

F. S. Schoucair carried out SPICE simulations on 6H-SiC NMOS operational amplifiers working from room temperature up to 450 °C [66].

Table 9.6 SiC operational amplifier gain bandwidth versus temperature (Brown *et al.* [64])

$T(°C)$	Gain (dB)	Bandwidth (KHz)
25	49	724
160	54	501
210	54	447
250	53	380
300	53	269

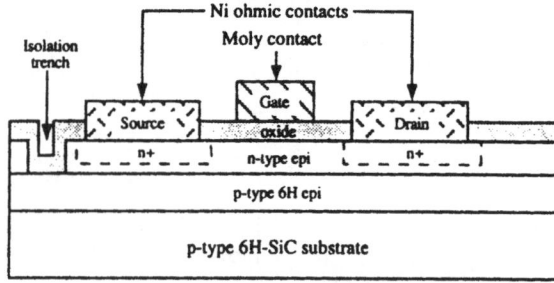

Figure 9.6 Depletion mode 6H-SiC MOSFET used for the flame detector circuits. (From Palmour *et al.* [63])

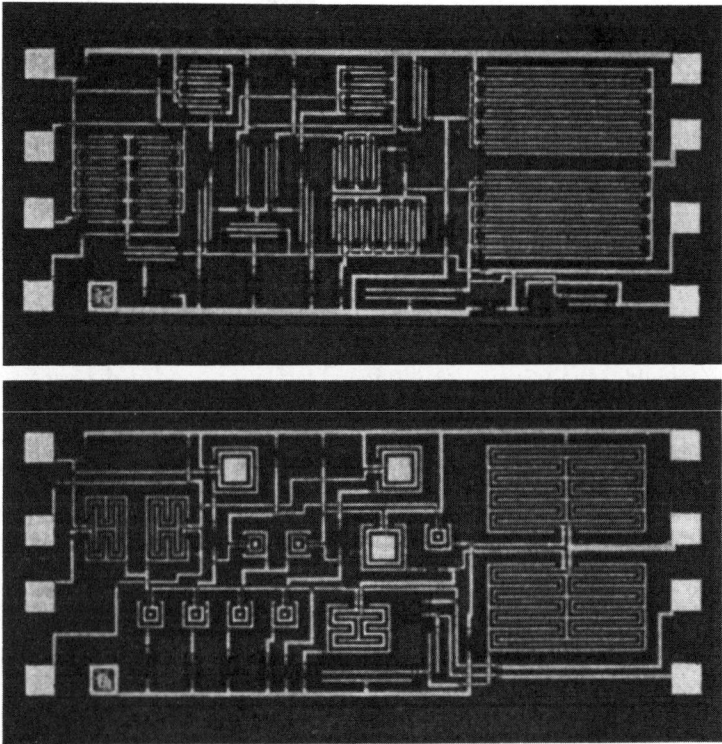

Figure 9.7 Two photomicrographs of SiC MOSFET operational amplifier chips: linear gate MOSFETs (above) and annular gate MOSFETs (below). (From Brown *et al.* [64])

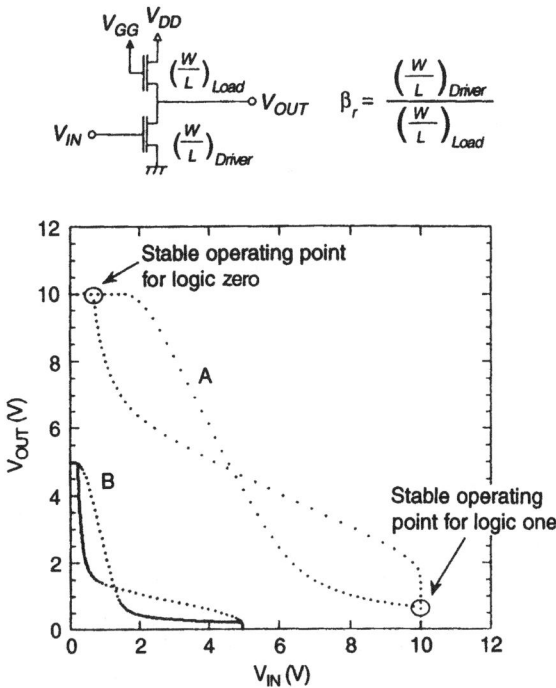

Figure 9.8 Circuit diagram of the SiC NMOS inverter and input–output characteristics at room temperature and 304 °C. Stable operating points at each temperature are determined by the intersections of the normal and reversed transfer curves. (From Xie *et al.* [67])

Several figures of merit were obtained: output voltage swing, unity-gain frequency, low-frequency small-signal voltage gain, phase margin, power dissipation, input common-mode voltage range, slew rate, setting time, systematic input offset voltage, DC common mode rejection ratio. The main result of the work was that circuits in Si MOS technology work up to 200 °C. Above this temperature the leakage current in the silicon technology starts to degrade the circuit performance. Temperatures between 200 °C and 450 °C require an integrated circuit made of SiC, even if the low mobility slows down its speed.

Digital circuits based on NOR gates have also been demonstrated with submicron depletion mode MOSFETs; an example is shown in Fig. 9.8. The first monolithic integrated circuits made in 6H-SiC with enhancement mode NMOS were demonstrated in 1994 by Xie *et al.* from Purdue University [67]. XNOR gates, D-latches, RS flip-flops, binary counters and half adders worked properly from room temperature to over 300 °C.

A class of gas sensors called field effect sensors includes MOS capacitors and MOS transistors. The gas-sensing principle is that molecules are dissociated at the catalytic metal surface. The ions diffuse through the metal film and form a dipole layer at the metal–insulator interface. The dipole layer contributes to a voltage, producing a shift of the $C-V$ curve of a MOS capacitor [68]. Field-effect gas sensors based on silicon work up to a maximum of 275 °C [69]. Silicon carbide MOS devices have been demonstrated to work up to 800 °C [70]. High temperature operation of gas sensors is desirable for several reasons [71]:

1. The use of gas sensors in a hot environment, e.g. close to a combustion process in a car.
2. Dissociation of gas molecules on a catalytic surface must overcome the dissociation barrier. Saturated hydrocarbons are not dissociated to any measurable degree below 300 °C.
3. It is possible to use the temperature as a parameter for distinguishing between different molecules as they are dissociated, and therefore detected, at different temperatures.
4. Heating in oxygen is an effective means of reactivating a poisoned catalytic surface. Capacitance–voltage measurements on 6H-SiC MOS capacitors with a platinum gate have demonstrated operation up to 800 °C. Hydrogen, methane, ethane, butane and propane were detected [70]. The conclusion is that 6H-SiC is stable at high temperatures, but the main problem is the mechanical cracking of the catalytic metal.

9.5.2 Schottky diode devices

It is easier to fabricate shorter gate lengths in metal–semiconductor field effect transistors (MESFETs) than in MOSFETs and junction field effect transistors (JFETs). Due to the low mobility but high saturation velocity for SiC, this is an important aspect when designing high speed and high temperature devices.

Some 3C-SiC MESFETs were fabricated during the mid 1980s by Yoshida *et al.* [72]. This work also pointed out the problem with aluminum electrodes at temperatures up to 400 °C. Other researchers, Kong *et al.* [75] and Kelner *et al.* [76], have also fabricated and tested 3C-SiC MESFETs. Silicon substrates were used in all cases. Both Palmour *et al.* [77] and Kelner *et al.* [78] later fabricated and tested 3C-SiC MESFETs grown on 6H-SiC substrates. They found much better characteristics due to fewer defects at the interface between the substrate and the film.

The first 6H-SiC MESFET was fabricated by Muench in 1977 [79]. In the early 1990s Palmour *et al.* produced a 6H-SiC MESFET which successfully operated up to 500 °C [28].

Comprehensive simulations of microwave power MESFETs have been done by Trew and coworkers [80]. Figure 9.9 shows the output power as a function of frequency for diamond, GaAs, 6H-SiC and 4H-SiC at different

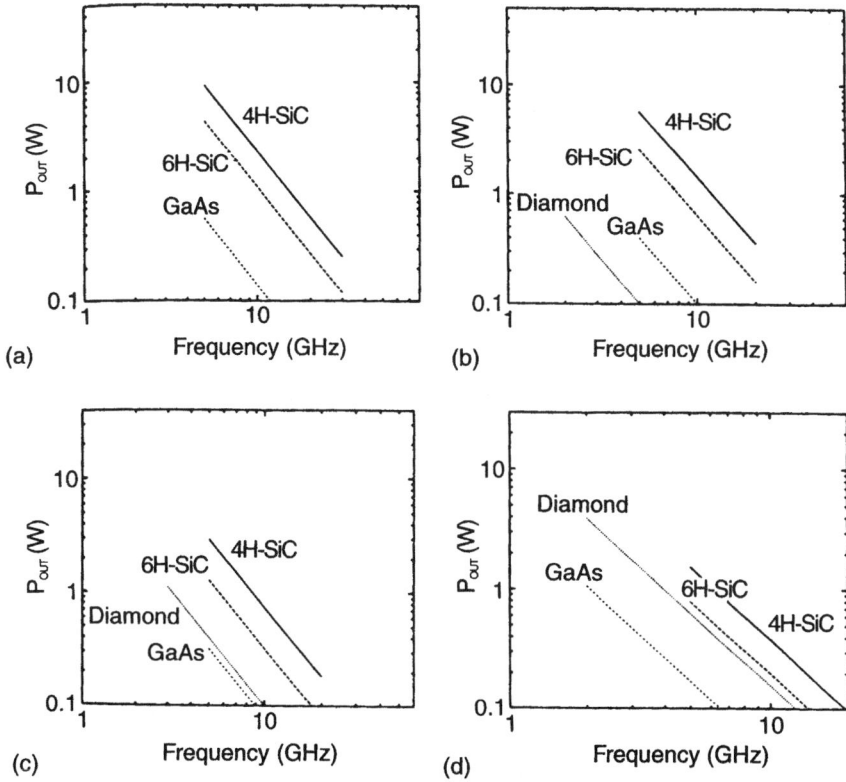

Figure 9.9 Comparison of the output power for diamond, GaAs, 6H-SiC and 4H-SiC MESFETs as a function of operating frequency at (a) 300 K, (b) 423 K, (c) 573 K and (d) 773 K. (From Shin *et al.* [80])

temperatures. The MESFET devices had a gate width of 1 mm and were operated in a class A. Room temperature performance is for a drain bias voltage of 40 V in the diamond and SiC MESFETs, a drain bias voltage of of 8 V in the GaAs devices. Figure 9.9 shows the superior output power at different frequencies for 4H-SiC MESFETs compared with the two other transistors. Note that, of the different polytypes of silicon carbide, 4H-SiC is regarded as the most promising due to its high mobility.

Schottky MIS diodes can also be used for high temperature gas sensing. On average a thin 0.15–0.2 nm silicon dioxide is used only to avoid pinning the barrier height. The principal mechanism is the same as in the MOS structure used for gas sensing. At high temperatures the molecules are dissociated at the catalytic metal, and the gas atoms form dipoles which then change the barrier height. Both capacitance–voltage and current–voltage can therefore be used to detect gases. Several catalytic metals, Ti, TaSi$_x$, Pd, Pt and Cr, were investigated; some of them showed Schottky diode properties up to 700 °C and several catalytic metals appear to become ohmic after heating at temperatures to 300 °C [81].

Finally, Schottky diodes were used in the first integrated circuit design for flame detection. The Schottky diode was used as a detector which worked up to 380 °C [63].

9.5.3 p–n Diode devices

JFETs are more suitable for high temperature applications than MOSFETs, due to the absence of gate oxides, thus producing higher reliability at elevated temperatures. The decreasing channel mobility, which reduces the current drive capability at high temperatures, and the long-term stability of the ohmic contacts are the two major problems for JFETs in high temperature applications. Dmitriev et $al.$ have fabricated 6H-JFETs with long channels ($L_g = 20 \mu$m), operating in the temperature range 20–500 °C, with a variation of g_m from 0.25 to 0.5 mS mm^{-1} [82]. Kelner et $al.$ have also fabricated similar devices ($L_g = 39 \mu$m) and characterized them over the range 20–350 °C [83]. Ivanov et $al.$ demonstrated very good JFET characteristics for temperatures in the range 27–427 °C and with $L_g = 10 \mu$m [84].

Good high temperature characteristics were recently found in buried gate JFETs [85]. Excellent low leakage current was shown up to 600 °C. The metal contacts, sputtered 90 nm Si/70 nm Pt, were stable in air up to 30 h at 600 °C. One application for these devices is in aerospace systems. Similarly good results for power and high temperature applications were demonstrated by Dohnke et $al.$ [86].

A breadboard solution of JFETs for the design of a DC-coupled differential voltage operational amplifier operating at 300 °C has also been realized [87]. Only n-channel depletion mode transistors were realized,

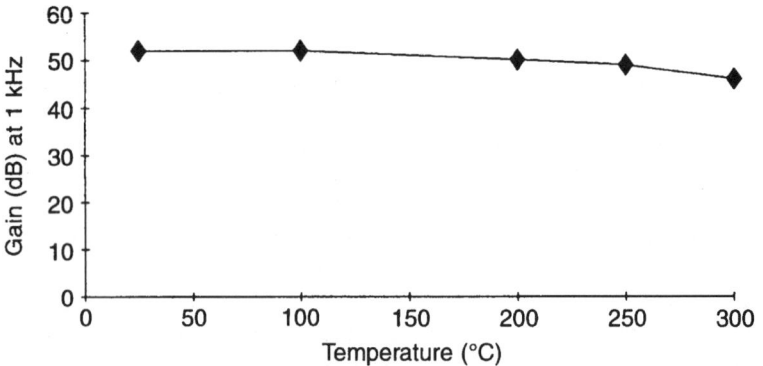

Figure 9.10 Open-loop gain versus temperature (at 1 kHz) for a JFET amplifier (breadboard configuration). (From Blackburn et $al.$ [87])

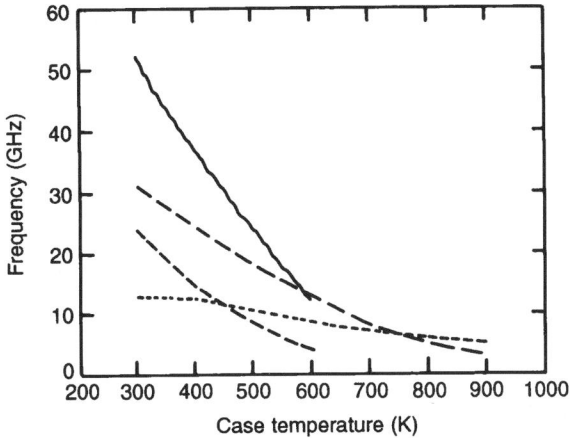

Figure 9.11 Calculated peak values of f_T and f_{max} at case temperatures from 300 K to 900 K for the 6H/3C-SiC HBT at $V_{cb} = 10$ V and the AlGaAs/GaAs HBT at $V_{cb} = 0$ V. (—) GaAs, f_T; (— — —) SiC, f_T; (— — —) GaAs, f_{max}; and ($\cdots$) SiC, f_{max}. (From Gao *et al.* [90])

matched pairs of transistors were not possible. However, the amplifier showed a gain which was relatively stable up to 300 °C (Fig. 9.10). From SPICE simulations it was found that the main reason for the decrease of gain as a function of temperature is the increase of leakage current with temperature [88].

There are several reasons why SiC bipolar transistors are less commonly realized. However, 6H-SiC bipolar transistors have been produced and operated with a current gain of 10.2 at 400 °C [89]. Good results have also been obtained by Ivanov and Chelnokov [94].

Since it is possible to lattice-match 6H-SiC and 3C-SiC (see sections 9.5.1 and 9.5.2), a simulation comparison between AlGaAs/GaAs HBTs and 6H-SiC/3C-SiC HBTs has been performed by Gao *et al.* [90]. The results are shown in Fig. 9.11.

SiC materials have been used in realizing blue light-emitting diodes (LEDs), but the indirect bandgap gives a low quantum efficiency of the order 0.02–0.03%. Several groups have produced blue light-emitting 6H-SiC LEDs (Siemens, Cree and the Ioffe Institute). The main light-emitting mechanism is a donor–acceptor pair recombination. The group from the Ioffe Institute operated a blue light-emitting diode up to 450 °C [91]. And Palmour *et al.* produced a blue light-emitting 6H-SiC LED in their flame detector circuit which worked up to 350 °C [63].

SiC is a suitable material for p–n diode UV detection at high temperatures, due to the larger optical absorption in the depletion region (absorption depth varies from 0.1 μm to 1000 μm for $\lambda = 200$–400 nm) as well as low leakage current. This allows UV detection against a red background. Temperature dependence of the 6H-SiC response in the visible range has been measured by Edmond *et al.* [92]. When the temperature changed from 22 °C to 400 °C the long-wavelength boundary shifts from 400 nm to 450 nm due to the variation of the bandgap with temperature. When the temperature increased from -50 °C to 350 °C, responsivity increased by a factor of around 1.5.

Finally, perhaps the most common application of p–n diodes in the future will be for power circuits at high temperatures. SiC diode rectifiers have shown a reverse leakage current as low as $5 \times 10^{-4}\,A\,cm^{-2}$ at 710 V and a temperature of 350 °C [92].

ACKNOWLEDGMENT

M. Willander thanks his colleague, Y. Mamontov, for his help in typing the text.

REFERENCES

1. Davis, R.F., Sitar, Z., Williams, B.E., Kong, H.S., Kim, H.J., Palmour, J.W., Edmond, J.A., Ryu, J. and Carter, C.H. Jr (1988) Critical evaluation of the status of the areas for future research regarding the wide bandgap semiconductors diamond, gallium nitride, and silicon carbide. *Mater. Sci. Eng. B*, **1**, 77–104.
2. Edgar, J.H. (1992) Prospects for device implementation of wide bandgap semiconductors. *J. Mater. Res.*, **7**, 235–252.
3. Davis, R.F. (1993) Deposition, characterization, and device development in diamond, silicon carbide, and gallium nitride thin films. *J. Vac. Sci. Technol. A*, **11**, 829–837.
4. Tong, L., Mehregany, M. and Matus, L.C. (1992) Mechanical properties of 3C silicon carbide. *Appl. Phys. Lett.*, **60**, 2992–2994.
5. Palmour, J.W., Kong, H.S. and Davis, R.F. (1987) High-temperature depletion mode metal–oxide–semiconductor field effect transistor in SiC thin films. *Appl. Phys. Lett.*, **51**, 2028–2030.
6. Dubey, M., Shanker Ram, U., Nath Rai, K. and Singh, G. (1973) The present state of structural stability and the growth of silicon carbide polytypes. *Phys. Stat. Sol. (a)*, **18**, 689–698.
7. Pandey, D. and Krishna, P. (1983) The origin of polytype structures. *Progr. Cryst. Growth Charact.*, **7**, 213–258.
8. Tairov, Y.M. and Tsvetkov, V.F. (1983) Progress in controlling the growth of polytypic crystals. *Progr. Cryst. Growth Charact.*, **7**, 111–162.
9. Verma, A.R. and Krishna, P. (1966) *Polymorphism and Polytypism in Crystals*, Wiley, New York.
10. Tairov, Y.M. and Tsvetkov, V.F. (1989) Controlled growth of polytypes and their importance for science and technology, in *Proc. Int. Conf. on Advanced Methods in X-ray and Neutron Structure Analysis of Materials* (ed. J. Hasek), Plenum, New York, pp. 331–339.
11. Koga, K., Nakata, T., Ueda, T., Matsushita, Y. and Fujikawa, Y. (1989) Polytype control of 4H/6H type SiC in the sublimation process and its application, in *Ext. Abstr. 3rd Int. Conf. on Amorphous and Crystalline Silicon Carbide and Related Materials*, pp. 689–690.
12. Kanaya, M., Takahashi, J., Fujiwara, Y. and Moritani, A. (1991) Controlled sublimation growth of single crystalline 4H-SiC and 6H-SiC and identification of polytypes by X-ray diffraction. *Appl. Phys. Lett.*, **58**, 56–58.

13. Lely, J.A. (1955) Darstellung von Einkristallen von Silizium Karbid und Beherrschung von Art und Menge der eingebauten Verunreinigungen. *Ber. Deutsch. Keram. Ges.*, **32**, 229.

14. Tairov, Y.M. and Tsvetkov, V.F. (1978) Investigation of growth processes of ingots of silicon carbide single crystals. *J. Cryst. Growth*, **43**, 209–219.

15. Tairov, Y.M. and Tsvetkov, V.F. (1981) General principles of growing large-size single crystals of various silicon carbide polytypes. *J. Cryst. Growth*, **52**, 146–150.

16. Ziegler, G., Lanig, P., Theis, D. and Weyrich, C. (1983) Single crystal growth of SiC substrate material for blue-light emitting diodes. *IEEE Trans. Electron Dev.*, **30**, 277–281.

17. Koga, K., Ueda, Y., Nakata, T., Yamaguchi, T. and Niina, T. (1987) Single crystal growth of 6H-SiC by a vacuum sublimation process and its application. *J. Vacuum Soc. Japan*, **30**, 886–892.

18. Carter, C.H. Jr, Tang, L. and Davis, R.F. (1987) Growth of single crystal boules of (6H)-SiC, in *Proc. 4th National Review Meeting on the Growth and Characterization of SiC, Raleigh NC.*

19. Barret, D.L., McHugh, J.P., Hopgood, H.M., Hopkins, R.H., McMullin, P.G. and Clarke, R.C. (1993) Growth of large SiC single crystals, *J. Cryst. Growth*, **128**, 352–362.

20. Mokhov, E.N., Shulpina, I.L., Tregubova, A.S. and Vodakov, Y.A. (1981) Epitaxial growth of SiC layers by sublimation 'sandwich method'. *Cryst. Res. Technol.*, **16**, 879–886.

21. Suzuki, A., Ikeda, M., Matsunami, H. and Tanaka, T. (1975) Liquid-phase epitaxial growth of 6H-SiC by vertical dipping technique. *J. Electrochem. Soc.*, **122**, 1741–1742.

22. Matsunami, H., Ikeda, M., Suzuki, A. and Tanaka, T. (1977) SiC blue LEDs by liquid-phase epitaxy. *IEEE Trans. Electron Dev.*, **24**, 958–961.

23. von Muench, W. and Kurzinger, W. (1978) Silicon carbide blue-emitting diodes produced by liquid-phase epitaxy. *Solid-State Electronics*, **21**, 1129–1132.

24. Hoffman, L., Ziegler, G., Theis, D. and Weyrich, C. (1982) Silicon carbide blue-light emitting diodes with improved external quantum efficiency. *J. Appl. Phys.*, **53**, 6962–6967.

25. Ikeda, M., Hayakawa, T., Yamagiwa, S. *et al.* (1979) Fabrication of 6H-SiC light emitting diodes by a rotation dipping technique: electroluminiscence mechanisms. *J. Appl. Phys.*, **50**, 8215–8225.

26. Matsunami, H. (1993) Growth and application of cubic SiC. *Diamond Rel. Mater.*, **2**, 1043–1050.

27. Nishino, S., Powell, J.A. and Will, H.A. (1983) Production of large-area single crystal wafers of cubic SiC for semiconductor devices. *Appl. Phys. Lett.*, **45**, 460–462.

28. Davis, R.F., Kelner, G., Shur, M., Palmour, J. and Edmond, J.A. (1991) Thin film deposition and microelectronic and optoelectronic device fabrication and characterization in monocrystalline alpha and beta silicon carbide. *Proc. IEEE*, **75**, 677–701.

29. Davis, R.F., Palmour, J. and Edmond, J.A. (1992) Epitaxial thin film growth, characterization and device development in monocrystalline silicon carbide. *Diamond Rel. Mater.*, **1**, 109–120.

30. Kong, H.S., Glass, J.T. and Davis, R.F. (1986) Epitaxial growth of beta-SiC thin films on 6H alpha-SiC substrates via chemical vapor deposition. *Appl. Phys. Lett.*, **49**, 1074–1077.

31. Kong, H.S., Glass, J.T. and Davis, R.F. (1988) Chemical vapor deposition and characterization of 6H-SiC thin films on off-axis 6H-SiC substrates. *Appl. Phys. Lett.*, **64**, 2672–2679.

32. Tairov, Y.M. and Vodakov, Y.A. (1977) Group IV materials (mainly SiC). *Topics Appl. Phys.*, **17**, 31–61.

33. Pensl, G. and Choyke, W.J. (1993) Electrical and optical characterization of SiC. *Physica B*, **185**, 264–283.

34. Goetz, W., Schoener, A., Pensl, G., Suttrop, W., Choyke, W.J., Stein, R. and Leibenzeder, S. (1993) Nitrogen donors in 4H-silicon carbide. *J. Appl. Phys.*, **73**, 3332–3338.

35. Ghezzo, M., Brown, D.M., Downey, E., Kretchmer, J., Hennessy, W., Polla, D.L. and Bakhru, H. (1992) Nitrogen-implanted SiC diodes using high-temperature implantation. *Electron Dev. Lett.*, **13**, 639–641.

36. Ghezzo, M., Brown, D.M., Downey, E. and Kretchmer, J. (1993) Boron-implanted 6H-SiC diodes. *Appl. Phys. Lett.*, **63**, 1206–1208.

37. Kim, H.J. and Davis, R.F. (1986) Theoretical and empirical studies of impurity incorporation into SiC thin films during epitaxial growth. *J. Electrochem. Soc.*, **133**, 2350–2357.

38. Wang, Y.C., Davis, R.F. and Edmond, J.A. (1991) *In situ* incorporation of Al and N and p–n junction diode fabrication in alpha(6H)-SiC thin films. *Electronic Mater.*, **20**, 289–294.

39. Palmour, J.W., Williams, B.E., Astell-Burt, P. and Davis, R.F. (1989) Crystallographic etching phenomenon during plasma etching of SiC (1 0 0) thin films in SF_6. *J. Electrochem. Soc.*, **136**, 491–495.

40. Yih, P.H. and Steckl, A.J. (1993) Effects of hydrogen additive on obtaining residue-free reactive ion etching of β-SiC in fluorinated plasmas. *J. Electrochem. Soc.*, **140**, 1813–1824.

41. Edmond, J.A., Ryu, J., Glass, J.T. and Davis, R.F. (1988) Electrical contacts to beta-silicon carbide thin films. *J. Electrochem. Soc.*, **135**, 356–362.

42. Waldrop, J.R. and Grant, R.W. (1990) Formation and Schottky barrier height of metal contacts to β-SiC. *Appl. Phys. Lett.*, **56**, 557–559.

43. Waldrop, J.R., Grant, R.W., Wang, Y.C. and Davis, R.F. (1992) Metal Schottky barrier contacts to alpha 6H-SiC. *J. Appl. Phys.*, **72**, 4757–4760.

44. Waldrop, J.R., Grant, R.W., Wang, Y.C. and Davis, R.F. (1993) Schottky barrier height and interface chemistry of annealed metal contacts to alpha 6H-SiC: crystal face dependence. *Appl. Phys. Lett.*, **62**, 2685–2687.

45. Bhatnagar, M., McLarty, P.K. and Baliga, B.J. (1992) Silicon carbide high-voltage (400 V) Schottky barrier diodes. *IEEE Electron Dev. Lett.*, **13**, 501–503.

46. Kimoto, T., Urushidani, T., Kobayashi, S. and Matsunami, H. (1993) High-voltage (> 1000 V) SiC Schottky barrier diodes with low on-resistance. *IEEE Electron Dev. Lett.*, **14**, 548–550.

47. Brown, D.M., Downey, E.T., Ghezzo, M., Kretchmer, J.W., Saia, R.J., Liu, Y.S., Edmond, J.A., Gati, G., Pimbley, J.M. and Schneider, W.E. (1993) Silicon carbide UV photodiodes. *IEEE Trans. Electron Dev.*, **40**, 325–333.

48. Kelner, G., Shur, M.S., Binari, S., Sleger, K.J. and Kong, H.S. (1989) High-transconductance β-SiC buried-gate JFETs. *IEEE Trans. Electron Dev.*, **36**, 1045–1049.

49. Matus, L.G., Powell, J.A. and Salupo, C.S. (1991) High-voltage 6H-SiC p–n junction diodes. *Appl. Phys. Lett.*, **59**, 1770–1772.

50. Shinohara, M., Yamanaka, M., Misawa, S., Onumura, H. and Yoshida, S. (1991) $C-V$ characteristics of MOS structures fabricated of Al doped p-type 3C-SiC epilayers grown on Si by chemical vapor deposition. *Jpn. J. Appl. Phys.*, **30**, 240–243.

51. Zaima, S., Onoda, K., Koide, Y. and Yasuda, Y. (1990) Effects of oxidation conditions on electrical properties of $SiC-SiO_2$ interfaces. *J. Appl. Phys.*, **68**, 6304–6306.

52. Wahab, Q., Sardela, M.R. Jr, Hultman, L., Henry, A., Willander, M., Janzen, E. and Sundgren, J.E. (1994) Growth of high-quality 3C-SiC epitaxial films on off-axis Si (0 0 1) substrates at 850 °C by reactive magnetron sputtering. *Appl. Phys. Lett.*, **65**, 725–727.

53. Palmour, J.W., Kong, H.S. and Davis, R.F. (1988) Characterization of device parameters in high-temperature metal–oxide–semiconductor field-effect transistors in beta-SiC thin films. *J. Appl. Phys.*, **64**, 2168–2177.

54. Nishino, S., Haguki, Y., Matsunami, H. and Tanaka, T. (1990) Chemical vapor deposition of single crystalline β-SiC films on silicon substrates with sputtered SiC intermediate layer. *J. Electrochem. Soc.*, **127**, 2674–2680.

55. Shibahara, K., Saito, T., Nishino, S. and Matsunami, H. (1986) Inversion-type n-channel MOSFET using antiphase-domain free cubic-SiC grown mode, in *Extended Abstracts of the 18th International Conference on Solid State Devices and Materials*, pp. 717–718.

56. Fuma, H., Misura, A., Tadano, H., Sugiyama, S. and Takigawa, M. (1988) Beta-SiC enhancement mode MOSFET, in *Extended Abstracts of the 2nd Conference on Solid State Devices and Materials*, pp. 13–16.

57. Furukawa, K., Hatano, A., Uemoto, A., Fugii, Y., Nakanishi, K., Shigeta, M., Suzuki, A. and Nakajima, S. (1987) Insulated-gate and junction-gate FETs of CVD-grown β-SiC. *IEEE Electron Dev. Lett.*, **8**, 48–49.

58. Kondo, Y., Takahasi, T., Ishi, K., Hayashi, Y., Sakuma, E., Misawa, S., Daimon, H., Yamaraka, M. and Yoshia, S. (1986) Experimental 3C-SiC MOSFET. *IEEE Electron Dev. Lett.*, **7**, 404–406.

59. Brown, D.M., Ghezzo, M., Kretchmer, J., Downey, E., Pimbley, J. and Palmour, J. (1994) SiC MOS interface characteristics. *IEEE Trans. Electron Devices*, **41**, 618–620.

60. Rebello, N.S., Shoucair, F.S. and Palmour, J.W. (1994) High temperature electrical characteristics of 6H silicon carbide MOSFETs (25–500 °C), in *Trans. 2nd Int. Conf. on High Temperature Electronics*, Vol. 1, pp. IV27–IV32.

61. Palmour, J.W., Edmond, J.A., Kong, H.S. and Carter, C.H. Jr (1994) Vertical power devices in silicon carbide, in *Proc. 1993 International Conf. on SiC and Related Materials* (eds M. Spencer and G. Harris), International Proceedings in Physics.

62. Palmour, J.W. and Lipkin, L. (1994) High temperature power devices in silicon carbide, in *Trans. 2nd Int. Conf. on High Temperature Electronics*, Vol. 1, pp. XI3–XI9.

63. Palmour, J.W., Waltz, D.G., Edmond, J.A., Carter, C.H. Jr, Gati, G. and Przybylko, S.J. (1994) High temperature silicon carbide flame detector circuit, in *Trans. 2nd Int. Conf. on High Temperature Electronics*, Vol. 1, pp. VII27–VII32.

64. Brown, D.M., Ghezzo, M., Kretchmer, J., Krishnamurthy, V., Michon, G. and Gati, G. (1994) High temperature silicon carbide planar IC technology and first monolithic silicon carbide operational amplifier IC, in *Trans. 2nd Int. Conf. on High Temperature Electronics*, Vol. 1, pp. XI17–XI22.

65. Siergiej, R., Agarwal, A., Burk, A., Clark, C., McDonald Hobgood, H., McMullin, P., Orphanos, P., Sriram, S., Smith, T. and Brandt, L. (1994) Novel silicon carbide MOSFETs for monolithic integrated circuits, in *Trans. 2nd Int. Conf. on High Temperature Electronics*, Vol. 1, pp. XI23–XI28.

66. Schoucair, F.S. (1994) Lecture Notes, High Temperature Electronics Summer School, North Carolina.

67. Xie, W., Cooper, J.A. and Melloch, M.R. (1994) Monolithic NMOS digital integrated circuits in 6H-SiC. *IEEE Electron Dev. Lett.*, **15**, 455–457.

68. Lundström, K.I., Shivaraman, M.S. and Svensson, C. (1975) A hydrogen sensitive Pd-gate MOS transistor. *J. Appl. Phys.*, **46**, 3876–3881.

69. Spetz, A., Armgardt, M. and Lundström, I. (1988) Comparison between platinum and iridium as gate in ammonia-sensitive metal–silicon dioxide–silicon structures. *Sensors and Materials*, **4**, 187–207.

70. Arbab, A., Spetz, A., Ul Wahab, Q., Willander, M. and Lundström, I. (1993) Chemical sensors for high temperature based on silicon carbide. *Sensors and Materials*, **4**, 173–185.

71. Baranzahi, A. (1995) Chemical sensors based on catalytic metal–oxide–silicon carbide, MOSiC, structures. *LiU-TEK-LIC-1994:54*.

72. Yoshida, S., Daimon, H., Yamanaka, M., Sakuma, E., Misawa, S. and Endo, K. (1986) Schottky-barrier field-effect transistors of 3C-SiC. *J. Appl. Phys.*, **60**, 2989–2992.

73. Daimon, H., Yamanaka, M., Shinohara, M., Sakuma, E., Misawa, S., Endo, K. and Yoshida, Y. (1987) Operation of Schottky-barrier field-effect transistors of 3C-SiC up to 400°C. *Appl. Phys. Lett.*, **51**, 2106–2108.

74. Yoshida, S., Endo, K., Sakuma, E., Misawa, S., Okumura, H., Daimon, H., Muneyama, E. and Yamanaka, M. (1987) Electrical properties of 3C-SiC and its application to FET, in *Materials Research Society 97* (eds E. Emin, T. Aselage and C. Woods), pp. 259–264.

75. Kong, H.S., Palmour, J., Glass, J.T. and Davis, R.F. (1987) Temperature dependence of the current–voltage characteristics of metal–semiconductor field effect transistors in n-type, beta-SiC grown via chemical vapor deposition. *Appl. Phys. Lett.*, **51**, 442–444.

76. Kelner, G., Binari, S., Sleger, K. and Kong, H. (1987) Beta-SiC MESFETs and buried-gate JFETs. *IEEE Electron Device Lett.*, **8**, 428–430.

77. See Palmour *et al.* [53] and Shibahara *et al.* [55].

78. Kelner, G., Shur, M., Binari, S., Sleger, K. and Kong, H.-S. (1989) High-transconductance β-SiC buried-gate JFETs. *IEEE Electron Dev.*, **36**, 1045–1049.

79. Muench, W., Hoeck, P. and Pettenpaul, E. (1977) Silicon carbide field-effect and bipolar transistors, in *IEDM Tech. Digest*, IEEE, New York, pp. 337–339.

80. Shin, M.W., Trew, R.J. and Bilbro, G.L. (1994) Large signal RF and DC performance of wide bandgap semiconductor, in *Trans. 2nd Int. Conf. on High Temperature Electronics*, Vol. 1, pp. IV15–IV19.

81. Karlsteen, M., Baranzahi, A., Lloyd Spetz, A., Willander, M. and Lundström, I. (1995) Electrical properties of inhomogeneous SiC MIS structures. *Journal of Electronic Materials*, **24**, 853–861.

82. Dmitriev, V.A., Ivanov, P.A., Strelchuk, A.M., Syrkin, A.L., Tsarenkov, B.V., Chelnokov, V.E. and Cherenkov, A.E. (1988) High-temperature SiC-6H field effect transistor with p–n gate. *Sov. Tech. Phys. Lett.*, **14**, 127–129.

83. Kelner, G., Binari, S., Shur, M., Sleger, K., Palmour, J. and Kong, H. (1990) AlGa–SiC buried-gate JFETs, in *Proc. 1990 Fall Meeting of the European Materials Research Society, Strasbourg, France.*

84. Anikin, M., Ivanov, P.A., Syrkin, A.L., Tsarenkov, B.V. and Chelnokov, V.E. (1989), in *Extended Abstracts of the 176th Meeting of the Electrochemical Society, Holywood FL*, p. 779.

85. Neudeck, P.G., Petit, J.B. and Salupo, C.S. (1994) Silicon carbide buried-gate junction field-effect transistors for high-temperature power electronic applications, in *Trans. 2nd Int. Conf. on High Temperature Electronics*, Vol. 1, pp. X23–X28.

86. Dohnke, K., Pebers, D., Rupp, R., Völk, J. and Stephani, D. (1994) High temperature junction field effect transistors for power applications, in *Trans. 2nd Int. Conf. on High Temperature Electronics*, Vol. 1, pp. IV15–IV20.

87. Blackburn, J., Scozzie, C.J., Tipton, W., Geil, B., DeLancey, M. and McGarrity, J.M. (1994) Silicon carbide junction field effect transistor amplifier operation at 300 °C, in *Trans. 2nd Int. Conf. on High Temperature Electronics*, pp. VII33–VII38.

88. Christopher, M., Winland, M., Davidsson, J., Massengill, L., Tipton, W. and Scozzi, S. (1994) High temperature circuit modeling for silicon carbide active electronic device technology, in *Trans. 2nd Int. Conf. on High Temperature Electronics*, pp. IV21–IV26.

89. Palmour, J.W., Edmond, J.A., Kong, H.S. and Carter, C.H. Jr (1993) 6H-silicon carbide devices and applications. *Physica B*, **185**, 461–465.

90. Gao, G., Sterner, J. and Morkoc, H. (1994) High frequency performance of SiC heterojunction bipolar transistors. *IEEE Trans. Electron Dev.*, **41**, 1092–1097.

91. Dmitriev, V.A., Linkov, I. Yu., Morozenko, Ya. v., Chelnokov, V.E. and Cheenkov, A.E. (1991) Blue SiC light-emitting diodes for operating temperature 500 °C, in *Trans. 1st Int. Conf. on High Temperature Electronics*, p. 506.

92. Edmond, J.A., Kong, H.S. and Carter, C.H. Jr (1993) Blue LEDs, UV photodiodes and high-temperature rectifiers in 6H-SiC. *Physica B*, **185**, 453–460.

93. Brown, D.M., Downey, E., Ghezzo, M., Kretchmer, J. and Liu, Y.S. (1991) Silicon carbide UV photodiodes, in *Trans. 1st Int. Conf. on High Temperature Electronics*, pp. 214–219.

94. Ivanov, P.A. and Chelnokov, V.E. (1992) Recent developments in SiC single-crystal electronics. *Semiconductor Science and Technology*, pp. 863–880.

GaN-based field effect transistors

10

M. Shur and M. A. Khan

10.1 INTRODUCTION

Excellent mechanical properties of GaN have attracted the attention of material and device scientists since the late 1920s and early 1930s. GaN is a direct wide band gap semiconductor, suitable for applications in visible and ultraviolet light emitters and detectors. Its good transport properties and its ability to form high quality heterostructures with AlGaN make the GaN-based material system very attractive for field effect transistors (FETs), especially for devices operating at high temperatures and/or in a harsh environment.

Within the past few years, GaN and AlGaN/GaN FETs have demonstrated tremendous progress, the best results almost challenging the GaAs-based material system. In this chapter we review the properties of GaN and related materials, such as crystal structure, mechanical properties, band structure and electron transport, discuss the available substrates, epitaxial material growth, etching and ion implantation and consider the important issues related to ohmic and Schottky contacts. We then discuss GaN–AlN–GaN semiconductor–insulator–semiconductor structures, review the results obtained for different types of GaN-based FETs (including GaN MESFETs, GaN MISFETs, GaN/SiN MISFETs, AlGaN/GaN heterostructure FETs (HFETs), AlGaN/GaN doped channel HFETs (DC-HFETs) and optoelectronic FETs) and compare the achieved and predicted microwave performance of these devices.

High Temperature Electronics. Edited by M. Willander and H.L. Hartnagel. Published in 1997 by Chapman & Hall, London. ISBN 0 412 62510 5.

10.2 PROPERTIES OF GaN AND RELATED MATERIALS

10.2.1 Crystal structure and mechanical properties

Just like Si, GaAs or SiC, GaN has a tetrahedral bond configuration. It crystallizes in one of two crystal structures – zinc blende (cubic) or wurtzite hexagonal (Table 10.1). The difference in these two structures is the relative angle between the atomic tetrahedra (Fig. 10.1). As seen from Table 10.1, the lattice constants of GaN and AlN are fairly close (but not as close as those for GaAs and AlAs, 0.5653 nm and 0.5661 nm, respectively). Figure 10.2 compares these lattice constants with the lattice constants of other wide band gap materials.

Table 10.2 lists some important material properties of wide bandgap semiconductors. As seen from Fig. 10.2 and Table 10.2, a variety of solid-state solutions and heterostructure material systems based on wide band gap semiconductors are possible. Among the wide band gap semiconductors SiC and GaN seem to be somewhat analogous to Si and GaAs, respectively. We will return to this analogy after comparing the transport properties of GaN with the transport properties of SiC.

Table 10.1 GaN and AlN lattice constants (from Strite and Morkoç [2])

Material	Lattice constants (nm)	Thermal expansion coefficient $(10^{-6} K^{-1})$
GaN zinc blende	0.450	3.17
GaN wurtzite	$a = 0.3189$ $c = 0.5185$	4.2
AlN wurtzite	$a = 0.3112$ $c = 0.4982$	5.3

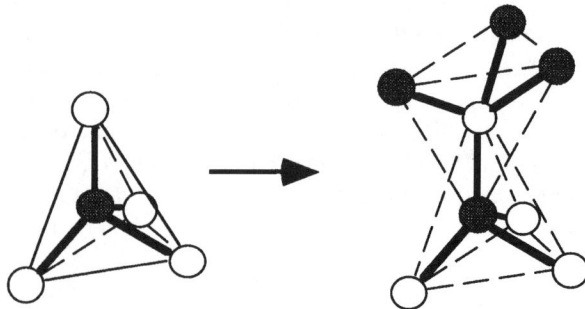

Figure 10.1 Bonding tetrahedron and two interpenetrating tetrahedra in the wurtzite structure. The bonding (with four nearest neighbors) is the same as for the diamond and zinc blende structures. (From Shur [1])

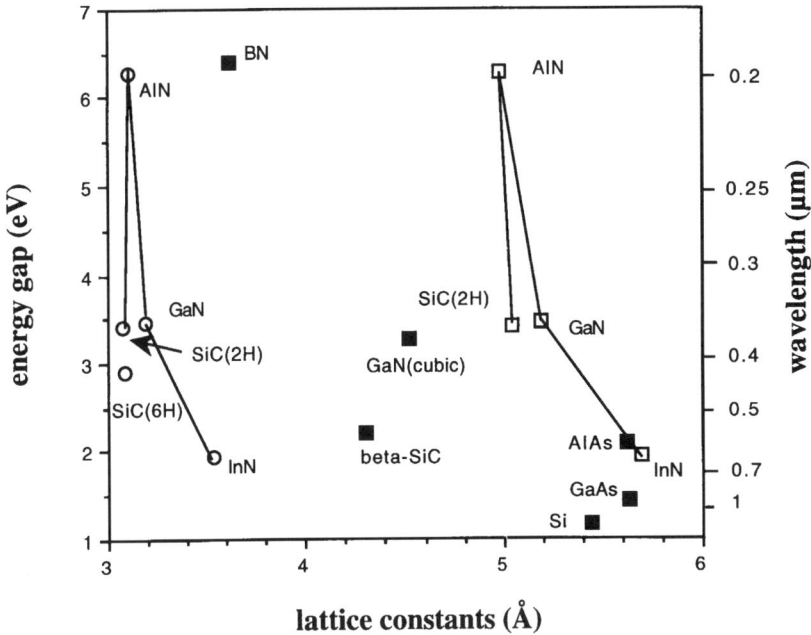

Figure 10.2 Energy gaps and lattice constants for wide band gap semiconductors: (○) *a* parameter, (□) *c* parameter and (■) cubic. Data for Si, GaAs and AlAs is included for comparison.

10.2.2 Transport properties

Monte Carlo simulations predict a high peak velocity, v_p, high saturation velocity, v_s, and a relatively high electron mobility, μ, in GaN ($v_p \approx 2.7 \times 10^5 \, \text{m s}^{-1}$, $v_s \approx 1.5 \times 10^5 \, \text{m s}^{-1}$ and $\mu \approx 1000 \, \text{cm}^2 \, \text{V}^{-1} \, \text{s}^{-1}$ at room temperature in GaN doped at $10^{17} \, \text{cm}^{-3}$) [3–7]. Table 10.3 lists typical parameters used in the Monte Carlo simulation of transport properties for GaN.

Figure 10.3a compares the velocity–field curve at room temperature for wurtzite GaN, α-SiC, Si and GaAs. Figure 10.3b shows the velocity–field characteristics of wurtzite GaN at different temperatures. GaN has a big advantage over other materials in terms of the electron peak velocity as well as in terms of the range of the electric fields over which the electron velocity remains really high. This should translate itself into a superior performance of GaN field effect transistors.

Even though typical values of the electron low field mobility in GaN at room temperature are on the order of 400–600 $\text{cm}^2 \, \text{V}^{-1} \, \text{s}^{-1}$, the best mobility values for the two-dimensional electron gas (2DEG) at the AlGaN/GaN heterointerface are more than $1000 \, \text{cm}^2 \, \text{V}^{-1} \, \text{s}^{-1}$, with even

Table 10.2 Estimated and measured material constants of α-SiC, β-SiC, α-GaN, β-GaN, AlN and InN important for device applications

Property	α-SiC (6H)	β-SiC	α-GaN	β-GaN	AlN	InN
Energy gap (eV)	2.9 (indirect)	2.2 (indirect)	3.4 (direct)	3.2 (direct)	6.2 (direct)	1.89 (direct)
Lattice constant, a (nm)	0.3081	0.4359	0.3189	0.452	0.311	0.354
Lattice constant, c (nm)	1.5117	–	0.5185	–	0.498	0.570
Density (g cm^{-3})	3.211	3.210	6.1		3.26	6.88
Static dielectric constant	9.66($\perp$) 10.03($\parallel$)	9.72	9.5 (8.9)		8.5	19.6
Dynamic dielectric constant	6.52	6.52	5.3		4.84	9.3
Electron mobility (cm^2 V^{-1} s^{-1})	330 (300 K) 70 (600 K)	1000 (300 K) 70 (600 K)	1200 (measured)			
Hole mobility (cm^2 V^{-1} s^{-1})	60	50	30	30	14	
Saturation velocity (m s^{-1})	2–2.5 × 10^5	2–2.5 × 10^5	2–2.5 × 10^5	2–2.5 × 10^5		
Electron effective mass ratio	0.25 trans. 1.5 longitud.	0.41	0.2	0.15	0.314	0.11
Light hole mass ratio	0.33	0.165	0.259	0.259	0.471	
Sound velocity (m s^{-1})	1.3 × 10^4	0.95 × 10^4	5 × 10^3	5 × 10^3	10^4	
Polar opt. phonon energy (eV)	0.104	0.121	0.092	0.092	0.11	
Acoustic deform potential (eV)	11.5		10 (21)	10 (21)	19	
Piezoelectric constant	$e_{33} = 0.2\,\mathrm{C\,m^{-2}}$		$e_{33} = 0.65\,\mathrm{C\,m^{-2}}$	$e_{33} = 0.56\,\mathrm{C\,m^{-2}}$	$d_{15} = 4 \times 10^{-8}\,\mathrm{m\,V^{-1}}$	
Thermal conductivity (W cm °C^{-1})	4.9 (10^{17} cm^{-3}) 3.2 (2 × 10^{18})	2.6 (10^{16} cm^{-3}) 1.3 (10^{20} cm^{-3})	1.5	1.5		
Melting temperature (°C)	3070	2830	>1700	>1700	3000	≈1100

Table 10.3 Typical parameters used for the Monte Carlo simulation of transport properties in GaN

	Zinc blende	Wurtzite
Lattice constants (nm)	0.452	$a = 0.3189$
		$c = 0.5185$
Energy gap (eV)	3.2	3.4
Dielectric constants	$\varepsilon_0 = 8.9$	$\varepsilon_0 = 8.9$
	$\varepsilon_\infty = 5.35$	$\varepsilon_\infty = 5.35$
Density $(g\,cm^{-3})$	6.095	6.095
Sound velocity $(m\,s^{-1})$	4.57×10^3	4.33×10^3
Effective mass in Γ valley	0.15	0.20
Acoustic deformation potential	10.1	10.1
Polar optical phonon energy (meV)	91.2	91.2

higher values possible in a very high quality material. Thus, as technology improves, the low field mobility of GaN can easily exceed that for Si FETs (in short n-channel Si FETs, where the surface scattering is dominant, the low field mobility is typically only $400–600\,cm^2\,V^{-1}\,s^{-1}$, i.e. much smaller than for bulk Si, where the mobility can be as high as $1500\,cm^2\,V^{-1}\,s^{-1}$). Even higher values of the electron mobility and electron velocity may be achieved in InGaN. Coupled with very high peak and saturation velocities and a semiinsulating substrate (plus a relatively low dielectric constant), this makes GaN-based materials extremely appealing for microwave and milli-meter wave applications.

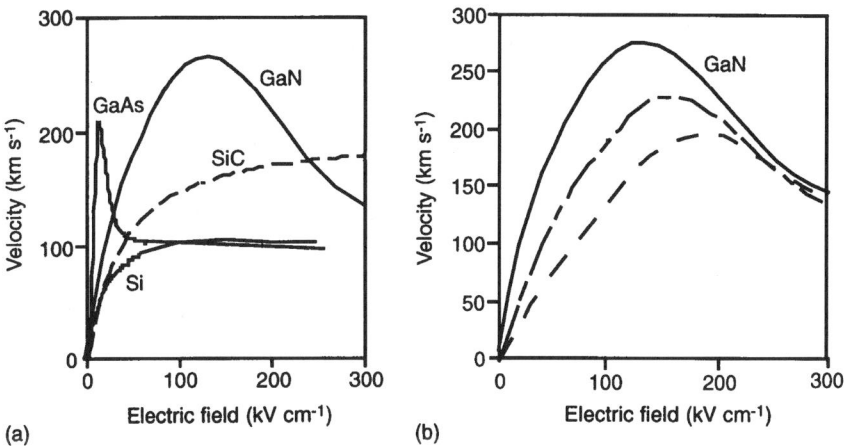

Figure 10.3 (a) Electron drift velocity at 300 K in GaN, SiC and GaAs. (b) Electron drift velocity in GaN at (—) 300 K, (---) 500 K and (---) 750 K [8].

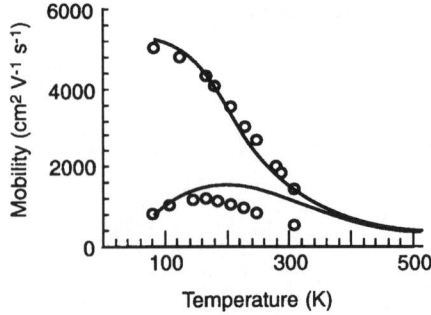

Figure 10.4 Measured and calculated Hall electron mobility in bulk GaN and in the two-dimensional electron gas [12].

The calculations of the low field mobility generally agree with the results of the Monte Carlo simulations and experimental data [3–12]. These calculations confirm that the electron mobility in GaN is smaller than the electron mobility in GaAs, comparable to the electron mobility in Si, and larger than the electron mobility in the most popular SiC polytypes.

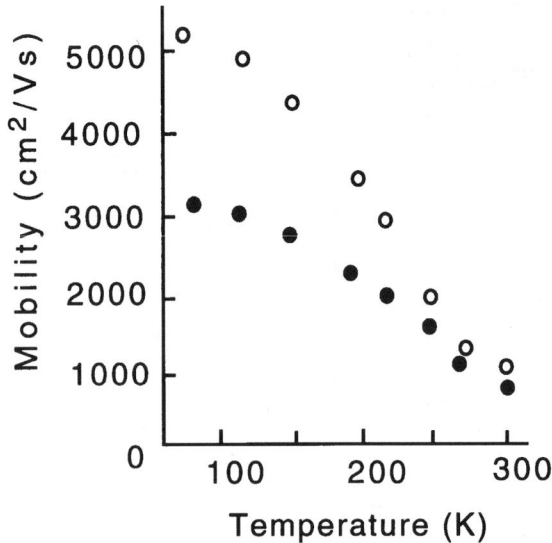

Figure 10.5 Temperature dependence of the electron mobility in AlGaN/GaN heterostructures with (○) nominally undoped channel ($n_s = 1 \times 10^{12}$ cm^{-3}) and (●) intentionally doped channel ($n_s = 5 \times 10^{12}$ cm^{-3}). Here n_s is the sheet carrier concentration in the channel at the heterointerface.

Shur *et al.* [12] compared the temperature dependences of the electron mobility in the two-dimensional electron gas (2DEG) in AlGaN/GaN heterostructures and in doped bulk GaN. Based on their experimental data and calculations, they demonstrated a large enhancement of the electron mobility in the 2DEG compared to bulk GaN (Fig. 10.4).

Khan *et al.* [13] reported on the measurements of the Hall mobility in doped channel GaN heterostructures. Their results (in agreement with the computed dependencies shown in Fig. 10.4) confirm that the electron mobility at room temperature hardly decreases with doping (Fig. 10.5). This important result led to the development of the AlGaN/GaN DC-HFETs, as discussed below.

Recent experimental data [14] show that the electron effective mass in the cubic phase of GaN is substantially smaller than in the hexagonal phase ($m_n = 0.15m_e$ for the cubic phase compared to $0.20m_e$ for the hexagonal phase). The energy gap is slightly smaller ($E_g = 3.2\,\text{eV}$ for the cubic phase compared to $3.4\,\text{eV}$ for the hexagonal phase). Figure 10.6 compares the Hall mobility for the hexagonal and cubic GaN; the mobility of cubic GaN should be considerably larger, in agreement with qualitative predictions made by Eastman [15]. Thus, in principle, the cubic GaN may be a superior material for field effect transistors if problems related to material growth are solved.

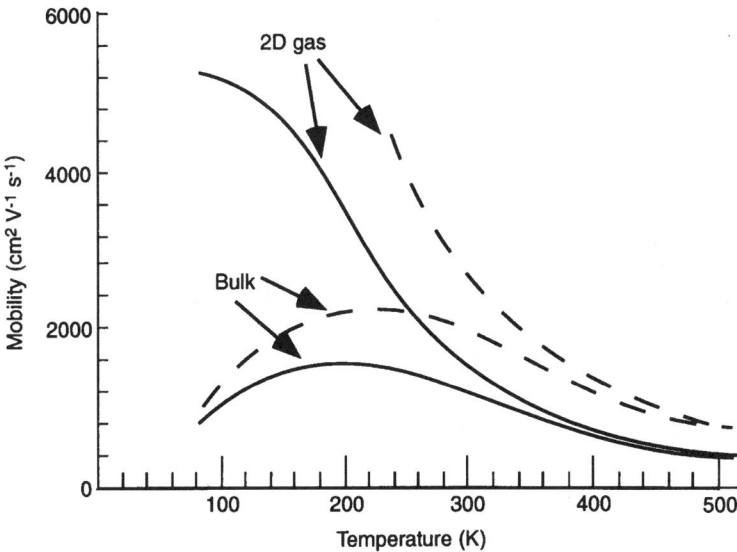

Figure 10.6 Electron Hall mobility in GaN versus temperature for two-dimensional electron gas and bulk GaN for (---) cubic GaN and (—) hexagonal GaN. For the bulk calculation, $n = 2 \times 10^{16}\,\text{cm}^{-3}$, $N_T = 2 \times 10^{17}\,\text{cm}^{-3}$; for the electron gas calculation, $n = 5 \times 10^{17}\,\text{cm}^{-3}$, $N_T = 6.5 \times 10^{16}\,\text{cm}^{-3}$. (From Shur *et al.* [12])

10.3 MATERIAL GROWTH AND DEVICE FABRICATION

GaN is now grown epitaxially on several different substrates: sapphire, 6H-SiC, 2H-SiC, Si and other materials. A typical technique of GaN growth is metallorganic vapor phase epitaxy (MOVPE). The review paper of Strite and Morkoç [2] discusses the different substrates and techniques for epitaxial growth. Electron cyclotron resonance (ECR) etching techniques have been used for FET fabrication [16]. Reactive ion etching and Cl-based etches have been explored as well [17]. Proton or helium bombardment has been used to isolate devices.

n-Type and p-type regions have been produced in GaN using Si^+ and Mg^+/P^+ implantation and subsequent anneal at $\sim 1100\,°C$ [18].

10.4 OHMIC AND SCHOTTKY CONTACTS

Just like for the GaAs-based material system, good ohmic contacts to GaN-based devices are crucial for obtaining good performance and for realizing reliable devices operating over a wide temperature range. Al and Au ohmic contacts to GaN yielded specific contact resistances of 10^{-4} and $10^{-3}\,\Omega\,cm^2$, respectively [19]. Khan et al. [20] reported on a systematic study of the contact resistance to doped layers of GaN. They demonstrated

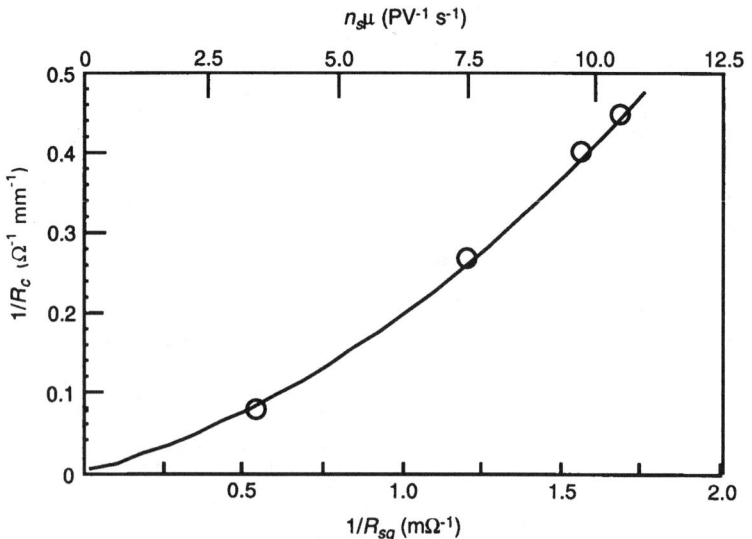

Figure 10.7 Inverse contact resistance versus inverse sheet resistance: (○) measured data and (—) analytical fit. (From Khan et al. [20])

the direct proportionality between the sheet resistance of the doped GaN layer and specific contact resistance. Binari [21] reported on the contact resistance in FETs as low as $0.5\,\Omega\,mm$ (comparable to good quality GaAs-based FETs).

Morkoç et al. [17] showed that TiN makes a good ohmic contact to GaN. Using a Ti/Al metallization system and rapid thermal anneal at $900\,°C$ for $30\,s$, they demonstrated a specific contact resistance of $8 \times 10^{-6}\,\Omega\,cm^2$. In this regard, Yoder [22] discussed the results by Tao et al. [23], who showed that TiN is free of electromigration at temperatures up to $408\,°C$ for a period of 10 years on silicon semiconductor material. As Yoder pointed out, this time could be even longer on refractory semiconductors such as III–Ns.

Typically, the contact resistance decreases with increasing conductivity of the device channel (Fig. 10.7).

Good quality Schottky contacts to GaN have been demonstrated by many groups. As pointed out by Morkoç et al. [17], the Schottky barrier to GaN depends on the metal work function, which means that the Fermi level is not pinned at the GaN–metal surface.

10.5 SEMICONDUCTOR–INSULATOR–SEMICONDUCTOR STRUCTURES

GaN and AlN have strongly pronounced piezoelectric properties due to large piezoelectric constants. For example, the piezoelectric constant e_{33} of AlN is 9.3 times larger than the piezoelectric constant e_{14} of GaAs.

In GaN–AlN–GaN semiconductor–insulator–semiconductor (SIS) structures with sufficiently thin AlN layers, the lattice constant mismatch is accommodated by internal strains instead of misfit dislocations. These strains induce electric fields and may strongly affect the carrier distributions near the interfaces. Bykhovski et al. calculated the strain-induced electric field and charge distribution in such structures [24]. They found that, in an SIS structure grown along a $(0\,0\,0\,1)$ crystallographic direction, the strain-induced electric fields can shift the flat band voltage, producing an accumulation region on one side and a depletion region on the other side of the AlN insulator. The surface charge density caused by the piezoelectric effect is on the order of $10^{12}\,cm^{-2}$. As a consequence of the asymmetry in the space charge distribution, the capacitance–voltage characteristics of the SIS structure become asymmetrical. The asymmetrical shift of the $C-V$ characteristics with respect to the origin is on the order of $1.5\,V$ for a $3\,nm$ AlN film. This asymmetry should vanish in a relaxed film. Hence, as mentioned in Bykhovski et al. [24] and depending on the film thickness, the capacitance–voltage measurements of the GaN–AlN–GaN SIS structures can be used for quantitative characterization of the degree of the AlN film relaxation.

The experimental $C-V$ characteristics of symmetrically doped SIS structures are strongly asymmetrical, in qualitative agreement with the theory [25]. The experimental results for 3.0 nm AlN film are in excellent agreement with the calculation for the totally unrelaxed structure. Thicker AlN films seem to be partially relaxed. Therefore, the starting point for the generation of misfit dislocation corresponds to $L \geqslant 3.0$ nm. This conclusion is in agreement with experimental studies of strained-layer GaN/AlN superlattices [26]. The upper limit for the complete relaxation has been estimated as 10 nm [26]. Other data show that the AlN film is not fully relaxed at this thickness [25]. Bykhovski *et al.* [25] believe that this indicates the high quality of their SIS structures. This also means that the piezoelectric effects play an important role in GaN/AlN/GaN SIS structures when the AlN thickness is less than or of the order of 10 nm [25].

Bykhovski *et al.* explained a partial relaxation of these films up to 10 nm of AlN thickness as follows. Some stacking faults and partial misfit dislocations may appear when the film thickness is larger than 3.0 nm. Their number gradually increases with the thickness up to $L = 10$ nm. Then, full misfit dislocations correspond to a smaller energy [27]. These results and further studies of the SIS structures indicate a possibility of obtaining thin, high quality AlN and AlGaN layers to provide effective gate isolation in AlGaN/GaN HFETs and DC-HFETs [28–30].

10.6   FIELD EFFECT TRANSISTORS

All basic processing and technological steps are available for fabrication of GaN and AlGaN/GaN field effects transistors of different types. GaN metal–semiconductor field effect transistors (MESFETs), GaN metal–insulator field effect transistors (MISFETs), doped channel AlGaN/GaN HFETs, inverted channel AlGaN/GaN HFETs, AlGaN/GaN DC-HFETs have all been developed (Fig. 10.8).

Khan *et al.* reported on the microwave performance of short channel (0.2 μm gate length) AlGaN/GaN heterostructure field effect transistors (HFETs) at room temperature [31] and elevated temperatures up to 110 °C with the cutoff frequency, f_T, up to 20 GHz and the maximum frequency of oscillations, f_{max}, up to 77 GHz [32]. Binari *et al.* [21, 33, 34] reported on the microwave performance of the AlGaN/GaN inverted HFETs, GaN MESFETs and Si_3N_4/GaN MISFETs with the gate length $L = 0.8$ μm at room temperature and obtained a smaller cutoff frequency but a larger $f_T L$ product. Binari *et al.* [21, 34] and Khan *et al.* [32, 35] also presented the results of the DC and microwave measurements, showing that these devices can operate at least up to 360 °C (Fig. 10.9).

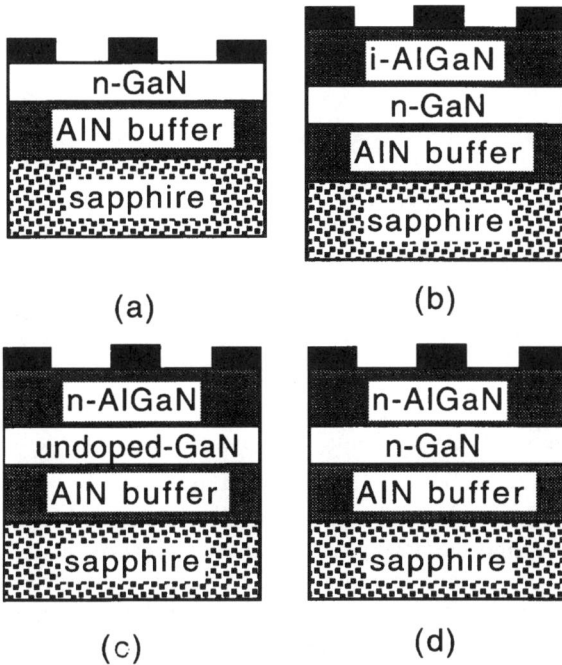

Figure 10.8 Schematic structure of different GaN and AlGaN/GaN FETs: (a) GaN MESFET, (b) GaN MISFET, (c) AlGaN/GaN HFET and (d) AlGaN/GaN DC-HFET. A GaN buffer layer may sometimes be used.

Let us now discuss a typical fabrication sequence for an AlGaN/GaN HFET. The device epilayer structure (Figs 10.8c and 10.10) is deposited over basal plane sapphire substrate using low pressure metallorganic chemical vapor deposition (MOCVD). A typical device structure may consist of a $0.5\,\mu m$ highly insulating GaN layer followed by a 10 nm thick conducting channel followed by a 3–10 nm thick $Al_{0.1}Ga_{0.9}N$ spacer layer. The unintentional doping in these layers is estimated to be around $5 \times 10^{17}\,cm^{-3}$, which results in a channel depletion at zero gate bias. The top n-type AlGaN layer controls the device threshold voltage.

HFET structures with varying gate lengths and widths and source–drain spacing are then fabricated on isolated mesas. These mesas are formed using reactive ion etching in a CCl_4 plasma and using photoresist as an etch mask. Ti/Al is often used as the source–drain metal and Ti and the Schottky barrier metal for the gate. For microwave applications, the device contact geometry is designed to allow for the wafer RF probing using cascade probes.

Figure 10.11 shows the measured transconductance in the saturation region at different temperatures. The transconductance is limited by a large

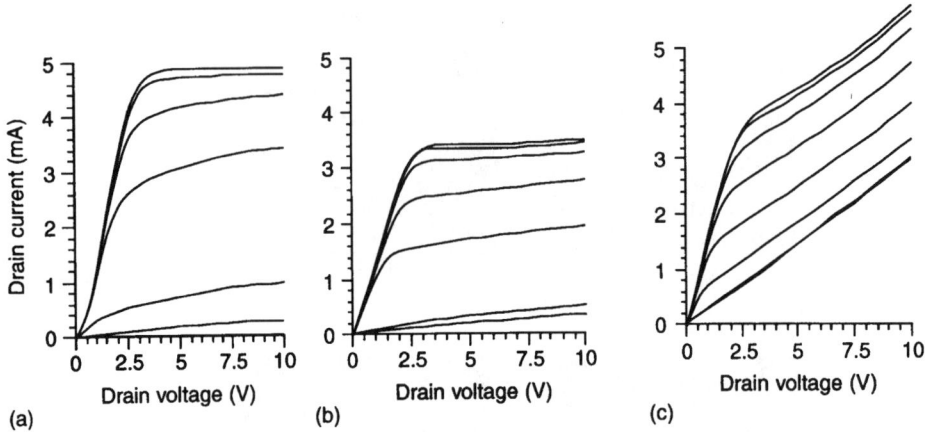

Figure 10.9 Current–voltage characteristics of AlGaN/GaN HFETs at (a) 25 °C, (b) 200 °C and (c) 300 °C.

series resistance, and the device performance can be drastically improved by making better ohmic contacts and more conductive channels.

Figure 10.12 shows the measured values of the cutoff frequency, f_T, and the maximum frequency of oscillations, f_{max}, as functions of temperature at a drain bias $V_{ds} = 20$ V. Except for temperature of 300 °C (for which the transistor characteristics were highly nonideal and which was at the very limit of our measurement system capabilities), the measured cutoff frequencies are roughly proportional to the extracted values of g_{mo}, in agreement

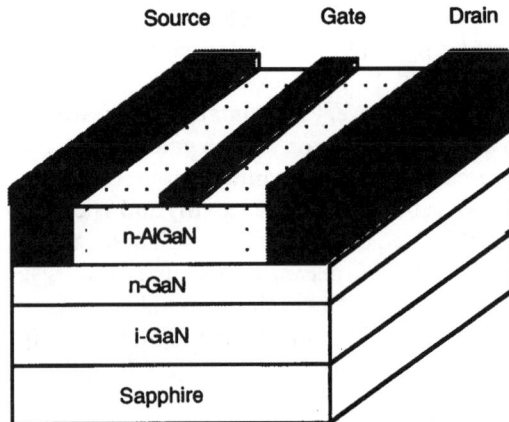

Figure 10.10 Device structure of AlGaN/GaN HFETs.

with the well-known relationship

$$f_T = \frac{g_{mi}}{2\pi(C_{gs} + C_{gd})} \tag{10.1}$$

where $g_{mi} = g_{mo}W$ is the device transconductance, W is the device width and C_{gs} and C_{gd} are the gate–source and gate–drain capacitances. In the saturation regime, $C_{gs} \approx 2\varepsilon_s WL/(3d_i) \gg C_{gd}$. Here $\varepsilon_s \approx 79.7\,\mathrm{pF\,m^{-1}}$ and d_i are the dielectric permittivity and thickness of the AlGaN barrier layer, respectively. (Our standard self-consistent quantum mechanical calculations of the electronic subbands for the two-dimensional electron gas in GaN showed that, for this semiconductor, the effective thickness of the two-dimensional electron gas, Δd, is much smaller than d_i, so Δd does not appreciably affect the channel capacitance.)

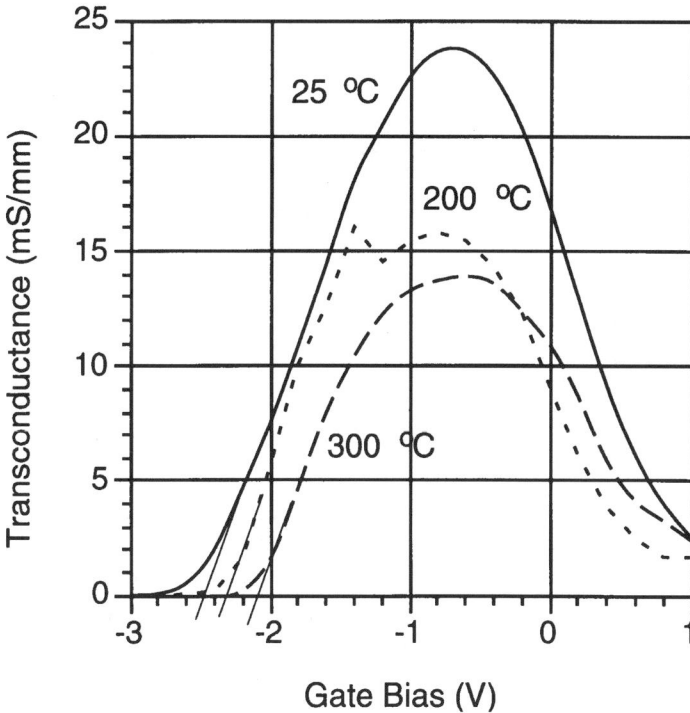

Figure 10.11 Transconductance at a drain–source bias (V_{ds}) of 10 V for temperatures of 25, 200 and 300 °C. The thin tangents near the bottom left show extrapolations of the transconductance curves used to determine the threshold voltages of approximately -2.5, -2.3 and -2.1 V at 25, 200 and 300 °C, respectively [32].

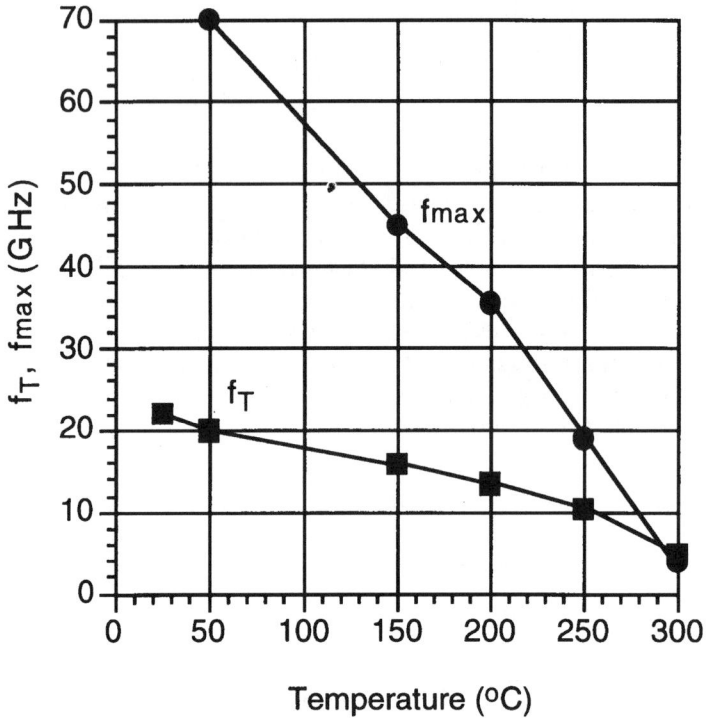

Figure 10.12 Values of f_T and f_{max} versus temperature: $V_{ds} = 20\,\text{V}$ and $V_{gs} = -0.8\,\text{V}$ [32].

Using equation (10.1), Khan *et al.* [32] extracted the barrier layer thickness

$$d_i \approx \frac{4\pi f_T \varepsilon_s L}{3g_{mo}} \approx 31.0\,\text{nm}$$

from the measured f_T data, in good agreement with the value of $d_i \approx 30.0\,\text{nm}$ estimated from their growth data.

The analysis of the data shown in Fig. 10.11 reveals that the device transconductance is limited by a large source series resistance, R_s, and by a limited sheet carrier concentration, n_s, of the two-dimensional electron gas (2DEG) in the device channel. In order to increase n_s, Khan *et al.* [20] proposed to use a doped channel approach, similar to that for doped channel AlGaAs/GaAs heterostructures [36]. Such an approach is even more advantageous in AlGaN/GaN structures because the additional impurity scattering has less of an effect in GaN than in GaAs, since electrons in GaN have a much higher effective mass. As we showed earlier [20], the channel doping also dramatically reduces the contact resistance. In doped channel structures, we have achieved very high values of the sheet electron concentration (higher than approximately $1.5 \times 10^{13} \, cm^{-2}$) and very high values of the sheet concentration–mobility products (up to approximately $10^{16} \, V^{-1} s^{-1}$). We recently designed and fabricated AlGaN/GaN DC-HFETs [13, 37, 38]. These devices exhibited transconductance up to $120 \, mS \, mm^{-1}$ and maximum drain–source saturation currents in excess of $600 \, mA \, mm^{-1}$ [13]. At high gate biases, the current–voltage characteristics exhibit negative differential resistance, related to device self-heating [39]. This means that the device characteristics should further improve with better heat sinking. The microwave measurements for this device show that it exhibited the highest product of cutoff frequency and gate length $(18.3 \, GHz \, \mu m)$ of any wide band gap semiconductor FET) [13].

We used the velocity saturation HFET model [40] in order to derive an expression linking f_T to the low field mobility μ, saturation velocity v_s, gate

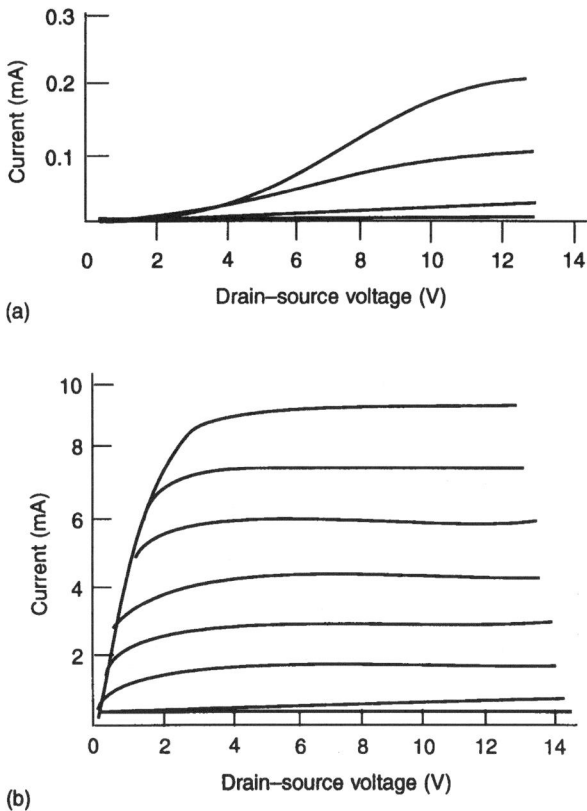

(a)

(b)

Figure 10.13 Measured current–voltage characteristics of the $Al_{0.15}Ga_{0.85}N/GaN$ HFETs at 25 °C: (a) in the dark and (b) under UV illumination [41].

length L and intrinsic gate voltage swing, $V_{GT} = V_{GS} - V_T$ where $V_{GS} = V_{gs} - I_{ds}R_s$, and V_{gs} is the applied gate–source voltage. Using the gradual channel approximation, we first calculated the velocity distribution in the channel, $v(x)$, at the saturation voltage (the voltage at which the electrons attain their saturation velocity at the drain). The analysis showed that f_T could be significantly improved by increasing the sheet carrier density in the channel; this provided us with the motivation of using doped channel devices in order to increase the cutoff frequency. We estimate that further improvements in the device design and a scaling down of the gate length should make it possible for AlGaN/GaN deep submicron devices to achieve cutoff frequencies of greater than 100 GHz.

The importance of a large electron sheet concentration in the channel for the improved FET performance is confirmed by a large transconductance increase in an $Al_{0.15}Ga_{0.85}N/GaN$ HFET under ultraviolet illumination

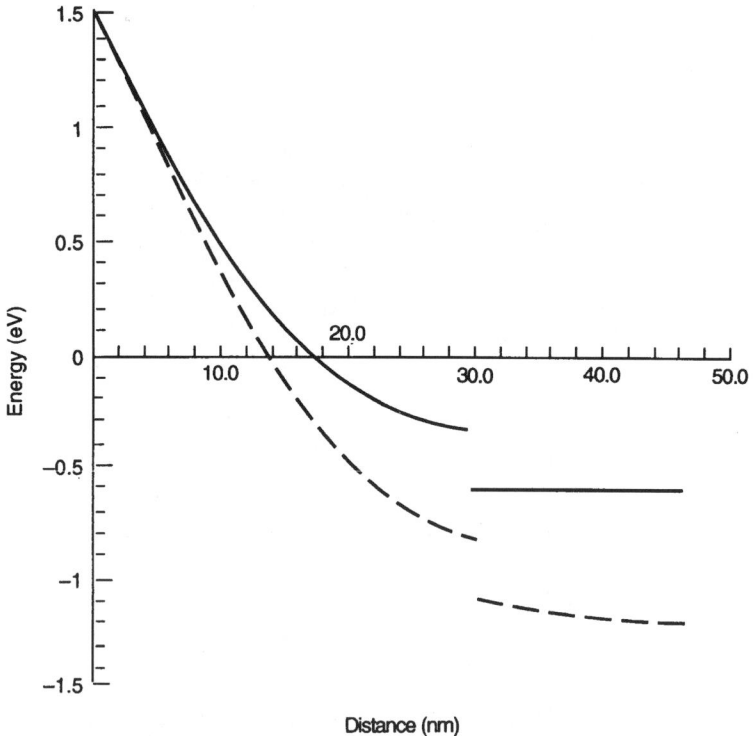

Figure 10.14 Band diagrams at the threshold (—) in the dark and (---) under illumination. Parameters used in the calculation: dielectric permittivity of AlGaN $\varepsilon_s = 7.9 \times 10^{-11}\,F\,m^{-1}$, Schottky barrier height $\Phi_b = 1.5\,eV$, ionized donor density in AlGaN layer $N_d = 2 \times 10^{18}\,cm^{-3}$, AlGaN layer thickness $x_n = 30\,nm$, conduction band discontinuity $E_c = 0.273\,eV$, surface positive charge trapped in the channel under illumination $N_t = 4 \times 10^{11}\,cm^{-3}$ [41].

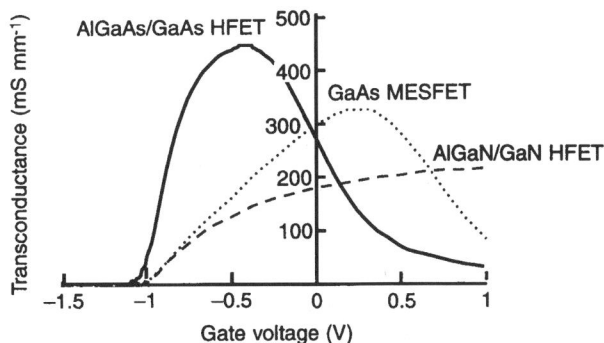

Figure 10.15 Computed device transconductance in the saturation region versus gate bias for three devices: $0.5\,\mu m$ gate GaAs MESFET, $0.5\,\mu m$ gate AlGaAs/GaAs HFET and $0.25\,\mu m$ AlGaN/GaN HFET [8].

[41]. The thickness of the $Al_{0.15}Ga_{0.85}N$ barrier layer in the devices described in Khan *et al.* [41] was approximately 30 nm. From the measured threshold voltage ($V_T \approx -0.6\,V$), we estimated the doping of the AlGaN layer to be on the order of $2 \times 10^{18}\,cm^{-3}$ (assuming the Schottky barrier height of 1.5 eV and $\Delta E_c \approx 0.27\,eV$). The GaN active layer was unintentionally doped. The nominal gate length, L, and the gate width, W, were $1\,\mu m$ and $150\,\mu m$, respectively.

Figure 10.13 shows the measured $I–V$ characteristics of the AlGaN/GaN HFETs at 25 °C in the dark and under illumination with an He–Cd laser (wavelength, $\lambda = 325\,nm$). We speculate that light creates electrons in the

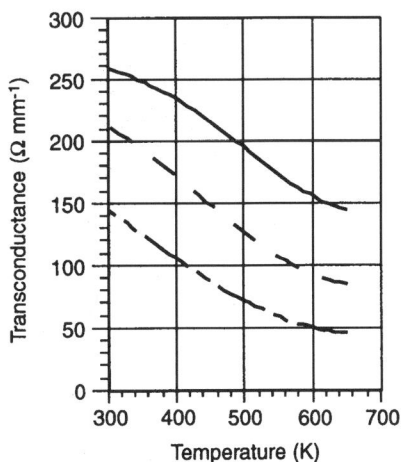

Figure 10.16 Computed transconductance versus temperature for (—) $0.5\,\mu m$, (– –) $1\,\mu m$ and (– - –) $2\,\mu m$ AlGaN/GaN HFETs [8].

GaN active layer, trapped holes in the undoped GaN layer and the active channel. The illumination acts as 'optical' channel underdoping. Figure 10.14 shows the band diagrams at the threshold in the dark (solid line) and under illumination (dashed line). A relatively modest positive trapped charge in the active channel ($4 \times 10^{11} \, cm^{-3}$) is sufficient to produce a considerable shift in V_T and is therefore sufficient to enhance the n_s value. (Even though much of the actual trapped positive charge in our devices may be located in the undoped GaN layer, this charge affects the device characteristics in a way similar to the charge trapped in the channel.)

The current under the 'optical channel doping' was fully modulated by the gate bias. The maximum g_m is on the order of $64 \, mS \, mm^{-1}$, nearly three times higher than we reported previously for similar devices with a comparable threshold voltage and many times larger than for the same device in the dark. This enhancement is explained by a large reduction in the source and drain series resistances, R_s and R_d, and by the enhancement of the n_s value. The calculations – based on a high electron mobility transistor (HEMT) model [40] – confirm that the g_m enhancement is caused by the reduction in R_s and by a larger n_s.

Our analysis also shows that transconductances of over $100 \, mS \, mm^{-1}$ should be achievable in submicron AlN/GaN HFETs with a relatively small

$$f_T = 36.1 \text{ GHz} \quad f_{max} = 70.8 \text{ GHz}$$

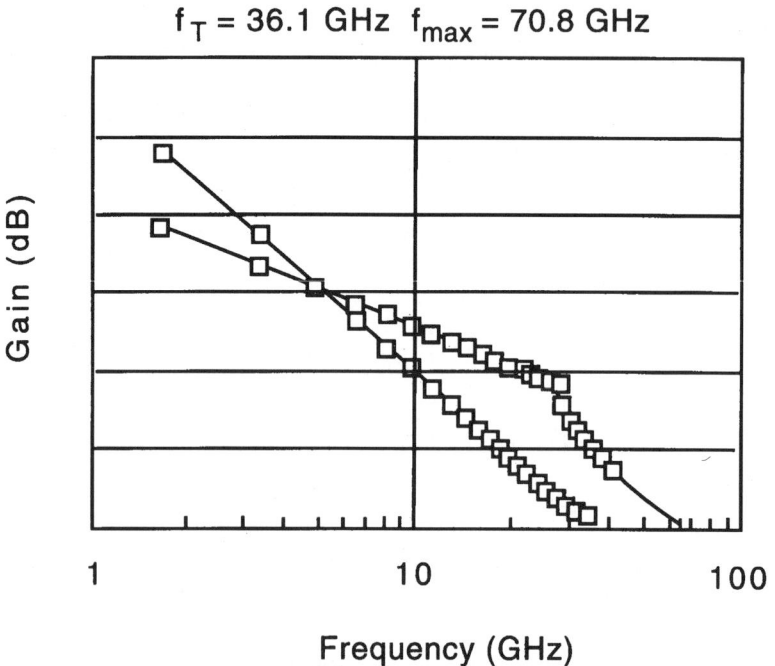

Figure 10.17 Gain versus frequency for 0.25 μm AlGaN/GaN DC-HFET [38].

drop at elevated temperatures (Figs 10.15 and 10.16). These results show that, in spite of a smaller thermal conductivity than for SiC, the excellent transport properties of GaN at elevated temperatures make these devices a viable alternative for high temperature microwave and digital applications compared to SiC. Our results obtained for 0.25 μm gate AlGaN/GaN DC-HFETs confirm a high potential of these devices for microwave operation (Fig. 10.17) [38].

10.7 OPTOELECTRONIC AlGaN/GaN FETs

Unique optical and electronic properties of the GaN/AlGaN material system open up numerous opportunities for visible-blind optoelectronic devices. These devices have a high sensitivity, a large gain–bandwidth product and can be integrated with GaN/AlGaN field effect transistors, which have already demonstrated an operation at microwave frequencies. A transparent sapphire substrate makes AlGaN/GaN HFETs well suited for optoelectronic applications.

The HFET photodetector [42] is based on a 0.2 μm gate AlGaN/GaN HFET and utilizes a shift in the threshold voltage caused by the light-generated carriers (Fig. 10.18).

The response of AlGaN/GaN heterostructure insulated gate field effect transistors (HIGFETs) to light is different than AlGaN/GaN HFETs (Fig. 10.19). The HIGFET respectively exhibits a low resistance state and a persistent high resistance state before and after the application of a high drain voltage. The device can be returned to the low resistance state by exposing it to optical radiation with sensitivity peaks at certain wavelengths (Figs 10.20, 10.21 and 10.22). A possible mechanism is electron trapping in the gate insulator near the drain edge of the gate.

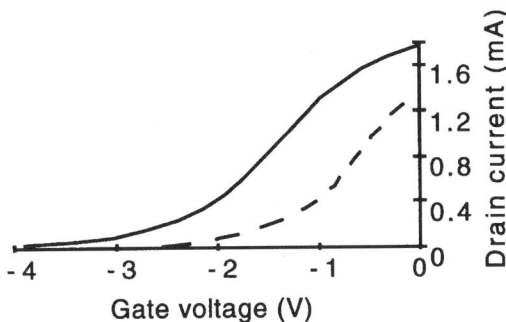

Figure 10.18 Current–voltage characteristics of AlGaN/GaN HFETs (---) in the dark and (—) under illumination [42].

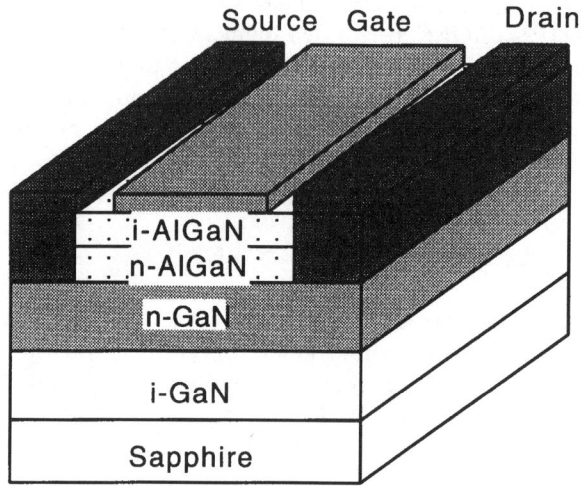

Figure 10.19 Schematic diagram of AlGaN/GaN HIGFET [43].

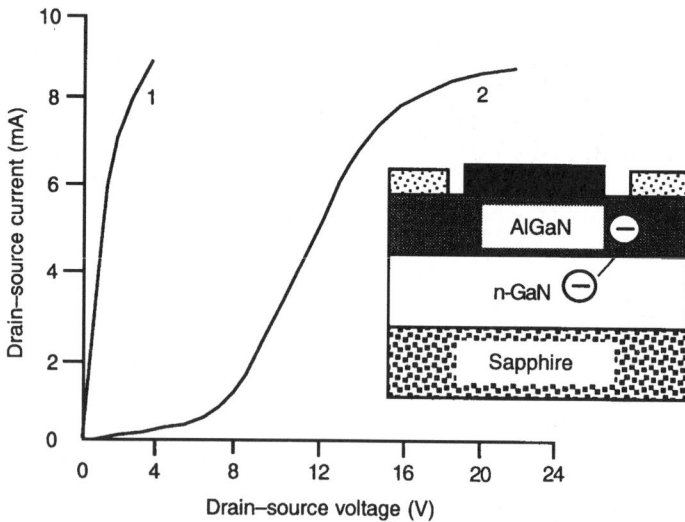

Figure 10.20 Current–voltage characteristic collapse. Curve 1 shows the on-state (before the application of high drain bias). Curve 2 shows the off-state (after the application of high drain bias). The inset illustrates the collapse mechanism–electron trapping in the barrier layer at the drain [43].

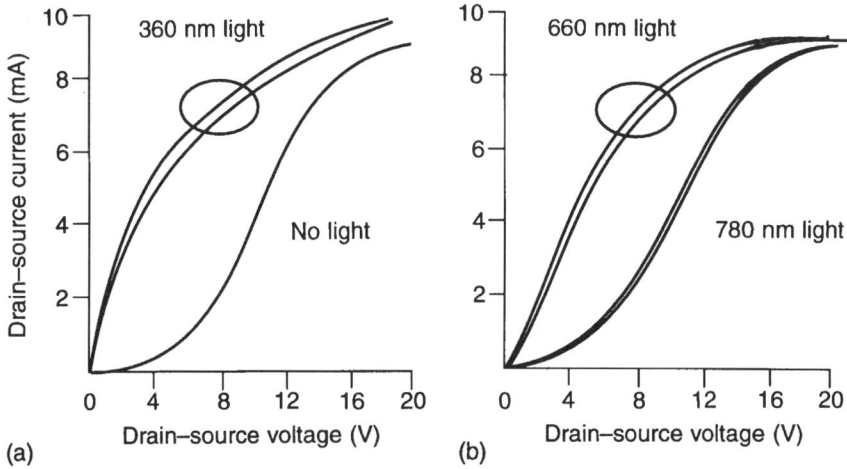

Figure 10.21 Current–voltage characteristics (a) in the dark and (b) under illumination [43].

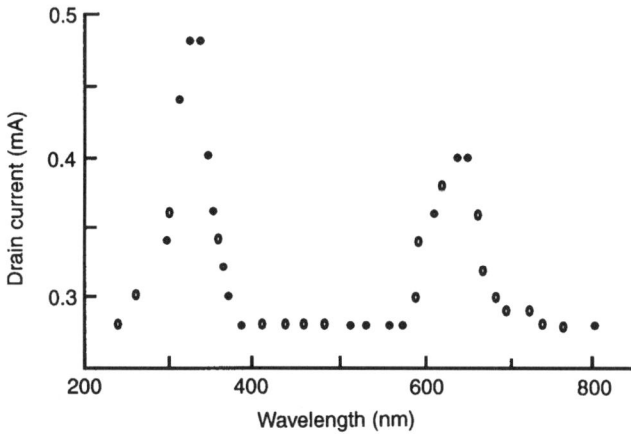

Figure 10.22 Dependence of the drain current on the wavelength of the optical radiation: $V_{ds} = 2\,\mathrm{V}$ (after $V_{ds} = 40\,\mathrm{V}$) [43].

10.8 CONCLUSIONS

GaN-based FETs have already demonstrated an impressive performance. Their excellent transport and optoelectronic properties should allow us to achieve excellent performance in harsh conditions and/or at elevated temperatures. Large band discontinuities in these materials should allow us to obtain very high densities of 2DEG (up to 2×10^{13} cm^{-2} for GaN); 2DEG should have much better transport properties than bulk electrons. This has already been confirmed by a good performance of AlGaN/GaN HFETs compared to other types of FETs. Further improvements in device technology and material quality should lead to the development of viable microwave and possibly millimeter wave devices, as well as optoelectronic integrated circuits based on AlGaN/GaN FETs.

RECENT DEVELOPMENTS

The field of wide band gap materials and devices is a rapidly developing field, and new important developments have occurred in five months prior to publication of this book. Among these developments are the demonstration of room temperature mobility in GaN of 1,500 cm^2 V^{-1} s^{-1} [44] and the development of ultra-low ohmic contacts to n-type GaN (with specific contact resistance of 8.9×10^{-8} ohm cm^{-2} [45]. As was reported in [45], these contacts are composite Ti/Al/Ni/Au (150 Å/2200 Å/400 Å/500 Å) on GaN preceded by reactive ion etching and followed with annealing at 900°C for 30 s.

Schottky barriers contacts with ideality factors close to unity have been fabricated [46]. Also, higher values of the GaN HFET transconductance have been achieved (up to 222 mS mm^{-1} under illumination and 210 mS mm^{-1} in the dark) [47]. The maximum frequency of oscillation for GaN-based HFETs approached 100 GHz [48].

ACKNOWLEDGMENTS

The authors are grateful to Dr A. D. Bykhovski for useful comments. This work has received support from the Office of Naval Research (monitored by Max Yoder) and the Army Research Office (monitored by John Zavada).

REFERENCES

1. Shur, M. (1996) *Introduction to Electronic Devices*, Wiley, New York.
2. Strite, S. and Morkoç, H. (1992) *J. Vac. Sci. Technol. B*, **10**(4), 1237–1266.
3. Littlejohn, M.A., Hauser, J.R. and Glisson, T.H. (1976) *Appl. Phys. Lett.*, **26**, 625.
4. Gelmont, B., Kim, K.S. and Shur, M. (1993) *J. Appl. Phys.*, **74**, 1818.
5. Shur, M., Gelmont, B., Saavedra-Munoz, C. and Kelner, G. (1994) in *Proc. 5th Conf. Silicon Carbide and Related Compounds*, Institute of Physics, Bristol and Philadelphia, Conference Series No. 137, p. 465.
6. Joshi, R.P. and Raha, P.K. (1994) in *Proc. 5th Conf. Silicon Carbide and Related Compounds*, Institute of Physics, Bristol and Philadelphia, Conference Series No. 137, p. 687.

7. Kolnik, J., Oguzman, I.H., Brennan, K.F., Wang, R., Ruden, P.P. and Wang, Y. (1995) *J. Appl. Phys.*, **78**(2), 1033.
8. Shur, M.S. and Khan, M.A. (1995) Electronic and optoelectronic AlGaN/GaN heterostructure field effect transistors, in *Proc. Symp. on Wide Band Gap Semiconductors and the 23rd State-of-the-Art Program on Compound Semiconductors (SOTAPOCS XXIII)* (eds F. Ren, D.N. Buckley, S.J. Pearton, P. Van Daele, G.C. Chi, T. Kamijoh and F. Schuermeyer), The Electrochemical Society, New Jersey, Vol. 95-21, pp. 128–135.
9. Chin, V.W.L., Tansley, T.L. and Osotchan, T. (1994) *J. Appl. Phys.*, **75**, 7365.
10. Rode, D.L. and Gaskill, D.K. (1995) *Appl. Phys. Lett.*, **66**, 1972.
11. Kim, J.G., Frenkel, A.C., Liu, H. and Park, R.M. (1994) *Appl. Phys. Lett.*, **65**, 91.
12. Shur, M., Gelmont, B. and Khan, A. (1996) High electron mobility in two-dimensional electron gas in AlGaN/GaN heterostructures, *J. Electronic Materials*, **17**, 325.
13. Khan, M.A., Chen, Q., Shur, M.S., Dermott, B.T. and Higgins, J.A. (1996) Microwave operation of GaN/AlGaN doped channel heterostructure field effect transistors, *IEEE Electron Device Lett.*, submitted.
14. Fanciulli, M., Lei, T. and Moustakas, T.D. (1993) *Phys. Rev. B*, **48**, 15144–15147.
15. Eastman, L. (1995) Paper presented at the Rump Session on Wide Band Gap Semiconductors, Device Research Conference, June 20.
16. Burm, J., Schaff, W.J. and Eastman, L.F. (1996) The fabrication of recessed gate GaN MODFETs, in *Proceedings of 1995 International Symposium on Compound Semiconductors*, Institute of Physics Conference Series No. 145, p. 605.
17. Morkoç, H., Strite, S., Gao, G.B., Lin, M.E., Sverdlov, B and Burns, M. (1994) *J. Appl. Phys.*, **76**(3), 1363–1398.
18. Pearton, S.J., Vartuli, C.B., Zolper, J.C., Yuan, C. and Stall, R.A. (1995) *Appl. Phys. Lett.*, **67**(10), 1435–1437.
19. Foresi, J.S. and Moustakos, T.D. (1993) *Appl. Phys. Lett.*, **62**, 2859.
20. Khan, M.A., Shur, M.S. and Chen, Q. (1996) Contact resistance in doped GaN/AlGaN heterostructures, *Appl. Phys. Lett.*, **68**, 3022.
21. Binari, S. (1995) in *Proc. Symp. on Wide Band Gap Semiconductors and the 23rd State-of-the-Art Program on Compound Semiconductors (SOTAPOCS XXIII)* (eds F. Ren, D.N. Buckley, S.J. Pearton, P. Van Daele, G.C. Chi, T. Kamijoh and F. Schuermeyer), The Electrochemical Society, New Jersey, Vol. 95-21, pp. 136–143.
22. Yoder, M. (1995) Private email communication, June 5.
23. Tao, J. *et al.* (1995) Electromigration characteristics of TiN barrier layer material. *IEEE Electron Device Trans.* **16**(6), 230.
24. Bykhovski, A., Gelmont, B. and Shur, M.S. (1993) The influence of the strain-induced electric field on the charge distribution in GaN–AlN–GaN SIS structure. *J. Appl. Phys.*, **74**(11), 6734.
25. Bykhovski, A., Gelmont, B., Shur, M.S. and Khan, A. (1994) Strain and charge distribution in GaN/AlN/GaN SIS structure, in *Proc. of 5th Conf. on Silicon Carbide and Related Materials*, Institute of Physics, Bristol and Philadelphia, Conference Series No. 137, pp. 691–694.
26. Sitar, Z., Paisley, M.J., Yan, B., Ruan, J., Choyke, W.J. and Davis, R.F. (1990) *J. Vac. Sci. Technol., B*, **8**, 316.
27. Roitburd, A.L. (1973) *Phys. Stat. Sol. (a)*, **16**, 329.
28. Bykhovski, A., Gelmont, B. and Shur, M.S. (1993) Strain and charge distribution in GaN–AlN–GaN SIS structure for arbitrary growth orientation, *Appl. Phys. Lett.*, **63**, 2243.
29. Bykhovski, A., Gelmont, B., Shur, M.S. and Khan, A. (1995) Current–voltage characteristics of strained piezoelectric structures, *J. Appl. Phys.*, **76**, 1616–1620.
30. Bykhovski, A., Gelmont, B. and Shur, M.S. (1995) Elastic strain relaxation in GaN–AlN–GaN semiconductor–insulator–semiconductor structures, *J. Appl. Phys.*, **78**(6), 3691–3696.
31. Khan, M.A., Kuznia, J.N., Olson, D.T., Schaff, W., Burm, G. and Shur, M.S. (1994) Microwave performance of 0.25 micron gate AlGaN/GaN heterostructure field effect transistor, *Appl. Phys. Lett.*, **65**(9), 1121–1123.
32. Khan, M.A., Shur, M.S., Kuznia, J.N., Burm, J. and Schaff, W. (1995) Temperature activated conductance in GaN/AlGaN heterostructure field effect transistors operating at temperatures up to 300 °C, *Appl. Phys. Lett.*, **66**(9), 1083–1085.

33. Binari, S., Rowland, L.B., Kelner, G., Kruppa, W., Dietrich, H.B., Doverspike, K. and Gatskill, D.K. (1994) *Electronics Lett.*, **30**(15), 1248.
34. Binari, S., Rowland, L.B., Kruppa, W., Kelner, G., Doverspike, K. and Gatskill, D.K. (1995) DC, microwave, and high-temperature characteristics of GaN FET structures, *Inst. Phys. Conf. Ser.*, **141**, 459–462.
35. Khan, M.A., Kuznia, J.N., Shur, M.S., Eppers, C., Burm, J. and Schaff, W. (1994) Paper presented at the 21st International Symposium on Compound Semiconductors, San Diego, September.
36. Ruden, P.P., Shur, M.S., Akinwande, A.I., Nohava, J., Grider, D. and Baek, J.H. (1990) AlGaAs–InGaAs–GaAs quantum well doped channel heterostructure field effect transistors. *IEEE Trans. Electron Devices*, **37**(10), 2171–2175.
37. Chen, Q., Khan, A., Yang, J.W., Sun, C.J., Shur, M.S. and Park, H. (1996) *Appl. Phys. Lett.*, **69**, 794.
38. Khan, M.A., Chen, Q., Shur, M.S., Dermott, B.T., Higgins, J.A., Burm, J., Schaff, W. and Eastman, L.F. (1996) Short channel of GaN/AlGaN doped channel heterostructure field effect transistors with 36.1 cutoff frequency, *Electronic Lett*, **32**(4), 357.
39. Shur, M.S. (1987) *GaAs Devices and Circuits*, Plenum, New York.
40. Lee, K., Shur, M.S., Fjeldly, T.A. and Ytterdal, T. (1993) *Semiconductor Device Modeling for VLSI*, Prentice-Hall, Englewood Cliffs NJ.
41. Khan, M.A., Shur, M. and Chen, Q. (1996) High transconductance AlGaN/GaN optoelectronic heterostructure field effect transistor, *Electronics Lett.*, **31**, 2130.
42. Khan, M.A., Shur, M.S., Chen, Q., Kuznia, J.N. and Sun, C.J. (1995) *Electronics Lett.*, **31**, 398.
43. Khan, M.A., Shur, M.S., Chen, Q.C. and Kuznia, J.N. (1994) *Electronics Lett.*, **30**, 2175.
44. Wu, Y.-F., Keller, B.P., Keller, S., Kapolnek, D., Kozodoy, P., Denbaars S.P. and Mishra, U.K. (1996) Very high breakdown voltage and large transconductance realized on GaN heterojunction field effect transistors, *Appl. Phys. Lett.*, **69**(10), 1438.
45. Fan, Z.-F., Mohammad, S.N., Kim, W., Aktas, Ö., Botchkarev, A.E. and Morkoç, H. (1996) *Appl. Phys. Lett.* **68**, 1672.
46. Mohammad, S.N., Fan, Z.-F., Kim, W., Aktas, Ö., Botchkarev, A.E., Salvador, A. and Morkoç, H. (1996) *Electron. Lett.* **32**, 598.
47. Mohammad, S.N., Fan, Z.-F., Salvador, A., Aktas, Ö., Botchkarev, A.E., Kim, W. and Morkoç, H. (1996) *Appl. Phys. Lett.* **69**(10).
48. Khan, M.A., Chen, Q., Shur, M.S., Dermott, B.T., Higgins, J.A., Burm, J., Schaff, W. and Eastman, L.F. (1996) CW Operation of Short Channel GaN/AlGaN Doped Channel Heterostructure Field Effect Transistors at 10 GHz and 15 GHz, *IEEE Electron Device Lett.*, **17**(12).

Index